Lactic Acid Bacteria: Latest Research

Lactic Acid Bacteria: Latest Research

Edited by **Sean Layman**

R Callisto Reference

New York

Published by Callisto Reference,
106 Park Avenue, Suite 200,
New York, NY 10016, USA
www.callistoreference.com

Lactic Acid Bacteria: Latest Research
Edited by Sean Layman

International Standard Book Number: 978-1-63239-442-2 (Hardback)

Contents

Preface

This book is a compilation of latest researches in Lactic Acid Bacteria (LAB). The ongoing scientific research in several parts of the globe on the proteomics, genomics and genetic engineering of lactic acid bacteria is enhancing our knowledge of their physiology, pushing further the boundaries for their potential applications. This book represents a collection of the excellent scientific research activities regarding the future applications of LAB. It is organized under three sections namely, Livestock Feed, Fish & Seafood Products, and New Fields of Application. It is intended for a great spectrum of readers including researchers, students, corporate R&D and academicians interested in this subject.

The researches compiled throughout the book are authentic and of high quality, combining several disciplines and from very diverse regions from around the world. Drawing on the contributions of many researchers from diverse countries, the book's objective is to provide the readers with the latest achievements in the area of research. This book will surely be a source of knowledge to all interested and researching the field.

In the end, I would like to express my deep sense of gratitude to all the authors for meeting the set deadlines in completing and submitting their research chapters. I would also like to thank the publisher for the support offered to us throughout the course of the book. Finally, I extend my sincere thanks to my family for being a constant source of inspiration and encouragement.

Editor

Livestock Feed

Ruminal Digestibility and Quality of Silage Conserved via Fermentation by Lactobacilli

Yang Cao, Yimin Cai and Toshiyoshi Takahashi

Additional information is available at the end of the chapter

1. Introduction

The utilization of whole crop rice (WCR) as an animal feed has proven economically viable, not only as a way of disposing of rice straw residues but also as a real alternative for feeding livestock in regions where rice is the main crop (Han et al., 1974). As a result, in Japan and other rice-producing countries, rice is no longer grown exclusively for human consumption but increasingly as a valuable forage crop. Forage rice is in fact believed to be an ideal alternative crop, not only in helping farmers adjust grain rice production but also in preserving the soil, leading to long-term utilization of the paddy field. Yet a major drawback of forage rice is that it yields low-quality silage, due to poor digestibility of nutrients, mostly crude proteins (Cai et al., 2003). Several processes have been developed to improve the fermentation and nutritional value of whole-crop silage from forage paddy rice. Breeding programs are carried out, and newly developed rice varieties with increased yield and amount of digestible nutrients are being grown and tested. Also, harvesting, preparation, and storage techniques are constantly being improved. However, WCR is usually insufficient in sugars and lactic acid bacteria (LAB), and may produce silages rich in ethanol rather than lactic acid and volatile fatty acid (VFA) (Cai et al., 2003). This could be attributed to the structure of the rice plant; the hollow stem may increase the air in a silo, facilitating yeast WCR is usually insufficient in sugars and lactic acid bacteria (LAB) (Cai et al., 2003), and may produce silages rich in ethanol rather than lactic acid and VFA (Yamamoto et al., 2004). This could be attributed to the structure of the rice plant; the hollow stem may increase the air in a silo, facilitating yeast growth especially in the early ensiling period. Furthermore, most of the processes used to date still rely on heavy chemical treatments with ammonia and sodium hydroxide and were reported to reduce the palatability of silage to ruminants (Cai et al., 2003; Enishi et al., 1998). Of the many factors that can affect silage fermentation, the type of microorganisms that dominate the process often dictates the final quality of the silage. For instance, homolactic fermentation by LAB is more desirable than other types of fermentation because the theoretical recoveries of dry matter and energy are greatest. During this type of fermentation, LAB utilizes water-soluble carbohydrates to produce lactic acid, the primary acid responsible for decreasing the pH in silage. In contrast, other fermentations are less efficient. Natural populations of LAB on plant material are often low in number and heterofermentative. Thus, the concept of using a microbial inoculant to silage involves adding fast-growing homofermentative LAB in order to dominate the fermentation, thereby producing higher-quality silage. Some of the commonly used homofermentative LAB in silage inoculants include *Lactobacillus plantarum*, *Lactobacillus acidophilus*, *Pediococcus acidilactici*, and *Enterococcus faecium*. Commercially available microbial inoculants contain one or more of these bacteria that have been selected for their ability to dominate the fermentation.

Food by-product such as tofu cake is high in crude protein and fatty acids (Xu et al. 2001). Not only could the by-products be utilized as a source of nutrients for ruminants, but using them to replace imported commercial feedstuffs could save energy in transportation, and possibly reduce the environmental impact of burning them as waste of burying them landfills. Preparing total mixed ration (TMR) silage is one practice whereby food by-products are stored and utilized as animal feeds in Japan (Imai, 2001). Our previous study (Cao et al., 2009b) showed that TMR silage with 30% dried tofu cake had the higher lactic acid content than that with rice bran or green tea waste. Alli et al. (1984) reported that as molasses can provide fermentable sugars for the production of organic acids, it has been used extensively as a fermentation aid, and that silage prepared with molasses may show a lower pH, higher residual water-soluble carbohydrates levels, greater quantities of lactic acid, lower levels of volatile basic nitrogen, and decreased dry matter (DM) loss compared to silage without molasses. Weinberg et al. (2003) also reported a high lactic acid content in silage ensiled with straw and molasses.

However, even if there is plenty of glucose as a substrate, if insufficient in lactic acid bacteria, to preparing good quality silage is difficult. Cai et al. (1999) reported that the factors involved in fermentation quality include chemical composition, particularly the water-soluble carbohydrates content of the silage material and the physiological properties of epiphytic bacteria. Conservation of forage crops by ensiling is based on natural fermentation in which epiphytic LAB convert sugars into lactic acid under anaerobic conditions. As a result, the pH decreases and the forage is preserved. The fermentation quality of silage is influenced by the size, diversity and activity of epiphytic LAB. The population density of LAB has been reported to range from 10–10^3 CFU/g fresh matter (FM) on standing forage crops to 10^3–10^7 CFU/g FM on chopped ones entering the silo. Generally, When LAB reaches at least 10^5 (CFU/g of FM), silage can be well preserved; if LAB values below 10^5 high-quality silage fermentation may need to be controlled using certain inoculants. *Lactobacillus plantarum* Chikuso-1 is shown to have great potential as an inoculant for WCR silage (Cai et al., 1999, 2003; Cai, 2001; Cao et al., 2011).

Furthermore, Commercial processing of vegetables results in many residues, but most are burned and dumped into landfills or used as compost, which is a waste of resources and leads to possible environmental problems due to unsuitable disposal. Demand for efficient use of food by-products is increasing due to economic and environmental concerns. Residues from vegetables such as white cabbage, Chinese cabbage, red cabbage, and lettuce are high in nutrients such as vitamins, minerals, and vegetable fiber, and large quantities of these vegetables are produced annually in many countries, including Japan. However, these vegetable residues perish easily because of their high moisture content. Technologies to create high-quality animal feed from vegetable residues and to provide long-term storage of the resulting silage need to be developed. In regions where vegetable residues are the main food by-product, the use of vegetable-residue silage as animal feed has proven economically viable, not only as a way of disposing of vegetable residues but also as an alternative livestock feed. To prepare fermented by-product mix as ruminant feed, it is important to investigate the digestive characteristics of these vegetable residues.

The purpose of this experiment was to examine the effects of lactic acid bacteria inoculant on the quality of fermentation and ruminal digestibility and fermentation characteristics of ruminant fed silage, and to determine the chemical and microorganism composition of vegetable residues, and the influence of lactic acid bacteria addition and moisture adjustment on fermentation quality and the in vitro ruminal fermentability of silages.

2. Materials and methods

Experiments were conducted with permission from the Committee of Animal Experimentation, and according to the animal care and use institutional guidelines for animal experiments at the Faculty of Agriculture, Yamagata University.

2.1. Silage preparation

Whole crop rice (Haenuki) was cultivated using transplant cultivating methods in a paddy field on an experimental farm at Yamagata University, Japan, and harvested at the full-ripe stage with a length of 2 cm. As shown in Table 1, TMR was prepared using compound feed (Kitanihon-Kumiai Feed, Yamagata, Japan), WCR, dried beet pulp, a vitamin-mineral supplement (Snow Brand Seed, Iwate, Japan), dried tofu cake (Zenno, Tsuruoka, Japan), molasses (sugarcane; Dai-Nippon Meiji Sugar Co., Tokyo, Japan) and LAB (*Lactobacillus plantarum* Chikuso-1, Snow Brand Seed, Sapporo, Japan). Experimental treatments included control silage without additive, or with molasses, (4% FM basis), LAB (5 mg kg^{-1} FM basis), and molasses+LAB. Moisture was adjusted with water to approximately 65%. Silages were prepared using a small-scale system of silage fermentation. Approximately 1 kg TMR was packed into plastic film bags (Hiryu BN-12 type, 270 mm × 400 mm, Asahikasei, Tokyo, Japan), and the bags were sealed with a vacuum sealer (SQ303, Sharp, Osaka, Japan). Three silos per treatment were prepared and stored in a room at 20–25°C for 60 days.

Residues of white cabbage (*Brassica campestris* L. var. *capitata*), Chinese cabbage (*Brassica rapa* L. *var. glabra Regel*), red cabbage (*Brassica oleracea* var. *capitata F. rubra*), and lettuce (*Lactuca sativa* L.) (three samples of each) were collected from a local commercial vegetable factory (Fujiyama factory, Matsuya Foods Company, Limited, Fujinomiya,

Shizuoka, Japan). The vegetable residue comprised the outside leaf part of white cabbage, Chinese cabbage, red cabbage, and lettuce with no added bacteria. Experimental treatments included control silage without additive or with LAB, beet pulp (DM, 90.7%; organic matter, 94.9%; crude protein, 8.4%; ether extract, 0.7%; acid detergent fiber, 25.6%; neutral detergent fiber, 52.1%; WSC, 2.1% of DM), and beet pulp+LAB. The strain FG1 (Lactobacillus plantarum Chikuso-1; Snow Brand Seed, Sapporo, Japan) isolated from a commercial inoculant was used. The de Man Rogosa Sharpe agar broth was inoculated with strain FG1 and incubated overnight. After incubation, the optical density of the suspension at 700 nm was adjusted to 0.42 using sterile 0.85% NaCl solution. The inoculum size of LAB was 1 mL of suspension per kilogram of FM. The inoculated LAB number was 1.0×10^5 CFU/g of FM. The addition ratio of beet pulp was 300 g per kg of FM. Silages were prepared using a small-scale system of silage fermentation (Cai et al., 1999). Approximately 100-g portions of forage material, chopped into about 20-mm length, were mixed well and packed into plastic bags (Hiryu KN type, 180 × 260 cm; Asahikasei, Tokyo, Japan). The bags were sealed with a vacuum sealer (BH 950; Matsushita, Tokyo, Japan). The plastic-bag silos were stored at a room at 25°C. There were ten bag silos per treatment. One bag of silo per treatment was opened on d 60. Samples were dried in a forced-air oven at 60°C for 48 h, ground to pass through a 1-mm screen with a Wiley mill (ZM200, Retch GmbH & Co. KG, Haan, Germany), and used for chemical analysis and in vitro digestibility measurements.

2.2. Chemical analyses

The TMR silages, vegetable and its silages were dried in a forced draft oven at 60°C for 48 h and ground into a 2-mm powder with a sample mill (Foss Tecator; Akutalstuku, Tokyo, Japan). Moisture, ash, crude protein, ether extract, and crude fiber contents were determined by general methods. Analyses of neutral detergent fiber and acid detergent fiber contents were made following Van Soest et al. (1991). Heat-stable amylase and sodium sulfite were used in the neutral detergent fiber procedure, and the results were expressed without residual ash. Nonfibous carbohydrate was calculated by the formula as: Nonfibous carbohydrate = organic matter − crude protein − nonfibous carbohydrate − ether extract. The fermentation products of silages were determined using cold-water extracts. Wet silage (50 g) was homogenized with 200 ml sterilized distilled water and stored at 4°C overnight (Cai et al. 1999). The pH of the silages was determined using a glass electrode pH meter (D-21, Horiba, Kyoto, Japan). Lactic acid was analyzed using the methods of Cai et al. (1999). Ammonia- N was determined as described by (Cai et al. 2003). To measure total VFA, silage and ruminal fluid were steam-distilled and titrated using sodium hydroxide. Dried VFA salt was separated and quantified using gas chromatography (G-5000A, Hitachi, Tokyo, Japan) equipped with a thermal conductivity detector and a stainless column (Unisole F-200, 3.2 mm × 2.1 m). The analytical conditions were as follows: column oven temperature, 140°C; injector temperature, 210°C; detector temperature, 250°C. V-score, which was used to assess silage quality, was determined from the proportion ammonia-N in the total nitrogen and VFA contents in the silage.

2.3. Cultures and incubations

Two adult wethers (average initial body weight, 78.5 kg) fitted with rumen cannulae were used as donors of ruminal fluid. The wethers were fed a basal diet of 50% reed canary grass (Phalaris arundinacea L.) hay and 50% commercial feed concentrate (Koushi-Ikusei-Special, Kitanihon-Kumiai-Feed, Miyagi, Japan) at maintenance energy level (2.0% DM of their body weight) and had free access to clean drinking water. They were fed once daily at 09:00 h. Wethers were cared for in accordance with the animal care and use institutional guidelines for animal experiments at the Faculty of Agriculture, Yamagata University (Tsuruoka, Japan).

Rumen fluid was collected through the rumen cannulae 2 h after feeding and diverted to plastic bottles. The fluid was filtered through four layers of cheesecloth and combines on an equal volume basis. The combined filtrate was mixed with CO_2-bubbled McDougall's artificial saliva (pH 6.8) at a ratio of 1:4 (vol/vol). Then 50 mL buffered rumen fluid was transferred to 128-mL serum bottles containing 0.5 g sample, and flushed with O_2-free CO_2. Tubes were capped with a butyl rubber stopper and sealed with an aluminum cap. Incubations were performed in triplicate at 39°C for 6 h (Mohammed et al., 2004) in a water bath with a reciprocal shaker (100 strokes/min).

2.4. Analysis of fermentation products

To terminate fermentation at the end of incubation, 25 μL of formaldehyde solution (35%) were injected into serum bottles, which were immediately sealed and cooled at room temperature. Gas samples were collected by air syringe from the serum bottles and injected into a gas chromatograph (GC323, GL Sciences, Tokyo, Japan) equipped with a thermal conductivity detector and a stainless steel column (WG-100 SUS, 1.8 m × 6.35 mm OD), and the methane production in each serum bottle was measured. The analytical conditions were as follows: column oven temperature at 50°C, injector temperature at 50°C, and detector temperature at 50°C.

2.5. *In vitro* DM digestibility, and methane and VFA production

Separate sub-samples of the supernatant were taken to determine the pH and VFA concentration. The bottles were rinsed with warm water to remove all solid residues, which were then oven-dried at 60°C and stored for further analyses. In total, 2 g of dried residue were oven-dried at 135°C and stored to determine DM digestibility.

2.6. Statistical analyses

Analyses were performed using the general linear model procedure (SAS institute, Cary, NC, USA). Data on fermentative characteristics, in vitro DM digestibility, and ruminal methane and VFA production of TMR silages were subjected to one-way analysis of variance Tukey's test was used to identify differences ($P < 0.05$) between means. Data on the fermentative characteristics of each vegetable silage opened on d 30 were analyzed by one-way analysis of variance. Data on chemical compositions, fermentative characteristics, in vitro ruminal DM digestibility, and fermentation products after 6-h incubation of silages opened on d 60 were analyzed using a completely randomized design with a 4 × 4 (vegetable residues × additive treatment) factorial treatment structure. The general linear model procedure of SAS version 9.0 (SAS Institute, Inc., Cary, NC) was used for the analysis, and the model included the main effects of vegetable residues and additive treatment, and their interactions. Sealing time and ensiling duration were excluded from the model because the 60-d silages, wastes processed, and silos were made only one time. The Tukey test was used to identify differences ($P < 0.05$) between means.

3. Results and discussion

3.1. Chemical composition of materials and silage

The contents of DM, crude protein, ether extract, nonfibous carbohydrate, ash, and neutral detergent fiber in molasses were 72.7, 4.3, 0.7, 83.6, 11.4, and 0%, respectively (Table 1). The contents of organic matter and crude protein in WCR were 86.5 and 5.3%, respectively. The contents of crude protein, nonfibous carbohydrate, and neutral detergent in the tofu cake were 30.1, 15.8, and 37.7%, respectively. And the chemical composition of vegetable residues is shown in Table 2. The DM of the four types of vegetable residues was less than 6%. Their OM contents were more than 70% of DM, and the crude protein and neutral detergent fiber contents were approximately 20% and 30% of DM, respectively. The water-soluble carbohydrates (including sucrose, glucose, and fructose) contents ranged from 8.4 to 21.7% of DM; the highest and lowest values were observed in white cabbage and lettuce residues, respectively.

Preparing TMR silage is one practice whereby food by-products are stored and utilized as ruminant feeds in Japan. It can avoid energy costs associated with drying, and may improve odors and flavors of unpalatable feed resources through fermentation in a silo (Cao et al., 2009a; Imai 2001; Wang & Nishino 2008). We have performed a number of experiments to investigate the effects of food by-products including tofu cake, rice bran, and

	WCR[1]	Concentrate[2]	Beet pulp	TC[3]	Molasses[4]
DM[5] (%)	36.0	88.2	90.7	91.3	72.7
CP[6] (% DM)	5.3	16.7	8.4	30.1	4.3
EE[7] (% DM)	2.2	3.8	0.7	12.2	0.7
NFC[8] (% DM)	32.1	60.1	33.7	15.8	83.6
Ash (% DM)	13.5	5.1	5.1	4.3	11.4
ADF[9] (% DM)	30.2	8.7	25.6	22.2	–
NDF[10] (% DM)	48.0	14.4	52.1	37.7	0

[1]Whole crop rice; [2]Formula feed ("Koushi Ikusei Special Mash" made by Zenno with 120g kg[-1] crude protein in fresh matter); [3]Tofu cake; [4]Wang & Goetsch (1998); [5]Dry matter; [6]Crude protein; [7]Ether extract; [8]Non-fibrous carbohydrate (100 – crude protein – ether extract – neutral detergent fiber – ash); [9]Acid detergent fiber; [10]Neutral detergent fiber.

Table 1. Chemical composition of WCR, concentrate, beet pulp and tofu cake used in total mixed ration silages (Cao et al., 2010b)

Item	White cabbage	Chinese cabbage	Red cabbage	Lettuce
DM, %	3.6 ± 0.02	2.2 ± 0.01	5.2 ± 0.02	2.0 ± 0.01
OM, % of DM	81.6 ± 0.30	70.8 ± 0.08	87.3 ± 0.24	75.4 ± 0.09
CP, % of DM	21.9 ± 0.01	20.6 ± 0.21	22.8 ± 0.32	21.1 ± 0.41
Ether extract, % of DM	2.8 ± 0.07	1.5 ± 0.02	1.7 ± 0.14	4.9 ± 0.07
ADF, % of DM	31.0 ± 0.20	26.5 ± 1.49	31.6 ± 0.42	29.4 ± 0.14
ADL, % of DM	2.3 ± 0.19	5.8 ± 0.02	3.2 ± 0.03	5.9 ± 0.30
NDF, % of DM	32.8 ± 0.56	27.7 ± 0.24	33.6 ± 0.40	31.4 ± 0.01
Sucrose, % of DM	12.1 ± 0.15	1.1 ± 0.02	6.3 ± 0.10	3.1 ± 0.13
Glucose, % of DM	4.3 ± 0.04	3.6 ± 0.07	2.5 ± 0.01	1.3 ± 0.04
Fructose, % of DM	5.3 ± 0.12	3.8 ± 0.05	3.7 ± 0.10	4.0 ± 0.18

[1]Values are means ± SD.

Table 2. Chemical composition1 of vegetables residues (Cao et al., 2011)

	Treatment				SEM[3]
	Control	M[1]	LAB[2]	M+LAB	
Ingredient					
LAB (mg kg^{-1} FM[4])	-	-	5	5	
M (% FM)	-	4	-	4	
Whole crop rice (% DM[5])	30	30	30	30	
Concentrate[6] (% DM)	25	25	25	25	
Vitamin-mineral supplement[7] (% DM)	1.5	1.5	1.5	1.5	
Dried beet pulp (% DM)	13.5	13.5	13.5	13.5	
Tofu cake (% DM)	30	30	30	30	
Chemical composition					
DM (%)	35.9	36.2	36.4	37.0	1.98
Organic matter (% DM)	92.6	92.5	92.6	92.3	0.25
Crude protein (% DM)	15.3	14.6	15.4	15.1	0.38
Ether extract (% DM)	5.1	5.3	5.5	5.6	0.24
Nitrogen free extract (% DM)	57.1	58.4	58.4	57.5	0.40
Crude fiber (% DM)	15.7	14.9	14.2	14.5	0.88
NFC[8] (% DM)	30.9	32.0	29.5	32.0	1.54
Crude ash (% DM)	7.4	7.5	7.4	7.7	0.25
Acid detergent fiber (% DM)	19.0	20.0	18.9	20.2	1.21
Neutral detergent fiber (% DM)	41.4	40.6	42.2	39.6	1.12

[1]Molassess; [2]Lactic acid bacteria (Lactobacillus plantarum); [3]Standard error of means; [4]Fresh matter; [5]Dry matter; [6]Formula feed ("Koushi Ikusei Special Mash" made by Zenno; total digestible nutrients, 70.0%; crude protein, 12.0% in fresh matter); [7]Commercial vitamin-mineral supplement product (Snow brand seed, Iwate, Japan); [8]Non-fibrous carbohydrate (100 – crude protein – ether extract – neutral detergent fiber – ash).

Table 3. Ingredient and chemical composition of total mixed ration silages (Cao et al., 2010b)

wet green tea waste on fermentation quality of WCR-containing TMR silage. Our previous study (Cao et al., 2009a, b) showed that silages with 30% tofu cake had higher lactic acid content, compared to those with rice bran and green tea waste. Therefore, we prepared TMR silage using tofu cake, and in order to investigate if adding LAB or molasses can further increase lactic acid content of the silages with tofu cake, LAB and molasses were added into these silages in this study. LAB can increase the lactic acid content of a silage (Cai, 2001; Cai et al., 2003), and was well used to prepare silage. Molasses is a fermentable carbohydrate (Maiga & Schingoethe, 1997) and many researchers (Alli et al., 1984) have reported its successful use with grass silage. In addition, molasses is a food by-product of sugar beet and sugarcane production. Molasses with high water-soluble carbohydrates is used as a major energy source for meat or milk production (Araba et al., 2002; Granzin & Dryden, 2005; Sahoo & Walli, 2008; Wang & gotsch, 1998).

3.2. Fermentation quality

As indicated by the low pH value (around 4.0) and ammonia-N/total N content (2.83–2.97%), high lactic acid content (2.49–2.87%), and V-score (99.8) for the silages, the four TMR silages were well preserved (Table 4). Although the levels of moisture, pH, acetic acid, propionic acid, butyric acid, and ammonia-N/total N and V-score did not differ significantly, lactic acid contents for the silages with LAB and molasses+LAB were higher ($P = 0.005$) than the control and molasses silages.

It is well established that LAB play an important role in silage fermentation, and LAB values have become a significant factor in predicting the adequacy of silage fermentation and in determining whether to apply bacterial inoculants to silage materials. Generally, when LAB reaches at least 10^5 (CFU/g of FM), silage can be well preserved (Cai 2001; Cai et al., 1999; Cai et al., 2003). However, the LAB values below 10^5 and aerobic bacteria values above 10^6 present in most WCR suggest that high-quality silage fermentation may need to be controlled using certain inoculants. The inoculant strain used in this study was Lactobacillus plantarum Chikuso-1; this strain promotes lactic acid fermentation and can grow well in low-pH environments. Therefore, silage prepared using this strain can promote the propagation of LAB, decrease pH, inhibit the growth of clostridia and aerobic bacteria, and improve silage quality (Cai et al., 1999).

	Treatment				SEM[3]	P-value
	Control	M[1]	LAB[2]	M+LAB		
pH	3.99	3.92	4.01	4.03	0.0391	0.585
Lactic acid (% FM[4])	2.49[a]	2.52[a]	2.84[b]	2.87[b]	0.0880	0.005
Acetic acid (% FM)	0.09	0.09	0.10	0.09	0.0032	0.934
Propionic acid (% FM)	0.003	0.003	0.001	0.001	0.0009	0.574
Butyric acid (% FM)	0.002	0.003	0.003	0.003	0.0001	0.776
ammonia-N/total-N (%)	2.97	2.91	2.92	2.83	0.1775	0.956
V-score	99.8	99.8	99.8	99.8	0.0090	0.970

[1]Molassess; [2]Lactic acid bacteria (Lactobacillus plantarum); [3]Standard error of means; [4]Fresh matter; Means within a row with different letters (a, b) differ ($P < 0.05$).

Table 4. Fermentative characteristics of total mixed ration silages (Cao et al., 2010b)

The pH, lactic acid, acetic acid, and ammonia-N were affected not only by vegetable, but also by addition and vegetable × addition. Comparison among the four types of vegetable silages revealed that the pH was the lowest ($P < 0.001$) in silage with white cabbage residue, followed by red cabbage, Chinese cabbage, and lettuce silages. The lactic acid content was highest in white cabbage silage ($P < 0.001$), whereas the acetic acid content was highest in lettuce silage ($P < 0.001$), followed by red cabbage, white cabbage, and Chinese cabbage silages. Propionic and butyric acids were not detected among the four types of vegetable silages; the ammonia-N concentration of white cabbage, red cabbage, and lettuce silages was lower ($P < 0.001$) than that of Chinese cabbage silage. Comparison of the treated silages showed that all silages treated with LAB or BP had lower ($P < 0.001$) pH values and ammonia-N concentrations, but higher ($P < 0.001$) lactic acid contents compared with the control silage.

Alli et al. (1984) assessed the effects of molasses on the fermentation of chopped whole-plant Leucaena. Silages were treated with molasses at a rate of 2.25% or 4.5% fresh weight at the time of ensiling, which led to increased rates of lactic acid production, lower pH, decreased DM loss, and reduced levels of ammonia-N compared to Leucaena to which no molasses was added. In the present experiment, WCR-containing TMR silages were treated with or without molasses at the rate of 4% fresh weight. Although the addition of molasses did not significantly influence the chemical composition, it increased DM and Non-fibrous carbohydrate contents by 3.06 and 3.56%, respectively. Adding molasses did not increase lactic acid content significantly, Adding LAB and molasses+LAB, however, increased lactic acid content significantly. This is probably because that, even if no molasses, there was enough fermentable sugars in TMR silage, and the LAB may have converted more fermentable sugars to lactic acid (Cai 2001; Cai et al., 2003). This study showed the advantage of LAB over molasses. Alli et al. (1984) reported that adding molassesf reduced the ammonia-N of silage, although in present study, ammonia-N did not differ among the four silages.

Furthermore, our previous study (Cao et al., 2011) has determined the effect of LAB inoculant and beet pulp addition on silage fermentation quality and *in vitro* ruminal DM digestion of vegetable residues. The silage treated with LAB or beet pulp had a lower pH and a higher lactic acid content than the control silage (Table 4). After 6 h of incubation, the LAB-inoculated silage had the highest DM digestibility and the lowest methane production. Weinberg et al. (2003) and Filya et al. (2007) reported that LAB inoculants affected the in vitro digestibility of alfalfa hay and corn silage, respectively, after 48 h incubation. In the present study, LAB inoculants not only increased ($P < 0.01$) the silage DM digestibility after 6 h in vitro incubation but also decreased ruminal methane production, which decreases the energy loss of feed (Cao et al., 2010a). Furthermore, LAB inoculants improve the fermentation quality of vegetable silage, which might decrease the degradation of crude protein in the silage; therefore, LAB-treated silage had a high concentration of ruminal ammonia-N. Adding beet pulp to vegetable silage did not affect the DM digestibility after 6 h in vitro incubation, but did increase the production of acetic acid, propionic acid, and even methane, while decreasing the production of butyric acid and ammonia-N. We cannot explain the mechanism of these effects. More research is needed to elucidate the probiotic effect of adding LAB or beet pulp to vegetable silage in ruminants.

Item	Moisture	pH	Lactic acid	Acetic acid	Ammonia-N
	%			g/kg of FM	
Vegetable residue means					
White cabbage	82.4	3.59c	20.9a	1.5c	0.26b
Chinses cabbage	82.9	4.26a	10.8b	1.3c	0.41a
Red cabbage	80.8	3.74b	12.7b	2.0b	0.26b
Lettuce	83.9	4.27a	10.4b	3.3a	0.28b
Additive treatment means					
Control	95.4a	4.46a	8.2c	2.4a	0.52a
LAB	95.1a	3.95b	10.8b	1.9b	0.26b
BP	69.1b	3.77c	17.6a	2.5a	0.23bc
LAB+BP	70.3b	3.68c	18.3a	1.4b	0.20c
Significance of main effects and interactions					
Vegetable residues (V)	0.249	<0.001	<0.001	<0.001	<0.001
Additive treatment (A)	<0.001	<0.001	<0.001	<0.001	<0.001
V × A	0.841	<0.001	<0.001	<0.001	<0.001

[1]Means within columns with different letters (a-c) differ ($P < 0.05$); [2]Propionic and Butyric acids were not detected; [3]Lactobacillus plantarum (Chikuso-1, Snow Brand Seed, Sapporo, Japan); [3]Beet pulp.

Table 5. Fermentation profile of vegetable residue silages prepared with LAB and BP after 60 days of storage (Cao et al., 2011)

The pH, lactic acid, acetic acid, and ammonia-N were affected not only by vegetable, but also by addition and vegetable × addition（Table 5）. Comparison among the four types of vegetable silages revealed that the pH was the lowest ($P < 0.001$) in silage with white cabbage residue, followed by red cabbage, Chinese cabbage, and lettuce silages. The lactic acid content was highest in white cabbage silage ($P < 0.001$), whereas the acetic acid content was highest in lettuce silage ($P < 0.001$), followed by red cabbage, white cabbage, and Chinese cabbage silages. Propionic and butyric acids were not detected among the four types of vegetable silages; the ammonia-N concentration of white cabbage, red cabbage, and lettuce silages was lower ($P < 0.001$) than that of Chinese cabbage silage. Comparison of the treated silages showed that all silages treated with LAB or BP had lower ($P < 0.001$) pH values and ammonia-N concentrations, but higher ($P < 0.001$) lactic acid contents compared with the control silage.

The factors involved in fermentation quality include not only the physiological properties of epiphytic bacteria but also the chemical composition of the silage material (Cai et al., 1999). In this study, the four types of vegetable residues had relatively high water-soluble carbohydrates contents; the epiphytic LAB transformed water-soluble carbohydrates into organic acids during the ensiling process, and as a result, the pH was reduced, which inhibited the growth of some microorganisms, such as bacilli, coliform bacteria, aerobic bacteria, yeasts, and molds. When silage was treated with LAB or BP, the fermentation tended to ensure rapid and vigorous results with the faster accumulation of lactic acid (Table 3) and lower pH values at earlier stages of ensiling, and it also inhibited the production of acetic acid and ammonia-N during silage fermentation and thus improved vegetable-residue conservation. The transitional behavior of the VFA in the silage during fermentation indicated sharp decreases in pH, and corresponding increases in lactic acid contents at earlier stages (7 d of ensiling) were typical of a good fermentation process and were in agreement with previous studies. Subsequently, a steady reduction in pH depicted stability, while lactic acid contents gradually stabilized after a decrease during storage. Some studies (Cai, 2001; Cai et al., 1999; Cai et al., 2003) have shown that the development of LAB peaks in the first 7 d in parallel with the rise in lactic acid concentration of silage, and this is followed by decreases in LAB numbers; however, no apparent decrease in LAB numbers was observed in this study. The d 60 LAB-treated silage had higher organic matter and crude protein, but lower water-soluble carbohydrates than did control silages. During silage fermentation, LAB could effectively utilize water-soluble carbohydrates to produce sufficient lactic acid to reduce pH and inhibit the growth of harmful bacteria; therefore, the resulting silage was of good quality. Furthermore, the moisture content of silage material is also a major factor influencing silage fermentation (Cai et al., 2003). An intrinsic characteristic of vegetable residues is their very high moisture content (95 to 98% of FM), and this is a major limitation to its use as livestock feed. Although dried vegetable residues can easily be incorporated into rations, the energy cost associated with drying wet vegetable residues has been increasing. Moreover, the risk of effluent production is high because of the low DM content. Therefore, pressed vegetable residues have been preferred for adjusting moisture with other feed to ensile. Cai et al. (1999) reported that high-moisture silage is more beneficial to lactic acid fermentation and has less risk of heat damage than low-moisture silage. In this study, according to our preliminary experiment and taking into consideration the cost of feed, the moisture of BP-treated silage was adjusted to 70%, and most beet pulp-treated silages had lower pH and ammonia-N and higher lactic acid content compared with control silage. It is possible that this is because the addition of BP not only adjusted the moisture content of the vegetable residues but also increased the water-soluble carbohydrates content; therefore, silages with added BP could greatly contribute to better lactic acid fermentation. Furthermore, we used a small-scale system of silage fermentation; all silages stored well and maintained high quality without aerobic deterioration in this study.

3.3. *In vitro* DM digestibility, and methane and VFA production

After in vitro 6 h incubation, DM digestibility, total VFA, acetic acid, isovaleric acid, valeric acid, and the acetic to propionic acid ratio did not differ significantly among the treatments (Table 6). However, methane production for the LAB silage and the molasses silage tended ($P = 0.065$) to decrease and increase, respectively, propionic acid for the LAB silage tended ($P = 0.061$) to increase, and butyric acid for the control silage was higher ($P = 0.008$) than the other silages.

	Treatment				SEM[3]	P-value
	Control	M[1]	LAB[2]	M+LAB		
DM[4] digestibility (%)	42.2	44.5	44.8	44.5	1.03	0.313
Methane production (L kg^{-1} DDM[5])	10.5	11.2	9.6	10.2	0.30	0.065
Total VFA (mmol 100 ml^{-1})	5.3	5.8	5.9	5.7	0.16	0.340
Acetic acid (A) (mol %)	37.0	38.9	38.0	38.8	0.55	0.142
Propionic acid (P) (mol %)	39.9	40.4	41.8	40.5	0.35	0.061
Butyric acid (mol %)	19.9[a]	17.8[b]	17.0[b]	17.8[b]	0.34	0.008
Isovaleric acid (mol %)	0.4	0.3	0.3	0.3	0.06	0.844
Valeric acid (mol %)	2.8	2.5	2.5	2.6	0.10	0.416
A/P	0.9	1.0	0.9	1.0	0.02	0.271

[1]Molassess; [2]Lactic acid bacteria (Lactobacillus plantarum); [3]Standard error of means; [4]Dry matter; [5]Digestible dry matter; Means within a row with different letters (a, b) differ ($P < 0.05$).

Table 6. *In vitro* dry matter digestibility, methane production and volatile fatty acid concentration after 6 hours incubation of total mixed ration silages (Cao et al., 2010b)

DM digestibility, VFA, and ammonia-N concentrations of vegetable-residue silage after 6 h incubation in vitro are shown in Table 7. Although DM digestibility was not influenced by vegetable, it was influenced by addition and by vegetable × addition; VFA, and ammonia-N were influenced by vegetable, addition, and vegetable × addition. DM digestibility did not differ among silages. However, ruminal CH_4 production of white and Chinese cabbage silages was lower ($P < 0.001$) than that of red cabbage and lettuce silages, and the total VFA production of red cabbage and lettuce residue silages was higher ($P = 0.014$) than that of Chinese cabbage silage. The acetic acid production of the lettuce silage was higher ($P < 0.001$) than that of the white cabbage silage. The propionic acid production of white cabbage was the highest ($P < 0.001$) among the four vegetable residues, followed by lettuce, which showed higher propionic acid production ($P < 0.001$) than did red or Chinese cabbage; the last two silages did not differ in this regard. Red cabbage had higher ($P < 0.001$) butyric acid production than Chinese cabbage and lettuce, and butyric acid production was higher in white cabbage ($P < 0.001$) than in Chinese cabbage. The A:P ratio of white cabbage silages was the lowest ($P < 0.001$) among the four types of vegetable silages. The highest and lowest ($P < 0.001$) ammonia-N production was found in white cabbage and lettuce silages, respectively. The LAB-treated silage had a higher ($P < 0.001$) DM digestibility than BP- and beet pulp+LAB-treated silages; it also had the highest ammonia-N production. Together with the control silage, LAB treated silage had lower ($P < 0.001$) total VFA, acetic acid, and propionic acid production, but higher ($P < 0.001$) butyric acid production and acetic acid:propionic acid ratio ratio compared with beet pulp- and LAB+beet pulp-treated silages.

In vitro DM digestibility was higher in silage with LAB than without LAB because LAB reduces DM loss in silage fermentation (Cai 2001; Cai et al., 2003). Furthermore, although there are some reports that adding molasses has no effect on DM digestibility (Granzin & Dryden 2005; Wang & Goetsch 1998), many more studies (Shellito et al., 2006; Sahoo & Walli 2008) have reported that diets with molasses have higher ruminal DM digestibility.

In the present experiment, there was a non-significant increasing trend in DM digestibility with molasses, LAB and molasses+LAB. Ruminal methane production and the molar proportion of propionic acid for silage with LAB decreased by 8.6% and increased by 4.8%, respectively. These might be because that adding LAB increased lactic acid content in the silage, when the silage containing high lactic acid content was incubated in vitro, there are two known mechanisms for the conversion of lactic acid or pyruvic acid to propionic acid, and when lactate acid is secondarily fermented by lactate-utilizing bacteria such as Megasphaera elsdenii, Selenomonas ruminantium, and Veillonella parvula, propionate is generally produced as a major product (Dawson et al., 1997) and this can reduce methanogenesis because electrons are used during propionate formation. But adding molasses, which has a high sugar content, may augment methane production in the rumen (Hindrichsen et al., 2005), perhaps because of which, molasses per se canceled (compensated) the effect of lactic acid content on methane production. A further research is necessary about the effect of molasses and the complex effects of molasses and LAB concerning the methane production in TMR silage. Furthermore, it is not yet clear why adding molasses or LAB decreased in vitro ruminal butyric acid in this study.

Item	DM digestibility	Total VFA	Acetic acid	Propionic acid	A/P[1]	Ammonia-N
	%		mmol/L			mg/L
Vegetable residue means						
White cabbage	44.9	43.0[ab]	24.8[b]	9.1[a]	2.7[b]	112.7[a]
Chinese cabbage	38.6	41.9[b]	28.0[ab]	7.4[bc]	3.8[a]	88.3[b]
Red cabbage	44.3	44.8[a]	27.5[ab]	6.7[c]	4.3[a]	91.8[b]
Lettuce	41.6	44.7[a]	29.2[a]	7.6[b]	4.0[a]	75.1[c]
Additive treatment means						
Control	41.3[ab]	40.1[b]	24.1[b]	6.0[b]	4.2[a]	122.4[b]
LAB	47.5[a]	42.2[b]	25.7[b]	6.5[b]	4.1[a]	164.1[a]
BP	39.8[b]	45.6[a]	29.3[a]	9.1[a]	3.2[b]	40.3[c]
BP+LAB	40.6[b]	46.4[a]	30.4[a]	9.2[a]	3.4[b]	41.2[c]
Significance of main effects and interactions						
Vegetable residues (V)	0.056	0.014	0.014	<0.001	<0.001	<0.001
Additive treatment (A)	0.001	<0.001	<0.001	<0.001	<0.001	<0.001
V × A	<0.001	0.009	0.01	<0.001	<0.001	<0.001

Means within columns with different letters (a-c) differ (P < 0.05).
[1]Acetic acid/propionic acid ratio.
[2]Digestible dry matter.
[3]Lactobacillus plantarum (Chikuso-1, Snow Brand Seed, Sapporo, Japan).
[4]Beet pulp.

Table 7. Measurements of dry matter digestibility, methane production, VFA concentration and ammonia-N after 6-h in vitro incubation with rumen fluid of vegetable residue silage (Cao et al., 2011)

4. Conclusions

The results of the present study show that adding LAB increased the lactic acid content of silage, had the potential to increase DM digestibility and to decrease ruminal methane production.

Author details

Yang Cao
College of Animal Science & Veterinary Medicine, Heilongjiang Bayi Agricultural University, Daqing, China

Yimin Cai
Japan International Research Center for Agricultural Sciences, Tsukuba, Japan

Toshiyoshi Takahashi
Faculty of Agriculture, Yamagata University, Tsuruoka, Japan

5. References

Alli, I., Fairbairn, R., Noroozi, E. & Baker, B. E. (1984). The effects of molasses on the fermentation of chopped whole-plant leucaena. *Journal of the Science of Food and Agriculture*, Vol. 35, No. 3, (March 1984), pp. 285–289, ISSN 0022-5142

Araba, A., Byers, F. M. & Guessous, F. (2002). Patterns of rumen fermentation in bulls fed barley/molasses diets. *Animal Feed Science and Technology*, Vol. 97, No. 1, (May 2002), pp. 53–64, ISSN 0377-8401

Cai, Y. (2001). The role of lactic acid bacteria in the preparation of high fermentation quality. *Japanese Journal of Grassland Science*, Vol. 47. No.5, (December 2001), pp. 527–533, ISSN 0447-5933

Cai, Y., Benno, Y., Ogawa, M. & Kumai, S. (1999). Effect of applying lactic acid bacteria isolated from forage crops on fermentation characteristics and aerobic deterioration of silage. *Journal of Dairy Science*, Vol. 82. No.3, (March 1999), pp. 520–526, ISSN 0022-0302

Cai, Y., Fujita, Y., Murai, M., Ogawa, M., Yoshida, N., Kitamura, R. & Miura, T. (2003). Application of lactic acid bacteria (*Lactobacillus plantarum* Chikuso-1) for silage preparation of forage paddy rice. *Japanese Journal of Grassland Science*, Vol. 49. No. 5, (December 2003), pp. 477–485, ISSN 0447-5933

Cao, Y., Cai Y., Takahashi ,T., Yoshida, N., Tohno, M., Uegaki, R., Nonaka, K. & Terada, F. (2011). Effect of lactic acid bacteria inoculant and beet pulp addition on fermentation characteristics and in vitro ruminal digestion of vegetable residue silage. *Journal of Dairy Science*, Vol. 94, No. 8, (March 2011), pp. 3902–3912, ISSN 0022-0302

Cao, Y., Takahashi, T. & Horiguchi, K. (2009a). Effects of addition of food by-procucts on the fermentation quality of a total mixed ration with whole crop rice and its digestibility, preference, and rumen fermentation in sheep. *Animal Feed Science and Technology*, Vol. 151, No. 1-2, (May 2009), pp. 1–11, ISSN 0377-8401

Cao, Y., Takahashi, T. & Horiguchi, K. (2009b). Effect of food by-products and lactic acid bacteria on fermentation quality and in vitro dry matter digestibility, ruminal methane and volatile fatty acid production in total mixed ration silage with whole-crop rice silage. *Japanese Journal of Grassland Science*, Vol. 55, No. 1, (April 2009), pp. 1–8, ISSN 0477-5933

Cao, Y., Takahashi, T., Horiguchi, K., Yoshida, N. & Cai., Y. (2010a). Methane emissions from sheep fed fermented or non-fermented total mixed ration containing whole-crop rice and rice bran. *Animal. Feed Science and Technology*, Vol. 157, No. 1, (April 2010), pp. 72–78, ISSN 0377-8401

Cao, Y., Takahashi, T., Horiguchi, K. & Yoshida, N. (2010b). Effect of adding lactic acid bacteria and molasses on fermentation quality and in vitro ruminal digestion of total mixed ration silage prepared with whole crop rice. *Grassland Science*, Vol. 56, No. 1, (March 2010), pp. 19–25 ISSN 1744-6961

Dawson, K. A., Rasmussen, M. A. & Allison, M. J. (1997). Digestive disorgers and nutritional toxicity. In: *The Rumen Microbial Ecosystem*, P. N. Hobson, C. S. Stewart, (Ed). 633-660, Blackie Academic and Professional, ISBN10 0751403660, London

Enishi, O., & Shijimaya, K. (1998). Changes of chemical composition of plant parts and nutritive value of silage in male sterile rice plant (*Oryza sativa* L.). *Grassland Science*, Vol. 44, No. 3, (October 1998), pp. 260–265, ISSN 1744-6961

Filya, I., Muck, R. E. & Contreras-Govea, F. E. (2007). Inoculant effects on alfalfa silage: Fermentation products and nutritive value. *Journal of Dairy Science*, Vol. 90, No. 11, (November 2007), pp. 5108–5114, ISSN 0022-0302

Granzin, B. C. & Dryden, G. M. (2005). Monensin supplementation of lactating cows fed tropical grasses and cane molasses or grain. *Animal Feed Science and Technology*, Vol. 120, No. 1-2 (May 2005) pp. 1–16, ISSN 0377-8401

Han, Y. W., & Anderson, A. W. (1974). The problem of rice straw waste. A possible feed through fermentation. *Economic Botany*, 28, pp. 338–344, ISSN 0013-0001

Hindrichsen, I. K., Wettstein, H. R., Machmuller, A., Jörg, B. & Kreuzer, M. (2005). Effect of the carbohydrate composition of feed concentratates on methane emission from dairy cows and their slurry. *Environmental Monitoring and Assessment*, Vol. 107, No. 1-3, pp. 329–350, ISSN 0167-6369

Imai, A. (2001). Silage making and utilization of high moisture by-products. *Japanses Journal of Grassland Science*, Vol. 47, No. 3, (August 2001), pp. 307–310, ISSN 0447-5933

Maiga, H. A. & Schingoethe, D. J. (1997). Optimaizing the utilization of animal fat and ruminal bypass proteins in the diets of lactating dairy cows. *Journal of Dairy Science*, Vol. 80, No.2, (February 1997), pp. 343–352, ISSN 0022-0302

Sahoo, B. & Walli, T. K. (2008). Effects of formaldehyde treated mustard cake and molasses supplementation on nutrient utilization, microbial protein supply and feed efficiency in growing kids. *Animal Feed Science and Technology*, Vol. 142, No. 3-4, (May 2008), pp. 220–230, ISSN 0377-8401

Shellito, S. M., Ward, M. A., Lardy, G. P., Bauer ,M. L. & Caton, J. S. (2006). Effects of concentrated separator by-product (desugared molasses) on intake, ruminal fermentation, digestion, and microbial efficiency in beef steers fed grass hay. *Journal of Animal Science*, Vol. 84, No. 6, (Joune 2006), pp. 1535–1543, ISSN 0021-8812

Van Soest, P. J., Robertson, J. B. & Lewis, B. A. (1991). Methods for dietary fiber, neutral detergent fiber, and non-starch polysaccharides in relation to animal nutrition. *Journal of Dairy Science*, Vol. 74, No. 10, (October 1991), pp. 3583–3597, ISSN 0022-0302

Wang, Z. S. & Goetsch, A. L. (1998). Intake and digestion by Holstein steers consuming diets based on litter harvested after different numbers of broiler growing periods or with molasses addition before deep-stacking. *Journal of Animal Science*, Vol. 76, No. 3, (March 1998), pp. 880–887, ISSN 0021-8812

Wang, F. J. & Nishino, N. (2008). Effect of aerobic exposure after silo opening on feed intake and digestibility of total mixed ration silage containing wet brewers grains or soybean curd residue. *Grassland Science*, Vol. 54, No. 3, (September 2008), pp. 164–166, ISSN 1744-6961

Weinberg, Z. G., Ashbell, G. & Chen, Y. (2003). Stabilization of returned dairy products by ensiling with straw and molasses for animal feeding. *Journal of Dairy Science*, Vol. 86, No. 4, (April 2003), pp. 1325–1329, ISSN 0022-0302

Xu, C. C., Suzuki, H. & Toyokawa, K. (2001). Characteristics of ruminal fermentation of sheep fed tofu cake silage with ethanol. *Animal Science Journal*, Vol. 72, No. 4, (July 2001), pp. 299–305, ISSN 1344-3941

Yamamoto, Y., Deguchi, Y., Mizutani, M., Urakawa, S., Yamada, H,. Hiraoka, H., Inui, K., Kouno, S. & Goto, M. (2004). Improvement of fermentation quality and dry matter digestibility of rice whole crop silage treated with fermented juice of epipytic lactic acid bacteria and mechanical processing. *Japanese Journal of Grassland Science*, Vol. 49, No. 6, (February 2004), pp. 665–668, ISSN 0447-5933

Chapter 2

Lactic Acid Bacteria
in Tropical Grass Silages

Edson Mauro Santos, Thiago Carvalho da Silva,
Carlos Henrique Oliveira Macedo and Fleming Sena Campos

Additional information is available at the end of the chapter

1. Introduction

Lactic Acid Bacteria (LAB) have applications in many industrial areas and play an important role in the preservation process of moist forages for animal feeding (silage).

The basic principle silage is to store the surplus forage keeping its stability and nutritional value until it is required to feed the animals. This process takes place in anaerobic conditions, where the lactic acid produced by the LAB inhibits the proliferation of spoilage microorganisms, which are less tolerant to acidic conditions. Thus, as the pH values decline, the silage losses decline as well due to the greater conversion of plant soluble carbohydrates (the main substrate for LAB) in lactic acid, with 96.9% rate of energy recovery (Mc Donald et al., 1991). The major soluble carbohydrates present in forage crops are fructose, glucose, sucrose and frutosanas, according to Woolford, (1984), sucrose and frutosanas are rapidly hydrolyzed in their monomers at forage harvest.

Lactic fermentation produces lactic acid as the main product. Therefore, homofermentative bacteria such as *Lactobacillus plantarum* are desirable in the silage fermentation process, once 87% of their metabolites become lactic acid. On the other hand, in the heterofermentative process, additional substances like ethanol, acetate and CO_2, are formed. Microbial inoculants used as additives include homofermentative LAB, heterofermentative LAB, or both combined. The specificity between the forage specie and its epiphytic micro flora implicates the need for studies related with isolation and identification of the main microorganism groups present in the forage used for silage.

In this chapter we will discuss the characteristics of tropical grasses, the main LAB species found in these grasses and how the LAB's are used to improve the quality of tropical grass' silages.

2. Tropical grass characteristics

The forage characteristics that contribute to a good fermentation are: dry matter content, autochthonous plant microbiota and, most importantly, the quantity of soluble carbohydrates. Corn and sorghum are the most appropriate grasses to make silages due to their high soluble carbohydrate contents and dry matter production. However, some studies have shown that different grasses can be utilized if they are ensilage at the right developmental stage or if appropriate additives are used (Zanine et al., 2010).

The decline in pH values inhibit the spoilage microorganism proliferation, which allows the silage nutritive values to be preserved. Thus, the best silage forages are the ones with high soluble carbohydrates contents, which should be sufficient to promote the fermentation and produce enough acid to preserve the silage. According to Ferreira (2002), the minimum soluble carbohydrates contents recommended to ensure adequate fermentation of good silage, varies between 6% and 12% of the dry mass. McDonald et al. (1991) found that, since the soluble sugar level is adequate, dry mass contents higher than 25% are sufficient to ensure a good silage production. The buffering capacity is another factor affecting the silage final product. It reflects the capacity to resist change in the pH values, determined by buffering substances, represented in plants by inorganic bases such as potassium (K) and calcium (Ca), protein, ammonia (N-NH₃), organic salts (malate, citrate).

Several factors affect the fermentation pattern and consequently the silage quality, including dry matter content, amount of soluble carbohydrates readily available and initial LAB population (Pereira et al., 2006). These inherent plant characteristics may vary according to species and maturity stage. Corn (*Zea mays* L.) and sorghum (*Sorghum bicolor* L. Moench), followed by millet (*Pennisetum glaucum*) and sunflower (*Helianthus annuus*) seems to be the most adapted species for silage due to the high soluble carbohydrates content, low buffering capacity, satisfactory dry matter productivity and quality of the silage produced. Although, sorghum silage nutritional value is considered lower than that of corn, it has shown an important role in forage production in Brazil and in the world as well, standing out as a resistant species to adverse environmental factors, such as drought stress (Miranda et al., 2010). This grass provides silage at low costs and the plant regrowth can be used (Rezende et al., 2011), because they keep the root system active.

As corn and sorghum have ideal characteristics for silage, a factor that drew the researcher's attention was the ideal harvest moment, considering the maturity stage and silage quality. Faria Júnior et al. (2011), working with the effect of seven grain maturity stages on the quality of sorghum BRS 610 silage, observed that the most appropriate stage for ensiling is the milk and soft dough stages, due to its higher silage fermentation quality and nutritional value.

Pearl millet silage presents high crude protein content as an intrinsic characteristic, when compared with corn and sorghum silage. Crude protein values varying from 8.51% to 10.68% were observed by Amaral et al. (2008). The storage system efficiency must not be defined only by the silage nutritional value, but also include the losses that occur from the plant harvest to the animal feeding (Neumann et al., 2007).

Sugarcane (*Saccharum officinarum* L.) is an important grass due to its tolerance to drought periods and high production potential of dry matter and soluble carbohydrates per hectare. The sugarcane silage confection has been unusual, being used more for animal feeding in its natural form, after cutting and chopping, but it can be recommended when desires to store the sugarcane in its higher nutritional value stage (the dry season) for use throughout the year (Molina et al., 2002). However, according to Santos et al. (2006), sugar cane silage becomes justifiable only when there is a surplus or when accidental burning of sugar cane fields happen, always taking into account the difficulty of achieving a good fermentation pattern due to intense alcoholic fermentation (8% to 17% of dry matter of ethanol) caused by yeast (Kung Jr. & Stanley, 1982), leading to losses of up to 30% of dry matter (Ferreira et al., 2007), accumulation of cell wall components and reduction in the *in vitro* dry matter digestibility. Furthermore, sugar cane silage has low aerobic stability, as result of high residual carbohydrate and lactic acid contents (McDonald et al., 1991). On the other hand, the adoption of the silage method represents a chance to keep the sugarcane nutritional value and allows better logistics for their manufacture and use, what implies the hand labor rationalization, concentrating the sugar cane harvest process in a particular time of year or time period, resulting in easier daily farm handling and maximizing the machinery use.

Thus, there has been a growing number of research projects, especially in Brazil, seeking additives that inhibit yeast growth in sugar cane silages (Valeriano et al., 2009). Nevertheless, some studies have shown that grasses can also be stored if they are ensiled at the ideal stage of development, or if the suitable additives are applied (Zanine et al., 2010).

Tropical weather grasses have high production in favorable seasons and a sharp decline in the less favorable ones. In this context, the surplus silage can be an option to increase the dry matter supply to the animals in unfavorable times. Such examples of tropical forages with a potential for silage are: *Brachiaria brizantha* (cv. Marandu), *Brachiaria decumbens* (cv. Basilisk), *Brachiaria humidicula*, *Panicum maximum* Jacq. (Cv. Colonião, Tobiatã, Tanzânia, Mombaça, Vencedor, Centauro, Massai), *Pennisetum purpureum* Schum. (Cv. Napier, Taiwan, Merker, Porto Rico, Cameroon, Mott), *Cynodon dactylon* (Tifton) and the hybrid of *Cynodon dactylon* x *C. nlemfuensis* (Coastcross). (Patrizi et al., 2004; Santos et al., 2006; Ribeiro et al., 2008; Oliveira et al., 2007; Zopollatto et al., 2009; Lopes & Evangelista, 2010). When compared to the others, elephant grass stands out in silages researches because of present high productivity and higher soluble carbohydrates concentration.

According to Evangelista et al. (2004), the tropical grasses present low dry matter contents, high buffering capacity and low soluble carbohydrates in growth stages in which they present good nutritive values, endangering the conservation through ensilage, once secondary fermentations are possible to occur. Bacteria from the *Clostridium* genus are favored by humid environments with high pH values and temperature. These bacteria are responsible for large losses because they produce CO_2 and butyric acid instead of lactic acid.

The grasses are colonized by a large number of LAB. In the most of the cases different species occur simultaneously in the same culture (Daeschel et al., 1987). According to Pahlow et al. (2003), in literature review studies, the species more commonly found in plants are *Lactobacillus plantarum, Lactobacillus casei, Pediococcus acidilactici, Enterococcus faecium.* Some heterofermentative lactic bacteria species can also be found in plants.

The lactic acid bacteria from the autochthonous microbiota are essential for the silage fermentation. However, no bacteria group varies as much as this one regarding number, with a detection limit of 10^1 to 10^5 CFU g^{-1} in alfalfa forage, 10^6 in perennial grasses and 10^7 in corn and sorghum (Pahlow et al., 2003).

The Table 1 shows contents of dry matter, crude protein, soluble carbohydrates and LAB number of mombaça grass (*Panicum maximum*) and *Brachiaria decumbens* with different regrowth ages. It is observed that in none of regrowth ages, neither grass showed dry matter content exceeding 30% and only the grasses cuted over 50 days after regrowth presented LAB population greater than 5 log CFU/g. On the other hand, there is a sharp drop in crude protein content with increasing regrowth age.

Signal grass (Brachiaria decumbens.)				
AGE (days)	DM (%)	CP (%)	SC (%)	LAB (log CFU/g)
30	20.99	9.65	2.62	3.93
40	21.23	6.97	2.92	4.81
50	21.94	5.86	3.13	5.37
60	22.35	5.30	2.73	5.32
70	23.67	4.37	2.53	5.51
Mombaça grass(Panicum maximum Jacq. cv. Mombaça)				
AGE (days)	DM (%)	CP (%)	SC (%)	LAB (log CFU/g)
30	17.75	7.43	3.34	4.35
40	19.63	7.30	4.12	4.56
50	21.50	6.47	4.18	5.16
60	23.38	4.94	5.43	5.55

Table 1. Dry matter (DM), crude protein (CP) and soluble carbohydrates (SC) and number of lactic acid bacteria (LAB) in signal grass and mombaça grass silage with different regrowth ages (Sousa et al., 2006).

Santos et al. (2011) studying the regrowth age influence in the LAB population observed that silages made with older plants presented LAB populations higher than the silages made with younger plants. According to Knicky (2005), it can be attributed to the increase in soluble carbohydrates and dry matter content, as well as to the decrease of anionic substances such as salts of organic acids, nitrate, sulfates, and so on. Pereira et al. (2005) found an increase in LAB population in elephant grass with the increase in regrowth age.

Meeske et al. (1999) found population of approximately 1 log CFU/g of fresh forage in *Digitaria eriantha.* Cai et al. (1998), analyzing Guinea grass (*Panicum maximum*) indigenous microbiota, found values lower than 3 log CFU/g of fresh forage. Pereira et al. (2007) reported initial LAB population of 4.92 log CFU/g in elephant grass plants.

Table 2 presents a data compilation of chemical composition and other parameters considered determinants of tropical grass silages quality, such as buffering capacity, soluble carbohydrates and pH values.

	Corn	Sorghum	Pearl millet	Sugar Cane	Elephant grass	Buffel grass	*Brachiaria brizantha*	*Brachiaria decumbens*
n*	6	6	6	7	5	4	6	6
DM	30.68	30.20	31.21	25.25	20.75	37.15	38.36	30.9
OM	96.91	92.79	90.9	97.45	90.91	90.60	92.89	92.25
CP	7.22	8.04	11.09	2.80	7.81	5.03	9.67	7.01
MM	5.81	4.45	9.1	2.68	9.53	9.92	5.29	7.53
EE	2.16	-	-	0.82	3.33	1.8	1.16	2.51
NDF	50.32	61.36	60.64	46.88	72.44	73.94	70.05	75.47
ADF	26.57	37.27	35.68	28.24	44.11	50.60	38.64	38.26
NFC	32.49	-	-	44.21	9.99	14.05	8.74	14.12
LIGNIN	4.72	6.2	4.24	4.72	6.24	8.4	4.67	5.9
IVDMD	59.19	52.87	-	53.87	60.90	37.4	58.77	51.61
pH	5.60	5.93	3.62	4.76	5.6	-	-	-
N-NH3	0.785	-	1.28	1.20	-	-	-	-
ETHANOL	-	-	-	2.12	-	-	-	-
YEASTS	5.30	-	-	2.71	-	-	-	-
BUFFERING CAPACITY	-	19.98	-	10.80	-	-	-	-
STARCH	21.31	-	-	5.50	-	-	-	-

Table 2. Chemical characterization of tropical grass used for silage. *Number of researches; DM = dry matter (%); OM = organic matter (%); CP = crude protein (%); EE = ether extract (%); NDF = neutral detergent fiber (%); NFC = non-fibrous carbohydrates (%); IVDMD = *in vitro* dry matter digestibility (%); N-NH3= ammonia nitrogen (% TN); ADF = acid detergent fiber (%); MM = mineral matter (%).

(Pariz, C.M. et al., 2011; Silva, T.C. et al., 2011; Viana, M.C.M. et al., 2011; Hu, W. et al., 2009; Martinez , J.C. et al., 2009; Valeriano, A.R, 2009; Benett, C.G.S. 2008; Reis, J.A.G. et al., 2008; Ribeiro, J.L. et al., 2008; Moreira, J.N. et al., 2007; Pedroso, A.F. et al., 2007; Velho, J.P. et al., 2007; Valadares Filho, S.C. et al., 2006; Velho, J.P. et al., 2006; Kollet, J.L. et al., 2006; Aroeira, L.J.M. et al., 2005; Bernardino, F.S. 2005; Moraes, E.H.B.K. et al., 2005; Santos, G.R.A. et al., 2005; Silva, A.V. et al., 2005; Patrizi, W.L. et al., 2004; Dairy, J. et al., 2003; Santos, M.V.F. et al., 2003; Landell, M.G.A. et al., 2002; Neumann, M. et al., 2002; Rodrigues, P.H.M. et al., 2002).

It is observed that tropical grasses have characteristics influenced by several factors, ranging from species choice to maturity stage at harvest. These factors are primordial in silage confection, because if handled properly, they will favor the LAB development, resulting in higher quality silage.

To understand how the factors related to the grass management will influence the LAB population dynamics consequently the fermentation, it is necessary to know the characteristics related to metabolism and the main tropical grass species.

3. Characteristics of lactic acid bacteria (LAB) present in tropical grasses

Lactic acid bacteria are gram-positive. They are negative catalase, do not present motility and do not produce spores. The final fermentation product is lactic acid, however, some groups produce considerable amount of CO_2, ethanol and other metabolites, these being called heterofermentative. Particularly, *Lactobacillus plantarum* are the larger silage fermentative bacteria (Ohmomo et al., 2002). *Lactococcus, Streptococcus* and *Enterococcus* are very important in the fermentation initial stage, because they keep an acidic environment, which then becomes, predominantly colonized by Lactobacillus.

Fermentation can be considered the anaerobic decomposition of organic compounds to organic products, which may be metabolized by the cells without the oxygen intervention. Under anaerobiosis conditions, phosphorylation occurs at the substrate level in which an organic acid donates electrons to a NAD^+, so that in microorganisms the NAD^+ needs to be regenerated and it occurs through various oxidation-reduction pathways, involving pyruvate or its derivatives, like acetyl-CoA. Pyruvate is a key molecule of fermenting microorganisms, from that, it can be formed by several compounds such as: acetaldehyde (ethanol), acetyl-CoA, lactate, acetoacetate (butyrate, isopropanol), acetoin (2, 3-butanediol, diacetyl), acetate, oxaloacetate, succinate, and propionate.

The homofermentative LAB are characterized by a faster fermentation rate, reduced proteolysis, higher lactic acid concentrations, lower acetic and butyric acids contents, lower ethanol content, and higher energy and dry matter recovery. Heterofermentative bacteria utilize pentoses as substrate for acetic and propionic acids production, which are effective at controlling fungi, at low pH values. The facultative heterofermentative use the same hexoses pathway of homofermentative, but they are able to ferment pentoses, as they have aldolase and fosfocetolase enzymes. The facultative heterofermentative may produce lactic and acetic acids when the substrate is a pentose, or lactic acid, ethanol and CO_2 when hexose is the substrate, due to the need of oxidation of two NAD molecules produced in the glycolytic pathway (White, 2000).

Table 3 summarizes the main lactic acid bacteria found in silages including some *Lactobacillus* with heterofermentative metabolism and some *Leuconostoc* species which have heterofermentative metabolism also.

For species of *Lactobacillus* genus were defined three groups based on the presence or absence of aldolase and fosfocetolase enzymes (Kandler and Weiss, 1986). These groups are as follows:

Lactobacillus		Enterococcus	Leuconostoc	Pediococcus
L. plantarum	L. brevis	E. faecalis	L. dextranicum	P. acidilactici
L. casei	L. buchneri	E. faecium	L. citrovorum	P. pentosaceus
L. curvatus	L. fermentum	E. lactis	L. mesenteroides	P. cerevisae
L. acidophilus	L. viridescens			

Table 3. Main lactic acid bacteria found in silages. (Woolford, 1984)

Group 1: Homofermentative, which ferment hexoses homolacticly almost exclusively to lactic acid (>85%), however, they are unable to ferment pentoses, due to the fosfocetolase enzyme lack;

Homofermentative Lactobacillus	
1A. *Lactobacillus delbrueckii* subsp. *Delbrueckii*	9. *L. helveticus*
1B. *Lactobacillus delbrueckii* subsp. *lactis*	10. *L. jensenii*
1C. *Lactobacillus delbrueckii* subsp. *bulgaricus*	11. *L. ruminis*
2. *L. acidophilus*	12. *L. salivarius*
3. *L. amylophilus*	13. *L. sharpeae*
4. *L. amylovorus*	14. *L. vitulinus*
5. *L. animalis*	15. *L. yamanashiensis*
6. *L. crispatus*	
7. *L. farciminis*	
8. *L. gasseri*	

Group 2: Facultative heterofermentative that use the same hexoses pathway as the one of group 1, but are able to ferment pentoses, since they have aldolase and fosfocetolase enzymes;

Facultative heterofermentative Lactobacillus	
16. *L. agilis*	20b. *L. coryniformis* subsp. *Torquens*
17. *L. alimentarius*	21. *L. curvatus*
18. *L. bavaricus*	22. *L. homohiochii*
19a. *L. casei* subsp. *Casei*	23. *L. maltaromicus*
19b. *L. casei* subsp. *pseudo-plantarum*	24. *L. murinus*
19c. *L. casei* subsp. *rhamnosus*	25. *L. plantarum*
19d. *L. casei* subsp. *tolerans*	26. *L. sake*
20a. *L. coryniformis* subsp. *coryniforms*	

Group 3: Obligately heterofermentative, which ferment hexoses, forming lactic acid, ethanol (or acetic acid) and CO_2, being able to still ferment pentose to form lactic and acetic acids.

Mandatory heterofermentative *Lactobacillus*	
27. *L. bifermentans*	36. *L. halotolerans*
28. *L. brevis*	37. *L. hilgardii*
29. *L. buchneri*	38. *L. kandleri*
30. *L. collinoides*	39. *L. kefir*
31. *L. confusus*	40. *L. minor*
32. *L. divergens*	41. *L. reuteri*
33. *L. fermentum*	42. *L. sanfrancisco*
34. *L. fructivorans*	43. *L. vaccinostercus*
35. *L. fructosus*	44. *L viridescens*

The homofermentative LAB presence in silage is extremely necessary. CO_2 generation results in carbon loss, ie, nutrient losses in plant materials. Therefore, homofermentative bacteria such as *Lactobacillus plantarum*, are desirable in the fermentation of silage.

Several lactic acid bacteria have antimicrobial peptides known as bacteriocins which are responsible for inhibiting the growth of or related species which have similar nutritional requirements. The bacteriocins action mechanism involves interaction with specific receptors on the cell membrane to its insertion resulting in proton-motive force dissipation and pores formation, which may cause cell viability loss (Montville and Chen, 1998; Ennahar et al., 2000).

According Lücke (2000), gram-negative bacteria are less susceptible to the action of bacteriocins from lactic acid bacteria due to the presence of outer membrane, which limits the access of peptides to the target site. In addition, the gram-negative bacteria are more sensitive to organic acid produced by LAB compared with the gram-positive bacteria (Ennahar et al., 2000).

Table 4 presents the lactic acid bacteria percentages isolated from sorghum plant in a study conducted by Tjandraatmadja et al. (1991). Likewise, *Lactobacillus plantarum* was the predominant specie and it kept 100 days after ensiling. It was observed the presence of *Lactobacillus fermentum* and *Lactobacillus brevis* heterofermentative bacteria in large quantities at the end of the ensiling process. It demonstrates that these bacteria are active during the fermentation process.

Evaluating the microbiological composition of silages obtained from three different grass species, Tjandraatmadja et al. (1994) found that *Lactobacillus plantarum* and *Pediococcus spp.* are the predominant species, observing one more time the presence of significant amounts of *Lactobacillus brevis* and *Lactobacillus fermentum* (Table 5). Santos et al. (2006) observed that

Lactobacillus plantrum was the predominant species in mombaça grass (*Panicum maximum*) and signal grass (*Brachiaria decumbens*).

Species	Days after ensiling			
	0	4	8	100
Lactobacillus plantarum	35	84	87	44
Leuconostoc spp.	59	0	0	0
Lactobacillus fermentum	6	6	4	7
Lactobacillus brevis	0	10	9	49

Table 4. Percentage of lactic acid bacteria species isolated from sorghum silage (Tjandraatmadja et al., 1991).

Species	Days after ensiling		
	P. maximum	*D. decumbens*	*S. sphacelata*
Lactobacillus plantarum	21	39	47
Lactobacillus coryneformis	6	21	0
Leuconostoc spp.	27	12	0
Enterococcus faeceium	0	10	4
Enterococcus faecalis	3	0	3
Pediococcus spp.	30	12	31
Lactobacillus brevis	7	6	11
Lactobacillus fermentum	6	0	4

Table 5. Main lactic acid bacteria (%) isolated from grasses (*Panicum maximum* cv Hami; *Digitaria decumbens*; *Setaria sphacelata* cv Kazungula) (Tjandraatmadja et al., 1994).

It is evident that *Lactobacillus plantarum* and the species from the *Pediococcus* genus are prevalent in forage plants. The species from *Leuconostoc* genus are present in plants. However, according to Chunjian et al. (1992) and Tjandraatmadja et al. (1991) they disappear early in the ensiling process.

According Lücke (2000), gram-negative bacteria are less susceptible to the action of bacteriocins from lactic acid bacteria due to the presence of outer membrane, which limits the access of peptides to the target site. In addition, the gram-negative bacteria are more sensitive to organic acid produced by LAB compared with the gram-positive bacteria (Ennahar et al., 2000).

Santos et al. (2011) conducted a study aiming to characterize and quantify microbial populations in signal grass harvested at different regrowth ages. The six lactic acid bacteria strains isolated from signal grass were characterized according Gram staining, catalase enzyme reaction, and bacilli form, submitted to growth and identification tests. The microbial isolates identification was performed by carbohydrates fermentation in API 50 CH kit (BioMéurix - France).

Regarding the predominant bacteria identification in signal grass plants, it is observed in Table 6 that all isolates had the form of short bacilli with rounded ends, arranged in pairs or in short chains (3-4 cells). All of them showed negative reaction to the catalase enzyme test and were gram-positive. None of the strains grew at pH 9.6 and 6.5% NaCl, but all grew at pH 7.2 and 4% NaCl at 45°C.

	Isolated strain						
	EB1	EB2	EB3	EB4	EB5	EB6	*Lactobacillus plantarum*
Test							
form	bacillus	bacillus	bacillus	bacillus	bacillus	bacillus	bacillus
Arranjement	DB*	DB	DB	DB	DB	DB	DB
Gram	+	+	+	+	+	+	+
Catalasis	-	-	-	-	-	-	-
Growth at different pH							
7,2	+	+	+	+	+	+	+
9,6	-	-	-	-	-	-	-
Growth at different salt concentartion (NaCl)							
NaCl 4%	+	+	+	+	+	+	+
NaCl 6,5%	-	-	-	-	-	-	-
Growth at different temperatures (T °C)							
15 °C	+	+	+	+	+	+	+
45 °C	+	+	+	+	+	+	+

Table 6. Morphology and biochemical characteristics of the isolates EB1, EB2, EB3, EB4, EB5, EB6, signal grass plant (*Brachiaria decumbens* cv. Basiliski). *DB: diplobacillus. (Santos et al., 2011).

According with the carbohydrate fermentation pattern (Table 7), the isolates EB1, EB2, EB5 e EB6 were identified as *Lactobacillus plantarum* with 99.9% of similarity.

The *Lactobacillus plantarum* specie, identified as dominant in signal grass plants (*Brachiaria decumbens* cv. Basiliski) (Santos et al., 2011) has been isolated and characterized as major species in several cultures. Lin et al. (1992) evaluated the corn and alfalfa autochthonous microbiota and found that from the total lactic acid bacteria isolated, over 90% were homofermentative lactic bacteria, being *Lactobacillus plantarum* the predominant specie. Tjandraatmadja et al. (1994), in studies on tropical grasses silage, found *Lactobacillus plantarum* and *Pediococcus spp.* as the predominant species.

	Isolated strain				Lactobacillus plantarum
	EB1	EB2	EB5	EB6	
Glycerol	-	-	-	-	-
Erythritol	(+)	(+)	(+)	(+)	-
D-arabinose	-	-	-	-	-
L-arabinose	+	+	+	+	+
Ribose	+	+	+	+	+
D-xylose	-	-	-	-	-
L-xylose	-	-	-	-	-
Adonitol	-	-	-	-	-
β-methyl D-xyloside	-	-	-	-	-
Galactose	+	+	+	+	+
D-glucose	+	+	+	+	+
D-frutose	+	+	+	+	+
D-mannose	+	+	+	+	+
L-sorbose	-	-	-	+	-
Rhamnose	(+)	(+)	(+)	(+)	-
Dulcitol	-	-	-	-	-
Inositol	-	-	-	-	-
Mannitol	+	+	+	+	+
Sorbitol	+	+	+	+	+
α-methyl D-mannose	-	-	-	-	+
α-methyl D-glycoside	-	-	-	-	-
N-acetyl-glucosamine	+	+	+	+	+
Amygdaline	+	+	+	+	+
Arbulin	+	+	+	+	+
Esculin	+	+	+	+	+
Salicin	+	+	+	+	+
Cellobiose	+	+	+	+	+
Maltose	+	+	+	+	+
Lactose	+	+	+	+	+
Melibiose	+	+	+	+	+
Saccharose	+	+	+	+	+
Trehalose	+	+	+	+	+
Inulin	-	-	-	-	-
Melezitose	+	+	+	+	+
D-raffinose	+	+	+	+	+
Amidon	-	-	-	-	-
Glycogene	-	-	-	-	-
Xylitol	-	-	-	-	-
β-gentibiose	+	+	+	+	+
D-turanose	+	+	+	+	+

| | Isolated strain | | | | Lactobacillus |
	EB1	EB2	EB5	EB6	plantarum
L-lyxose	-	-	-	-	-
D-tagatose	-	-	-	-	-
D-fucose	-	-	-	-	-
L-fucose	-	-	-	-	-
D-arabitol	(+)	(+)	(+)	(+)	-
L-arabitol	-	-	-	-	-
Gluconate	+	+	+	+	+
2 Cetogluconate	-	-	-	-	-
5 Cetogluconate	-	-	-	-	-

Table 7. Carbohydrate fermentation pattern of the isolates EB1, EB2, EB5, and EB6, signal grass plants (*Brachiaria decumbens* cv. Basiliski). + Intense fermentation, - no fermentation; (+) less intense fermentation (Santos et al., 2011).

In another study, Rocha (2003), evaluating the lactic acid bacteria populations in elephant grass plants cv. Cameroon (*Pennisetum purpureum* Schum) identified the isolates as *Lactobacillus casei ssp. Pseudoplantarum*, using the carbohydrate fermentation profile as an identification criterion. Santos et al. (2011) observed the *Lactobacillus plantarum* as LAB predominant specie in signal grass (*Brachiaria decumbens* Stapf). Based on the reported above, it is observed that there were differences between the LAB dominant species among the cultures evaluated, however *Lactobacillus plantarum* has been identified as the predominant specie for most plants.

4. Lactic acid bacteria and their effects on silage fermentation

A suitable acidification is essential for the silage successful preservation, especially when the crop moisture is relatively high, condition which favors the proliferation of spoilage microorganisms. The acidity prevents the development of spoilage microorganisms because they are less tolerant to the acidic conditions than lactic acid bacteria (Woolford, 1984; McDonald et al., 1991).

Among the fermentation stages, aerobic remains during the filling and some hours after the silage closing. The growth of aerobic microorganisms such as yeasts, fungi and bacteria, favored by high concentrations of oxygen (O_2) with the plant respiration process, promotes the O_2 reduction, initiating the active fermentation process. Thus, occurs a sharp drop in silage pH due to the formation of organic acids from sugars, in which initially actuate the heterofermentative bacteria and enterobacteriaceae, that becomes, then, dominated by homofermentative until the pH falls to below 5.0.

In the stability phase, when only the lactic acid bacteria are active, the anaerobic and acidic pH conditions preserve the silage until the opening time. When the silo is opened, it typically happen the molds and yeasts growth. The inhibition of the fungi multiplication through the contact with O_2 is called aerobic stability (Santos et al., 2006).

According to Ohmomo et al. (2002) in the early fermentation stage, Lactococus species, such as Lactococcus lactis, Enterococcus faecalis, Pediococcus acidilactici, Leuconostoc mesenteroides, and Lactobacillus species such as Lactobacillus plantarum, Lactobacillus cellobioses grow together with aerobic microorganisms like yeasts, molds and aerobic bacteria, due to the presence of air between the plant particles. At the same time, it is the plant respiration process. To promote the fermentation, an anaerobic environment is formed making the population to become predominantly composed by LAB, basically Lactococcus and Lactobacillus.

At the final fermentation stage, *Lactobacillus* becomes prevalent, due to their tolerance to the acidity. However, the silage LAB is pretty well diversified, depending on plant material properties, silage technology and silo type. The LAB predominance change from *Lactococcus* to *Lactobacillus* usually occurs in the final fermentation stage. According to Langston et.al (1960), these chemical changes is resulted from bacterial or plant enzymes action making the conversion of carbohydrates into other components such as gas and organic acids, as well as the partial protein breakage resulting in formation of non-protein structures.

The LAB use as microbial inoculants have been widely documented in research (Penteado et al., 2007; Ávila et al., 2009a; Ávila et al., 2009b; Jalč et al., 2009; Reich & Kung Jr., 2010).

Zopollatto et al. (2009) in a meta-analysis study (1999-2009) found a data limitation on the effect of microbial additives in silage quality. They observed that the number of conduced studies is not enough to provide conclusive positions regarding the effects of additives, emphasizing also the data scarcity in certain areas, such as dairy cattle performance. The results documented by these authors show that the magnitude of the response, especially on animal performance, is low. Thus, the justification for the use of additives should be evaluated considering the losses reduction in silage and the higher plant nutritional value preservation. Furthermore, they found that the response intensity varies with plant species and microorganism studied, suggesting a specificity between these components.

However, studies conducted in the 1980s and 1990s had already shown that the fermentation responses differ between strains of the same species (Wooflford & Sawczyc 1984, Hill, 1989; Fitzsimons et al., 1992). Hill (1989) found that inoculating corn silage with two *Lactobacillus plantarum* strains isolated from corn and grass, the dominant strain after ensiling was the isolated from corn. The same was observed for the grass silage, where the dominant lactic bacteria strain of were the one isolated from grass.

Many inconclusive results observed in silage fermentation studies may be related to this principle, which must have been overlooked. The specificity between the forage specie and its epiphytic microflora implicates in the need for studies related with isolation and identification of the main microorganism groups present in the forage used for silage. Ávila et al. (2009b) isolated *Lactobacillus buchneri* strains from sugar cane (*Saccharum officinarum* L.) and found that *L. buchneri* UFLA SIL 72 addition reduced the fungi population and the ethanol concentration in silages. Santos et al. (2007) observed reduction in ammonia concentration and enterobacteria population in mombaça grass silage (*Panicum maximum*) inoculated with *Lactobacillus plantarum*, which were isolated from the epiphytic microflora.

Thereby the silage inoculants can facilitate or accelerate the ensiling process, but they do not replace the fundamental factors (plant maturity, dry matter content, oxygen exclusion), which are essential for producing good quality silage. Among these factors the regrowth age is the one that influences all the silage characteristics, from fermentation to the nutritional value, considering the losses.

Meeske & Basson (1998) evaluated the effect of inoculant containing *Lactobacillus acidophilus*, *Lactobacillus delbruekii ssp. bulgaricus* and *Lactobacillus plantarum* on corn silage and found no inoculants effect on pH values and the lactic acid production. According to the authors, the high LAB concentrations present in the plant before ensiling led to such results. Furthermore, the amount of bacteria from *Clostridium* genus present in greater numbers in the treatment without inoculants had no effect on the protein content decrease of the untreated silage. It was not detected the butyric acid formation.

The high residual soluble carbohydrates content in silage, mainly the ones made of corn, sorghum and sugarcane, favors the aerobic deterioration process by fungi and yeasts, causing losses after the silo opening. However, the organic acids produced by fermentation, mainly acetic acid, have fungicidal effect and can mitigate the deterioration, increasing silage aerobic stability (Ranjit & Kung Jr. 2000; Kung Jr. & Ranjit, 2001). Therefore, inoculants containing heterofermentative LAB (e.g. *Lactobacillus buchneri*) have been used to increase the silage aerobic stability.

Ávila et al. (2009a) evaluated the aerobic stability of mombaça grass silage (*Panicum maximum* Jacq. cv. Mombaça) inoculated with two *Lactobacillus buchneri* strains, one provinient from a commercial inoculant and another isolated from sugarcane (*Saccharum officinarum* L.) silage. It was observed an increase in dry matter content after silo opening, while the carbohydrate ratio did not change due to the low residual concentration, characteristic of grass silage. The ammonia (NH_3) concentrations were above the 12% of the total-N recommended by Molina et al. (2002) for good quality silage, indicating high proteolysis during fermentation, due to low soluble carbohydrates supply, what makes possible a rapid decline of pH values.

Table 8 present few studies evaluating the effect of LAB on the silage fermentation. It is observed that there is a pattern of responses, as discussed previously, and its effect depends of the crop used, the microorganism strain and its concentration at the inoculation time. Although significant, the effects are of low magnitude, which leads to reflect about the use of inoculants without the microbiological principles and characteristics of forage plants knowledge.

Kleinschimit and Kung Jr. (2006), in a meta-analysis study (43 experiments), evaluated the *Lactobacillus buchneri* effect on fermentation and aerobic stability of corn, grasses and small grains silages. In general, the inoculation reduced pH, lactic acid concentration and mold counts. At the same time increases in acetic acid concentrations and aerobic stability were detected in all silage types. The increase in aerobic stability was more pronounced in corn silage. Furthermore, it was observed an increase in the propionic acid and ethanol concentrations, on the other hand decreases in soluble carbohydrates concentrations were found in grass and small grains silages. It was observed correlation between acetic acid concentration and fungi population reduction.

Crop	Microrganism	pH[1]	NH3[2]	LA[3]	AA[4]	PA[5]	BA[6]	ET[7]	AE[8]	DML[9]	DMR[10]
			% total N			% DM			hours		%
Grass	LP	--	--	++	--		--	ns	ns	--	++
Corn	LB	++	ns	--	++	ns		ns	+		ns
Grass		--	ns	--	++	++		++	+		--
Corn	PA/ LP	--		--	--	--	--				++
Wheat	LB	++	ns	--	++				++	++	
	LP	ns	--	++	ns				ns	--	
	LP/ LB	ns	--	ns	++				++	--	
Sorghum	LB	ns	ns	--	++				++	++	
	LP	ns	--	ns	ns				--	--	
	LP/ LB	ns	--	--	++				++	--	
Wheat	LP/EF	--		++	ns		ns		ns		
	L.Pe	--		++	ns		ns		ns		
Wheat	LB	ns		ns	++				+	++	
	LP	ns		--	++				+	--	
Sugar cane	LB	ns		ns	++	++	ns	--	+		
Sunflowerl	SF/ PA/ LP	ns	ns	ns	ns	ns	ns		ns	ns	
	LP/ L.	ns	ns	ns	ns	ns	ns		ns	ns	
	SF/ LP	--	--	ns	--	++	ns		ns	ns	
Potato +	LB	--	--	++	++	ns	--		++		
WB*	LPa/ LL/ PA	--	--	++	--	--	--		--		

Table 8. Effect of inoculants with lactic acid bacteria on the fermentation of the silage. *Potato by-product + 30% of wheat bran; [1]lactic acid, [2]acetic acid, [3]propionic acid, [4]butyric acid, [5]ethanol, [6]aerobic stability, [7]dry matter losses, [8]dry matter recovery. ns = not significant, + = numerical increase, - = decreasing numbers; + + = significant increase (P <0.05) / - = significant decrease (P <0.05). (Filya et al., 2000; Rodrigues et al., 2001; Weinberg et al., 2002; Filya, 2003; Kleinschimit & Kung Jr., 2006; Rowghani & Zamiri, 2009; Ávila et al., 2009b; Nkosi et al., 2010; Santos et al., 2011). LP = *Lactobacillus plantarum,* EF = *Enterococcus faecium,* LPe = *Lactobacillus pentosus,* SF = *Streptococcus faecium,* PA = *Pediococcus acidilacti, L = Lactobacillus sp.,* LB = *Lactobacillus buchneri;* Pac = *Propionibacterium acidipropionici;* LPar = *Lactobacillus paracasei paracasei* LL = *Lactococcus lactis.*

In concluded studies, the inoculation with *Lactobacillus buchneri* changed silages fermentation pattern, decreasing the lactate/acetate ratio, without compromising the processes efficiency, because the dry matter values recovery remained above 90%, as the minimum value recommended for this variable in these plants. The authors also suggest the existence of culture-specific effect.

Evaluating barley silage inoculated with *Lactobacillus buchneri,* Taylor et al. (2002) observed a decrease in yeasts and molds number, contrasting with an increase in aerobic stability. Changes in dry matter consumption and milk production were not affected.

The homofermentative LAB are used in order to improve the fermentation of the silage by increasing the concentration of lactic acid, which reduces the ammonia and the loss of dry matter. The heterofermentative LAB, for its turn, promote improvements, especially after the opening of the silo, increasing the aerobic stability of silage by inhibiting the growth of molds and yeasts. Thus, many research papers have recommended the use of inoculant combining the above two groups of LAB, due to its greater efficiency compared to the isolated use.

5. Use of additives and management practices aimed at the development of lactic bacteria in tropical grass silages

For an appropriate fermentation process with lactic acid predominance, it is necessary to provide ideal conditions for the LAB to develop and predominate in the silage environment. In order to attend these conditions it is used some additives, which can absorb moisture or provide soluble carbohydrates, making this way a more propitious environment to the LAB growth. Some management practices may also be employed with the same purpose.

The key point in the management of grass for silage is undoubtedly the harvest time. Grass harvested in advanced maturity stage present high LAB population, however high tissues lignification is an intrinsic characteristic also, what reduces its nutritional value. In contrast, young grasses have good nutritional value, however it also have unfavorable characteristics to the fermentation process, such as high humidity, low LAB population and high buffering capacity. In case of young grasses it can be used various additives. In case of mature grasses it can be settled a point in which the dry matter content and the LAB populations are suitable and the nutritive value is not compromised.

Research conducted with tropical grasses, evaluating the addition of a wide variety of additives, show that the increase in forage dry matter content or soluble carbohydrates supply favors lactic fermentation and, in most cases, reduces the silage losses. Among many, it has been used wheat bran, corn, fruit pulp and biodiesel industry by-products, sugar cane molasses and even tropical fruits such as jackfruit (Zanine et al., 2006; Pardo et al., 2008; Santos et al., 2008; Rêgo et al., 2010; Andrade & Melotti, 2004; Zanine et al., 2010; Silva et al., 2011). It is important to remind that these additives should be used respecting the level recommended by the authors, otherwise the effects can endanger the fermentative process.

Andrade & Melotti (2004) evaluated the effect of 20 additives on the silage quality made of elephant grass with 80 days (Tables 9 and 10).

In this study, it is observed that cotton fiber, sweeping residue, corn meal, elephant grass hay and guandu hay were used as additives, absorbing moisture (90.91% of dry matter) .The sweeping residue and molasses were used to supply carbohydrates (97.65%).

Looking at N-NH$_3$ results, it seems that the use of urea, cotton fiber, elephant grass hay, guandu hay, corn meal and molasses with urea, resulted in increased protein degradation during fermentation process. However, no changes were observed in the lactic acid concentration.

Treatment	DM	pH	N-NH₃	Lactic acid	Acetic acid	Butyric acid
	%		% total N		%DM	
Control (without aditive)	15.58f	4.15b	12.39d	2.40a	0.30b	0.00b
Urea 0.5 %	15.49f	5.36a	35.76abc	1.05a	1.81a	0.57a
Cotton fiber (10%)	23.25b	5.33a	36.07ab	1.8a	0.66b	1.73a
Elephant grass hay (10%)	25.88a	4.26b	25.63bcd	2.48a	0.46b	0.12b
Guandu hay(10%)	25.78a	4.21b	8.33d	1.38a	0.58b	0.14b
Drying for 6 hours	19.84cd	4.08b	15.17d	1.81a	0.30b	0.02b
Sugar waste (2%)	16.50de	4.09b	13.68d	4.69a	0.66b	0.00b
Corn Meal (2%)	16.90de	4.00b	13.68d	2.47a	0.28b	0.00b
Corn Meal (4%)	20.39c	4.00b	12.94d	4.96a	1.15a	0.08b
Corn Meal (6%)	21.60c	4.04b	12.01d	4.41a	0.33b	0.00b
Corn Meal (2%) / Urea (0.5%)	17.96de	4.19b	36.67ab	5.31a	0.53b	0.04b
Corn Meal (4%)/ Urea (0.5%)	20.26c	4.29b	49.36a	1.96a	0.85b	0.05b
Corn Meal (6%) / Urea (0.5%)	20.43c	4.20b	46.86a	2.25a	0.38b	0.01b
Dried Molasses (1%)	16.95de	4.04b	10.52d	3.60a	0.22b	0.00b
Dried Molasses (2%)	17.58de	3.92b	10.27d	3.29a	0.23b	0.00b
Dried Molasses (3%)	16.67de	3.89b	9.43d	3.98a	0.35b	0.00b
Dried Molasses (1%) Urea (0.5%)	17.20de	4.18b	34.93abc	1.25a	0.46b	0.04b
Dried Molasses (2%) Urea (0.5%)	18.20de	4.09b	32.43abc	5.24a	0.44b	0.04b
Dried Molasses (3%) Urea (0.5%)	17.55ed	3.97b	11.50d	4.84a	0.36b	0.00b
Biosilo inoculant	15.88f	4.06b	15.24d	2.61a	0.50b	0.03b
CV (%)	7.04	5.55	34.87	50.62	62.54	137.65

Table 9. Dry matter (DM) content and fermentation pattern of elephant grass, Napier, ensiled with different additives (Andrade & Melotti, 2004). DM = dry matter (%), CP = crude protein (% DM), N-NH₃ = ammonia nitrogen/total nitrogen (%), lactic acids, acetic and butyric acids: values in % of the silage DM. Equal means in column do not differ (P>0.05): CV = coefficient of variation.

The lowest in vitro dry matter digestibility was obtained with the use of guandu hay. On the other hand the highest one was obtained using corn meal and urea (Table 10). Compared to the control treatment, only the urea and cotton fiber had higher dry matter loss (11.0 and 10.5%, respectively).

According to the authors, it is not recommended the inclusion of urea, hay and cotton fiber in elephant grass silage. Additives rich in nonstructural carbohydrates, such as corn meal and molasses can be used, however, further studies are required to establish suitable levels for better fermentation. The microbial inoculant 'Biosilo' does not benefit the elephant grass silage.

Treatment	IVDMD (%DM)	DML (%)
Control (without aditive)	41.62abcde	6.80b
Urea 0.5 %	34.47abcde	11.00a
Cotton fiber (10%)	27.62de	10.50a
Elephant grass hay (10%)	34.12abcde	9.80b
Guandu hay(10%)	26.36e	7.00b
Drying for 6 hours	41.71abcde	6.70b
Sugar waste (2%)	42.89abcd	6.85b
Corn Meal (2%)	41.36abcde	6.70b
Corn Meal (4%)	45.68abc	7.20b
Corn Meal (6%)	41.81abcde	5.70b
Corn Meal (2%) /Urea (0.5%)	50.30ab	6.60b
Corn Meal (4%)/ Urea (0.5%)	51.31a	7.10b
Corn Meal (6%) /Urea (0.5%)	41.82abcde	7.10b
Dried Molasses (1%)	40.03abcde	6.80b
Dried Molasses (2%)	46.84abc	6.65b
Dried Molasses (3%)	45.25abc	6.80b
Dried Molasses (1%) Urea (0.5%)	43.73abc	6.90b
Dried Molasses (2%) Urea (0.5%)	47.15bc	7.10b
Dried Molasses (3%) Urea (0.5%)	49.65ab	6.85b
Biosilo inoculant	32.52de	7.00b
CV (%)	13.70	18.5

Table 10. *In vitro* dry matter digestibility (IVDMD) and dry matter losses (DML) of elephant grass, Napier, ensiled with different additives (Andrade and Melotti, 2004). Equal means in column do not differ (P>0.05), CV = coefficient of variation.

In more recent studies, evaluating the effect of four additives in sugar cane silage (sugarcane with 1.5% of urea; 0.5% of urea + 4% of corn; 0.5% of urea + 4% of dried cassava, 1.5% of starea and sugar cane control), Lopes & Evangelista (2010) concluded that the additive 0.5% urea + 4% corn, provides better results to the sugar cane silage.

Ávila et al. (2006), using combinations of different additives types (citrus pulp, wheat bran, and corn meal) with various doses (3, 6, 9 and 12%), found that Tanzania grass has low soluble carbohydrates contents and citrus pulp was the additive which contributed to increase the forage carbohydrate concentration and to reduce the buffering capacity. It provides an increase in the relation soluble carbohydrate x buffering capacity and better conditions for the fermentation process, resulting in better quality silages.

Besides the additives, some management practices from the harvest time to the silo sealing can influence the LAB development. When the grass is chopped at harvest time, the LAB population tends to increase due to reactivation of dormant and non-culturable cells. Thus, as faster the time between cutting the grass and sealing the silo, better will be the fermentation conditions.

The well done compaction and sealing is one of the secrets for good silage. It serves to expel the air from inside the forage mass, considering that air presence affects the fermentation process, implicating in losses caused by undesirable microorganisms. According to Senger et al. (2005) the original material must present compression level exceeding 650 kg/m^3 of green matter, reducing the quality losses of the ensiled material.

Furthermore, the particle size influences the compression and consequently the silo density. Igarasi (2002) observed an inverse relationship between particle size and silage density, suggesting that as smaller the particle size greater the density, and thus there will be more oxygen remaining among the plant particles.

Neumann et al. (2007) evaluating the effect of particle size (small: 0.2 to 0.6 cm or large: 1.0 to 2.0 cm) and cutting height of corn plants (low: 15 cm or higher: 39 cm) on silage fermentation dynamics and opening period, found that small sized particles provide greater compression efficiency and consequently reduces temperature and pH gradients in the silo opening time. The temperature differential between silage and environment is greater on the top, what is related with the time that the silo remain opened and exposed to the external environment and also the lower compression efficiency. It causes an increase in ammoniac nitrogen content and elevation of silage pH values, indicating changes in silage nutritional value.

The plant moisture content and the particle size after chopping are directly related to the compression. Excessively wet forage provides favorable conditions for butyric fermentation and, favors nutrients losses through leaching, and proteins degradation. On the other hand, forage with high dry matter content hinders compaction and air expulsion in the ensiling process. Amaral et al. (2007) found that increase in compression of 100 to 160 kg MS/m^3 increased effluent production from 2.2 to 9.8 kg/t of green matter.

Summarizing, as faster and more efficient the process of harvest, chopping, compaction and sealing, greater is the amount of LAB present in silage, and thus lower the losses.

6. Conclusions

The increase in lactic acid fermentation is a big challenge for tropical grass silages confection, determining the success of this technology. It is really important to know the species of lactic acid bacteria prevalent in tropical grasses as well as their metabolism in order to obtain maximum use with its utilization.

The use of lactic acid bacteria as microbial inoculants in tropical grasses silage still shows some inconsistency in the results obtained in research works. More research that evaluates their effects on the fermentation parameters, dry matter losses and mainly on the quality, regarding nutrient intake and animal performance is required.

However, tropical grass silages represent a promising technology for livestock in areas threatened by periodic droughts. Furthermore, in tropical countries like Brazil, this practice has been quite taken by the producers.

Author details

Edson Mauro Santos and Carlos Henrique Oliveira Macedo
Department of Animal Science, Federal University of Paraiba, Areia, PB, Brazil

Thiago Carvalho da Silva
Department of Animal Science, Federal University of Viçosa, Viçosa, MG, Brazil

Fleming Sena Campos
Department of Animal Science, Federal University of Bahia, Salvador, BA, Brazil

7. References

Amaral, R.C., Bernardes, T.F., Siqueira, G.R. & Reis, R.A. (2007). Características fermentativas e químicas de silagens de capim-marandu produzidas com quatro pressões de compactação. *Revista Brasileira Zootecnia*, Vol.36, No.3, (May 2007), pp. 532-539, ISSN 1806-9290

Amaral, P.N.C., Evangelista, A.R., Salvador, F.M. & Pinto, J.C. (2008). Qualidade e valor nutritivo da silagem de três cultivares de milheto. *Ciência agrotécnica*, Vol.32, No.2, (March 2007), pp. 611-617, ISSN 1981-1829

Andrade, S.J.T. & Melotti, L. (2004). Efeito de alguns tratamentos sobre a qualidade da silagem de capim-elefante cultivar Napier (*Pennisetum purpureum*, Schum). *Brazilian Journ al of Veterinary Research and Animal Science*, Vol.41, No.6, (February 2004), pp. 409-415, ISSN 1678-4456

Ávila, C.L.S., Pinto, J.C., Figueiredo, H.C.P., Morais, A.R., Pereira, O.G. & Schwan, R.F. (2009a) Estabilidade aeróbia de silagens de capim-mombaça tratadas com *Lactobacillus buchneri*. *Revista Brasileira de Zootecnia*, Vol. 38, No. 5 (May 2009), pp.779-787, ISSN 1806-9290

Ávila, C.L.S., Pinto, J.C., Tavares, V.B. & Santos, I.P.A. (2006). Avaliação dos conteúdos de carboidratos solúveis do capim-tanzânia ensilado com aditivos. *Revista Brasileira de Zootecnia*, Vol.35, No.3, (Abril 2006), pp. 648-654, ISSN 1806-9290

Aroeira, L.J.M.; Paciullo, D.S.C.; Lopes, F.C.F.; Morenz, M.J.F.; Saliba, E.S.; Silva, J.J. & Ducatti , C. (2005). Disponibilidade, composição bromatológica e consumo de matéria seca em pastagem consorciada de *Brachiaria decumbens* com *Stylosanthes guianensis*. *Pesquisa agropecuária brasileira*, Vol.40, No. 4, (April 2005), pp.413-418, ISSN 0100-204X

Benett, C.G.S.; Buzetti, S.; Silva, K.S.; Bergamaschine, A.F. & Fabricio, J.A. (2008). Produtividade e composição bromatológica do capim-Marandu fontes e doses de nitrogênio. *Ciência e agrotecnologia.*, Vol. 32, No. 5, (September/October 2008), pp. 1629-1636, ISSN 1413-7054

Bernardino, F.S.; Garcia, R.; Rocha, F.C.; Souza, A.L. & Pereira, O.G. (2005). Produção e características do efluente e composição bromatológica da silagem de capim-elefante contendo diferentes níveis de casca de café. *Revista Brasileira de Zootecnia*, Vol.34, No.6, (November/December.2005), pp.2185-2191, ISSN 1806-9290

Cai, Y., Benno Y. Ogawa, M., Ohmomo, S., Kumai, S. & Nakase, T. (1998). Influence of Lactobacillus spp. from an inoculant and of weissella and *Leuconostoc* spp from forage crops on silage fermentation. *Applied and Environmental Microbiology*, Vol. 64, No. 8 (August 1998), pp. 2982-2987, ISSN 1098-5336

Chunjian, L., Bolsen, K.K., Brent, B. E. & Fung, D. Y. C. (1992). Epiphytic lactic acid bacteria succession during the pre-ensiling periods of alfafa and maize. *Journal of Applied Bacteriology*, Vol. 73, No. 5 (November 1992), pp. 375-387, ISSN 1364-5072

Daeschel, M.A., Anderson, R.E. & Fleming, H.P. (1987). Microbial ecology of fermenting plant materials. *FEMS Microbiology Reviews*, Vol. 46, No. 3 (September 1987), pp. 357-367, ISSN 0168-6445

Ennahar, S., Sashihara, T., Sonomoto, K., Ishizaki, A. (2000). Class IIa bacteriocins: biosynthesis. Structure and activity. *FEMS Microbiology Reviews*, Vol. 24, No.1 (January 2000), pp.85-106, 2000 ISSN 0168-6445

Evangelista, A.R., Abreu, J.G., Amaral, P.N.C., Salvador, F.M. & Santana, R.A.V. (2004). Produção de silagem de capim-marandu (*Brachiaria brizantha* Stapf cv. Marandu). *Ciência agrotécnica*, Vol.28, No.2, (July 2003), pp. 443-449, ISSN 1413-7054

Faria Júnior, W.G., Gonçalves, L.C., Ribeiro Júnior, G.O., Carvalho W.T.V., Maurício, R.M., Rodrigues, J.A.S., Faria, W.G., Saliba, E.O.S., Rodriguez, N.M. & Borges, A.L.C.C. (2011). Effect of grain maturity stage on the quality of sorghum BRS-610 silages. *Arquivo Brasileiro de Medicina Veterinária e Zootecnia*, Vol.63, No.5, (June 2011), pp. 1215-1223, ISSN 1678-4162

Ferreira, D.A., Gonçalves, L.C., Molina, L.R., Castro Neto, A.G. & Tomich, T.R. (2007). Características de fermentação da silagem de cana-de-açúcar tratada com uréia, zeólita, inoculante bacteriano e inoculante bacteriano/enzimático. *Arquivo Brasileiro de Medicina Veterinária e Zootecnia*, Vol.59, No.2, (January 2007), pp. 423-433, ISSN 0102-0935

Filya, I. (2003). The effect of *lactobacillus buchneri* and *Lactobacillus plantarum* on the fermentation, aerobic stability and ruminal degradability of low dry matter corn and sorgum silage. *Journal of Dairy Science*, Vol. 86, No.11 (November 2003), pp. 3575-3581, ISSN 1525-3198

Filya, I, Ashbell, G., Hen, Y. & Weinberg, Z.G. (2000). The effect of bacterial inoculants on the fermentation and aerobic stability of whole crop wheat silage. *Animal Feed Science and Technology*, Vol. 88, No. 1-2 (November 2000), pp.39–46, 0377-8401

Fitzsimons, A., Duffner, F., Curtin, D., Brophy, G., O'Kiely, P. & O'Connel, M. (1992). Assessment of *Pediococcus acidilactici* as a potential silage inoculant. *Applied and Environmental Microbiology*, Vol. 58, No. 9 (September 1992), pp. 3047-3052, ISSN 1098-5336

Hu, W.; Schmidt, R.J.; Mcdonell, E.E.; Klingerman, C.M. & Kung Jr., L. (2009). The effect of *Lactobacillus buchneri* 40788 or *Lactobacillus plantarum* MTD-1 on the fermentation and aerobic stability of corn silages ensiled at two dry matter contents. *Journal of Dairy Science*, Vol. 92, No. 8, (August 2009), pp.526-535, ISSN 0022-0302

Hill, H. A. (1989). Microbial ecology of lactobacilli in silage. *Proceedings of the 2nd Forage Symposium*, Pioneer Hi-Bred International, Johnston, IA, pp. 47-64

Igarasi, M.S. (2002). Controle de perdas na ensilagem de capim Tanzânia (*Panicum maximum* Jacq. Cv. Tanzânia) sob os efeitos do teor de matéria seca, do tamanho de partícula, da estação do ano e da presença de inoculante microbiano. *Dissertação* (Mestrado em Ciência Animal e Pastagens) - Escola Superior de Agricultura Luiz de Queiroz, Piracicaba, pp. 152

Jalč1, D., Laukova, A., Simonova, M., Váradyová, Z. & Homolka, P. (2009). The use of bacterial inoculants for grass silage: their effects on nutrient composition and fermentation parameters in grass silages. *Czech Journal of Animal Science*, Vol. 54, No. 2 (February), pp.84-91

Kandler, O. & Weiss, N. (1986). Lactobacillus. In: *Bergey's manual of systematic bacteriology*. Sneath, P.H.A., Mair, N.S., Sharpe, M. E. &Holt, J. G. Baltimore: Williams and Wilkins.

Knicky, M. (2005). Possibilities to improve silage conservation. In: http://pub.epsilon.slu.se/834/1/Thesis_for_epsilon2.pdf. (Consultado em 21/11/2005).

Kleinschimit, D.H. & Kung Jr., L. (2006). A meta-analysis of the effects of Lactobacillus buchneri on the fermentation and aerovic stability of corn and grass and small-grains silages. *Journal of Dairy Science*, Vol. 89, No. 10 (October 2006), pp. 4005-4013, ISSN 1525-3198

Kollet, J.L.; Diogo, J. M.S. & Leite, G.G. (2006). Rendimento forrageiro e composição bromatológica de variedades de milheto (*Pennisetum glaucum* (L.) R. BR.). *Revista Brasileira de Zootecnia*, Vol.35, No.4, (July/August 2006), pp.1308-1315, ISSN 1806-9290.

Kung Jr., L. & Ranjit N.K. (2001).The effect of *Lactobacillus buchneri*and other additives on the fermentation and aerobic stability of barley silage. *Journal of Dairy Science*, Vol. 84, No. 5 (May 2001), pp.1149-1155, ISSN 1525-3198

Kung Jr., L., Stanley, R.W. (1982). Effect of stage of maturity on the nutritive value of whole-plant sugarcane preserved as silage. *Journal of Animal Science*, Vol. 54, No. 4 (April 1982), pp.689-696

Landell, M.G.A.; Campana, M.P.; Rodrigues, A.A. et al. (2002). A variedade IAC86-2480 como nova opção de cana-de-açúcar para fins forrageiros: manejo de produção de uso na alimentação animal. Campinas: Instituto Agronômico, 2002. 39p. (Série Tecnologia APTA, *boletim técnico* IAC; 193).

Langston, C.W. & BOUMA, C. (1960). A study of the microorganisms from grass silage. II. The lactobacilli. Applied Microbiology., Vol. 8, No. 4 (July 1960), 223-234, ISSN 1098-5336

Lin, C., Bolsen, K.K., Brent, B.E., Hart, R.A., Dickerson, J.T., Feyerherm, A.M., Aimutis, W.R. (1992). Epiphytic microflora on alfafa and whole-plant corn. *Journal of Dairy Science*, Vol. 75, No. 9 (September 1992), pp. 2484-2493, ISSN 1525-3198

Lopes, J. & Evangelista, A.R. (2010). Características bromatológicas, fermentativas e população de leveduras de silagens de cana-de-açúcar acrescidas de ureia e aditivos absorventes de umidade. *Revista Brasileira de Zootecnia*, Vol.39, No.5, (May 2009), pp. 984-991, ISSN 1806-9290

Lücke, F. K. (2000). Utilization of microbes to process and preserve meat. *Meat Science*, Vol. 56, No. 2 (October 2000), pp. 105-115, ISSN 0309-1740

Martinez , J.C. (2009). Efeito de *Lactobacillus* na Fermentação e Estabilidade Aeróbica de Silagem de Milho. *Revista inter rural*, Vol. 3, No. 27, (November 2009), pp. 21-22, ISSN 0103-9458

McDonald, P.J., Henderson, A.R. & Heron, S.J.E. (1991). *The biochemistry of silage* (2ª Ed.) Mallow Chalcombe Publications, ISBN 0948617225

Meeske, R., Basson, H.M. & Cruywagen, C. W. (1999). The effect of a lactic acid bacterial inoculant with enzymes on the fermentation dynamics, intake and digestibility of *Digitaria eriantha* silage. *Animal Feed Science Technolog*, Vol. 81, No. 3 (October 1999), pp.237-248, ISSN 0377-8401

Meeske, R., Basson, H. M. (1998). The effect of a lactic acid bacterial inoculant on maize silage. *Animal Feed Science Technology*, Vol. 70, No. 3 (February 1998), pp.239-274, ISSN 0377-8401

Miranda, N.O., Góes, G.B., Andrade Neto, R.A. & Lima, A.S. (2010). Sorgo forrageiro em sucessão a adubos verdes na região de Mossoró, RN. *Revista Brasileira de Ciências Agrárias*, Vol.5, No.2, (June 2010), pp. 202-206, ISSN 1981-0997

Molina, L. R., Ferreira, D. A., Gonçalves, L. C., Castro Neto, A. G. & Rodrigues, N. M. (2002). Padrão de fermentação da silagem de cana-de-açúcar (*Saccharum officinarum* L.) submetida a diferentes tratamentos. *Anais da Reunião da sociedade brasileira de zootecnia*, Recife- PE, July, 2002

Montville, T.J. & Chen, Y. (1998). Mechanistic action of pediocin and nisin: recent progress and unresolved questions. *Applied Microbiology Biotechnology*, Vol. 50, No. 5 (November 1998), pp. 511-519, ISSN 1432-0614

Moreira, J.N.; Lira, M.A.; Santos, M.V.F. & Araújo, G.G.L. (2007). Potencial de produção de capim buffel na época seca no semiárido pernambucano. *Revista Caatinga*, Vol.20, No.3, (July/September 2007), pp.22-29, ISSN 0100-316X

Moraes, E.H.B.K.; Paulino, M.F.; Zervoudakis, J.T.; Valadares Filho, S.C. & Moraes, K.A.K. (2005). Avaliação qualitativa da pastagem diferida de *Brachiaria decumbens* Stapf., sob pastejo, no período da seca, por intermédio de três métodos de amostragem. *Revista Brasileira de Zootecnia*, Vol.34, No.1, (January/February 2005), pp. 30-35, ISSN: 1806-9290

Neumann, M., Muhlbach, P.R.F., Nornberg, J.L., Ost, P.R. & Lustosa, S.B.C. (2007). Efeito do tamanho de partícula e da altura de corte de plantas de milho na dinâmica do processo fermentativo da silagem e no período de desensilagem. *Revista Brasileira de Zootecnia*, Vol.36, No.5, (March 2007), pp.1603-1613, ISSN 1806-9290

Neumann, M.; Restle, J.; Alves Filho, D.C.; Brondani, I.L.; Pellegrini, L.G. & Freitas, A.K. (2002). Avaliação do valor nutritivo da planta e da silagem de diferentes híbridos de sorgo (*Sorghum bicolor*, L. Moench). *Revista Brasileira de Zootecnia*, Vol.31, No.1, (January/February 2002), pp.293-301, ISSN 1806-9290

Nkosi, B.D., Meeske, R., van der Merwe, H.J., Groenewald, I.B. (2010). Effects of homofermentative and heterofermentative bacterial silage inoculants on potato hash silage fermentation and digestibility in rams. *Animal Feed Science and Technology*, Vol. 157, No. 3 (May 2010), p.195–200, ISSN 0377-8401

Ohmomo, S., Tanaka, O., KItamoto, H. K. & Cai, Y. (2002). Silage and microbial performance, old history but new problem. *JARQ*, Vol. 40, No. 2 (April 2002), pp. 59-71, ISSN

Oliveira, J.S., Santos, E.M., Zanine, A.M., Mantovani, H.C., Pereira, O.G. & Rosa, L.O. (2007). Populações microbianas e composição química de silagem de capim-mombaça (Panicum maximum) inoculado com *Streptococcus bovis* isolado de rúmen. *Archives of Veterinary Science*, Vol. 12, No. 2, pp.35-40, ISSN 1517-784X

Oliveira, J.S., Ferreira, R.P., Cruz, C.D., Pereira, A.V., Botrel, M.A., Von Pinho, R.G., Rodrigues, J.A.S., Lopes, F.C.F. & Miranda, J.E.C. (2002). Adaptabilidade e Estabilidade em Cultivares de Sorgo. *Revista Brasileira Zootecnia*, Vol.31, No.2, (August 2002), pp. 883-889, ISSN 1806-9290

Pariz, C.M.; Azenha, M.V.; Andreotti, M.; Araújo, F.C.M.; Ulian, N.A. & Bergamaschine, A.F. (2011). Produção e composição bromatológica de forrageiras em sistema de integração lavoura-pecuária em diferentes épocas de semeadura. *Pesquisa agropecuária brasileira*, Vol.46, No.10, (October 2011), pp. 1392-1400, ISSN 0102-0935

Pardo R.P., Castello Branco van Cleef, E H, da Silva Filho J C, Castro Neto, P, & Neiva Júnior, A.P. (2008). Diferentes níveis de torta de nabo forrageiro (*Raphanus sativus*) como aditivo na silagem de capim elefante. *Livestock Research for Rural Development*, Vol. 20, No. 10 (October 2008), ISSN 0121-3784

Patrizi, W.L.; Madruga Jr.; C.R.F.; Minetto, T.P.; Nogueira, E. & Morais, M.G. (2004). Efeito de aditivos biológicos comerciais na silagem de capim-elefante (*Pennisetum purpureum* Schum). *Arquivo Brasileiro de Medicina Veterinária e Zootecnia*, Vol.56, No. 3, (June. 2004), pp. 392-397, ISSN 0102-0935

Pahlow, G., Muck, R.E. & Driehuis, F. (2003). Microbiology of ensiling. In: Silage Science and Technology. Madison. *Proceedings…* Madison: ASCSSA-SSSA, Agronomy 42, pp.31-93

Pedroso, A.F.; Nussio, L.G.; Loures, D.R.S.; Paziani, S.F.; Igarasi, M.S.; Coelho, R.M.; Horii, J. & Rodrigues, A.A. (2007). Efeito do tratamento com aditivos químicos e inoculantes bacterianos nas perdas e na qualidade de silagens de cana-de-açúcar. *Revista Brasileira de Zootecnia*, Vol.36, No.3 (May/June 2007), pp.558-564, ISSN 1806-9290

Penteado, D. C. S., Santos, E. M., Carvalho, G. G. P., Oliveira, J. S., Zanine, A. M., Pereira, O. G. & Ferreira, C. L. L. F. (2007). Inoculação com Lactobacillus plantarum da microbiota em silagem de capimmombaça. *Archivos de Zootecnia*, Vol.56, No.214, (July 2007), pp. 191-202, ISSN 0004-0592

Pereira, O.G., Rocha, K.D. & Ferreira, C.L.LF. (2007). Composição química, caracterização e quantificação da população de microrganismos em capim-elefante cv. Cameroon (*Pennisetum purpureum*, Schum.) e suas silagens. *Revista Brasileira de Zootecnia*, Vol.36, No.6, (agosto 2007), pp. 1742-1750, ISSN 1806-9290

Pereira, O. G., Santos, E. M., Ferreira, C. L. L. F., Mantovani, H. C. & Penteado, D. C. S. (2006). Populações microbianas em silagem de capim-mombaça de diferentes idades de rebrotação. *Anais da XLIII Reunião anual da sociedade brasileira de zootecnia*. João Pessoa-PB, julho 2006

Pereira, O.G., Sousa, L.O. & Penteado, C.S. (2005). Populações microbianas, pH e relação nitrogênio amoniacal/N total em silagens de capim-elefante com diferentes idades de rebrotação. *Anais da XLII Reunião anual da sociedade brasileira de zootecnia*. UFG-Goiânia, July 2005

Ranjit, N. K. & Kung Jr, L. (2000). The effect of *Lactobacillus buchneri*, *Lactobacillus plantarum*, or a chemical preservative on the fermentation and aerobic stability of corn silage. *Journal of Dairy Science*. Vol.83, No.5, (agosto 2000), pp.526-535, ISSN 2131-2144

Reich, L.J. & Kung Jr, L. (2010). Effects of combining *Lactobacillus buchneri* 40788 with various lactic acid bacteria on the fermentation and aerobic stability of corn silage. *Animal Feed Science and Technology*. Vol.159, No.34, (october 2010), pp.105-109, ISSN 0377-8401

Reis, J.A.G.; Reis, W.; Macedo, V.P. & Sousa, M.M. (2008). Diferentes níveis de uréia adicionados à cana-de-açúcar (*Saccharum officinarum* L.) no momento de sua hidrólise alcalina. *PUBVET*, Vol.2, No.4, (January. 2008), pp. 1- 12, ISSN 1982-1263

Rezende, P. M., Alcantara, H.P., Passos, A.M.A., Carvalho, E.R., Baliza, D.P. & Oliveira, G.T.M. (2011). Rendimento forrageiro da rebrota do sorgo em sistema de produção consorciado com soja. *Revista Brasileira de Ciências Agrárias*, Vol.6, No.2, (April 2011), pp. 362-368, ISSN 1981-0997

Ribeiro, J.L.; Nussio, L.G.; Mourão, G.B.; Mari, L.J.; Zopollatto, M. & Paziani, S.F. (2008). Valor nutritivo de silagens de capim-Marandu submetidas ao efeito de umidade, inoculação bacteriana e estação do ano. *Revista Brasileira de Zootecnia*, Vol. 37, No.7, (July 2008), pp. 1176-1184, ISSN 1806-9290

Rocha, K. D. (2009). Silagens de capim-elefante cv. Cameroon, de milho e de sorgo produzidas com inoculantes ênzimo-bacterianos: populações microbianas, consumo e digestibilidade. *Dissertação* (Mestrado em zootecnia). Universidade Federal de Viçosa-MG, pp. 93

Rowghani, E. & Zamiri, M.J. (2009). The effects of a microbial inoculant and formic acid as silage additives on chemical composition, ruminal degradability and nutrient digestibility of corn silage in sheep. *Iranian Journal of Veterinary Research*. Vol.10, No.2, (September 2009), pp.110-118, ISSN 1728-1997

Rodrigues, P.H.M.; Senatore, A.L.; Andrade, S.J.T.; Ruzante, J.M.; Lucci, C.S. & Lima, F.R. (2002). Efeito da adição de inoculantes microbianos sobre a composição bromatológica e perfil fermentativo da silagem de sorgo produzida em silos experimentais. *Revista Brasileira de Zootecnia*, Vol.31, No.6, (November/Decemper 2002), pp. 2373-2379, ISSN 1806-9290

Santos, E. M., Pereira, O. G., Rasmo, G., Ferreira, C. L. L. F. ; Oliveira, J.S., Silva, T.C. & Rosa, L.O. (2011). Microbial populations, fermentation profile and chemical composition of signalgrass harvsted of different rgrowth ages. *Revista Brasileira de Zootecnia*. Vol.40, No.4, (October 2010), pp.747-755, ISSN 1806-9290

Santos, E.M., Zanine, A.M., Dantas, P.A.S., Dórea, J.R.R., Silva, T.C., Pereira, O.G., Lana, R.P. & Costa, R.G. (2008). Composição bromatológica, perdas e perfil fermentativo de silagens de capim-elefante com níveis de jaca. *Revista Brasileira de Saúde e Produção Animal*, Vol.9, No.1, (March 2008), pp. 71-80, ISSN 1519-9940

Santos, E.M. (2007). Populações microbianas e perfil fermentativo em silagens de capins tropicais e desempenho de bovinos de corte alimentados com dietas contendo silagens de capim Mombaça. *Tese* (Doutorado) Universidade Federal de Viçosa. pp. 126

Santos, E.M., Zanine, A.M. & Oliveira, J.S. (2006). Produção de silagem de gramíneas tropicais. *Revista Electrónica de Veterinaria*, Vol.7, No.7, (July 2006), pp., ISSN 1695-7504

Santos, E. M., Pereira, O. G., Ferreira, C. L. L. F., Mantovani, H. C., Penteado, D. C. S., Oliveira, J.S. & Sousa, L.O. (2006). Isolamento, identificação e caracterização de *Lactobacillus* predominantes em gramíneas tropicais. *Anais da XLIII Reunião anual da sociedade brasileira de zootecnia*. João Pessoa-PB, July 2006

Santos, G.R.A.; Guim, A.; Santos, M.V.F.; Ferreira, M.A.; Lira, M.A.; Dubeux Jr, J.C.B. & Silva, M.J. (2005). Caracterização do pasto de capim-buffel diferido e da dieta de bovinos, durante o período seco no sertão de Pernambuco. *Revista Brasileira de Zootecnia*, Vol.34, No.2, (March/April 2005), pp.454-463, ISSN 1806-9290

Santos, M.V.F.; Dubeux Jr, J.C.B.; Silva, M.C.; Santos, S.F.; Ferreira, R.L.C.; Mello, A.C.L.; Farias, I. & Freitas, E.V. (2003). Produtividade e composição química de gramíneas tropicais na Zona da Mata de Pernambuco. *Revista Brasileira de Zootecnia*, Vol.32, No.4, (July/August 2003), pp.821-827, ISSN 1806-9290

Senger, C.C.D., Muhlbach, P.R.F., Sánchez, L.M.B., Peres Netto, D. & Lima, L.D (2005). Composição química e digestibilidade 'in vitro' de silagens de milho com distintos teores de umidade e níveis de compactação. *Ciência Rural*, Vol.35, N°.6, (December 2005), pp. 1393-1399, ISSN 0103-8478

Silva, T.C.; Edvan, R.L.; Macedo, C.H.O.; Santos, E.M.; Silva, D.S. & Andrade, A.P. (2011). Características morfológicas e composição bromatológica do capim buffel sob diferentes alturas de corte e resíduo. *Revista Trópica*. Vol. 5, No. 2, (July 2011), pp. 30-39, ISSN 1982-4831

Silva, T.C., Dorea, J. R.R., Dantas, P.A.S., Santos, E.M., Zanine, A.M. & Pereira, O.G. (2011). Populações microbianas, perfil fermentativo e composição bromatológica de silagens de capim-elefante com níveis de jaca. *Archivos de Zootecnia*. Vol.60, No.4, (Setember 2009), pp.247-255, ISSN 0004-0592

Silva, A.V.; Pereira, O.G.; Garcia, R.; Valadares Filho, S.C.; Cecon, P.R. & Ferreira, C.L.L.F. (2005). Composição bromatológica e digestibilidade *in Vitro* da matéria seca de silagens de milho e sorgo tratadas com inoculantes microbianos. *Revista Brasileira de Zootecnia*, Vol.34, No.6, (November/December 2005), pp.1881-1890, ISSN 1806-9290

Sousa, L.O., Santos, E.M., Penteado, D.C.S., Pereira, O.G., Carvalho, G.G.P. & Oliveira, J.S. (2006). Composição bromatológica de silagem de capim-mombaça inoculada com *lactobacilus plantarum* da microbiota epifítica. *Anais do VI Congresso nacional de zootecnia – Zootec*. Recife-PE, November, 2006

Taylor, C.C., Ranjit, N.J. & Mills, J.A. (2002). The effect of treating whole-plant barley with *Lactobacillus buchneri* 40788 on silage fermentation, aerobic stability, and nutritive value for dairy cows. *Journal of Dairy Science*, Vol.85, No.18, (November 2002), pp. 1793-1800, ISSN 1839–1854

Tjandraatmadja, M., Norton, B.W. & Macrae, I.C. (1994). Ensilage characteristics of three tropical grasses as influenced by stage of growth and addition of molasses. *World Journal of Microbiology and Biotechnology.* Vol.10, No.9, (August 1994), pp.74-81, ISSN 0959-3993

Tjandraatmadja, M., Norton, B. W. & Macrae, I. C. (1991). Fermentation patterns of forage sorghum ensiled under different environmental conditions. *World Journal of Microbiology and Biotechnology.* Vol.7, No.4, (July 1991), pp.206-218, ISSN 0959-3993

Valadares Filho, S.C.; Magalhães, K.A. & Rocha Jr, V.R. (2006). *Tabelas brasileiras de composição de alimentos para bovinos.* 2.ed. UFV, ISBN 859060413-6,Viçosa, Minas Gerais

Valeriano, A.R.; Pinto, J.C.; Ávila, C.L.S.; Evangelista, A.R.; Tavares, V.B. & Schwan, R.F. (2009). Efeito da adição de *Lactobacillus* sp. na ensilagem da cana-de-açúcar. *Revista Brasileira de Zootecnia,* Vol.38, No.6, (June 2009), pp. 1009-1017, ISSN 1806-9290

Velho, J.P.; Mühlbach, P.R.F.; Genro, T.C.M.; Velho, I.M.P.H.; Nörnberg, J.L.; Orqis, M.G. & Kessler, J.D. (2006). Alterações bromatológicas nas frações dos carboidratos de silagens de milho "safrinha" sob diferentes tempos de exposição ao ar antes da ensilagem. *Revista Brasileira de Zootecnia,* Vol.35, No.4, (July/August 2006), pp.1621-1628, ISSN 1806-9290

Velho, J.P.; Mühlbach, P.R.F.; Nörnberg, J.L.; Velho, I.M.P.H.; Genro, T.C.M. & Kessler, J.D. (2007). Composição bromatológica de silagens de milho produzidas com diferentes densidades de compactação. *Revista Brasileira de Zootecnia,* Vol.36 No.5, (September/October 2007), pp. 1532-1538, ISSN 1806-9290

Viana, M.C.M.; Freire, F.M.; Ferreira, J.J.; Macêdo, G.A.R.; Cantarutti, R.B. & Mascarenhas, M.H.T. (2011). Adubação nitrogenada na produção e composição química do capim Braquiária sob pastejo rotacionado. *Revista Brasileira de Zootecnia*, Vol.40, No.7, (July 2011), pp.1497-1503, ISSN 1806-9290

Zanine, A.M., Santos, E.M., Dorea, J.R.R., Dantas, P.A.S., Silva, T.C. & Pereira, O. G. (2010). Evaluation of elephant grass with adition of cassava scrapings. *Revista Brasileira de Zootecnia.* Vol.39, No.12, (April 2010), pp.2611-2616, ISSN 1806-9290

Zanine, A.M.; Santos, E.D.; Ferreira, D.J., Pereira, O.G., & Almeida, J.C.C. (2006) Efeito do farelo de trigo sobre as perdas, recuperação da matéria seca e composição bromatológica da silagem de capim mombaça. *Brazilian Journal of Veterinary Research and Animal Science.* Vol.53, No.6, (February 2006), pp.803-809, ISSN 1413-9596

Zhang, T., LI, L., Wang, X., Zeng, Z., Hu, Y. & Cui, Z. (2009). Effects of Lactobacillus buchneri and Lactobacillus plantarum on fermentation, aerobic stability, bacteria diversity and ruminal degradability of alfalfa silage. *World Journal of Microbiology and Biotechnology.* Vol.25, (January 2009), pp.965-971, ISSN 0959-3993

Zopollatto, M., Daniel, J.L.P. & Nussio, L.G. (2009). Aditivos microbiológicos em silagens no Brasil: revisão dos aspectos da ensilagem e do desempenho de animais. *Revista Brasileira de Zootecnia*. Vol.38, No.spe, (February 2006), pp.170-189, ISSN 1806-9290

Weinberg, z.g., Ashbell, G., Hen, Y., Azrieli, A., Szakacs, G. & Filya, I. (2002). Ensiling whole-crop wheat and corn in large containers with *Lactobacillus plantarum* and *Lactobacillus buchneri*. *Journal of Industrial Microbiology & Biotechnology*. Vol.28, No.19, (February 2002), pp.7-11, ISSN 1476-5535

White, D. (2000). *The physiology and biochemistry of prokaryotes* (2). Oxford University Press, ISBN 0195125797, USA

Woolford, M. K. (1984). *The silage fermentation*. Marcel Dekker, ISBN 0824770390, New York

Fish & Seafood Products

Selection of *Lactobacillus* Species from Intestinal Microbiota of Fish for Their Potential Use as Biopreservatives

Mahdi Ghanbari, Masoud Rezaei and Mansoureh Jami

Additional information is available at the end of the chapter

1. Introduction

Despite recent advances in seafood production, seafood safety is still an important public health issue. It is clear that indigenous bacteria present in marine environment as well as resulting from post contamination during processing are responsible for many cases of illnesses [1-3]. In the last years, traditional processes applied to seafood like salting, smoking and canning have decreased in favor of mild technologies involving lower salt content, lower heating temperature and vacuum (VP) or modified atmosphere packing (MAP, 3-5). Most of these treatments are usually not sufficient to destroy microorganisms and in some cases psychrotolerant pathogenic such as *Listeria monocytogenes* or spoilage causing bacteria can develop during prolonged shelf-life of these products [2,5,6]. As several of these products are eaten raw, it is therefore essential that adequate precautious and preservation technologies are applied to maintain their safety and quality. Among alternative preservation technologies, particular attention has been paid to biopreservation to extend the shelf-life and to enhance the hygienic quality of perishable food products such as seafood, thereby minimizing the impact on nutritional and organoleptic properties [1,7,8]. In this context, lactic acid bacteria (LAB) possess a major potential in biopreservation strategies, since they are safe to consume, and during storage they naturally dominate the microbiota of many foods [7-11]. Lactic acid bacteria are gram-positive, non-sporulating and catalase negative rods or cocci that ferment various carbohydrates mainly to lactate and acetate [12]. Accordingly, they are commonly associated with nutritious environments like foods, decaying material and the mucosal surfaces of the gastrointestinal and urogenital tract [12- 14], where they enhance the host protection against pathogens [13]. Their antagonistic and inhibitory properties are due to the competition for nutrients and the production of one or more antimicrobially active metabolites such as organic acids (lactic and acetic acid), hydrogen peroxide, and antimicrobial peptides like bacteriocins [8-11,15-17]. Bacteriocins are ribosomally synthesized peptides that exert their antimicrobial activity against either strains of the same species as the bacteriocin producer (narrow range), or to more distantly related species (broad range) [7,15,18]. An important reason for research on LAB based bacteriocins is due to their activity at nanomolar concentrations against number of bacterial pathogens [1,3,5,6,19,20]. Some bacteriocins even exhibit their activities against multidrug-resistant nosocomial pathogens such as methicillin-resistant *Staphylococcus aureus* (MRSA) and vancomycin-resistant enterococci [VRE, 17, 21]. Thus they also may have some big potential in medical and veterinary applications. Fermented food and plant material have been a well-known source for bacteriocin-producing LAB, but isolates from the intestinal of animals and humans has become an increasingly important source for such strains due to an increased awareness of their importance as probiotics. In fish the presence of LAB is meanwhile well documented and the bio-protective potential of some strains and/or their bacteriocin has been highlighted in the last years [4-6,16,18,22-26]. Kvasnikov et al. [12] described the presence of lactic acid bacteria, including *Lactobacillus* in the intestines of various fish species at larval, fry and fingerling stages inhabiting ponds in Ukraine. They give information on the changes in their composition as a function of the season of the year and life-stage of the fish. However, it was discussed that some human activities like artificial feeding in ponds would have had an effect on the bacterial composition and load in some fish, like carp (*Cyprinus carpio*) which showed the highest content of lactic acid bacteria in the intestines. Cai et al. [27] described the lactic acid bacteria in *Cyprinus carpio* collected from the Thajin river in Thailand. They reported the presence of *Enterococcus* spp. and the dominance of *Lactococcus garviae*, an emerging zoonotic pathogen, in *Cyprinus carpio*. Bucio Galindo et al. [23] studied the distribution of lactobacilli in the intestinal content of river fish and reported that various species of lactobacilli were present in relatively high numbers in the intestines of edible freshwater fish from the river, especially in warm season but in low numbers in cold season. There are no reports on the presence of *Lactobacillus* in the intestines of sturgeon fish inhabiting Caspian Sea, whereas other groups of bacteria have been studied in more details. In comparison with other food products of dairy or meat origin, only few bacteriocinogenic LAB strains have been recovered from seafood. The present study focuses on the characterization of antimicrobial compounds produced by the lactobacilli isolates, in addition, their ability to inhibit the growth of relevant food borne pathogens as well as of spoilage bacteria and last but not least, of contaminants in aquaculture.

2. Materials and methods

2.1. Fish intestine samples

Two species of Persian sturgeon (*Acipenser persicus*) and Beluga (*Huso huso*) were collected from the south coast of Caspian Sea in Iran. Twenty two individuals of these fish in adult stage were selected. The weight and length of the fish were measured before dissection. The fish were sacrificed by physical destruction of the brain, and the number of incidental organisms was reduced by washing the fish skin with 70% ethanol. Then, the ventral surface was opened with sterile scissors. After dissecting the fish, 1 g of the intestinal tract content of each fish was removed under aseptic condition and placed into previously weighed flasks containing storage medium.

2.2. Media and culture condition

Intestinal content was homogenized in a storage medium using a vortex mixer. One milliliter was transferred to reduced neutralized bacterial peptone (NBP, Oxoid L34, Hampshire, England) 0.5 g/L, NaCl 8 g/L, cysteine.HCl 0.5 g/L, pH adjusted to 6.7 [29]. Afterwards serial dilutions were spread on plates of selective media and incubated at the following conditions. Columbia blood agar (CAB, Oxoid CM 331) was used as a selective medium to make an estimation of the cultivable total anaerobic counts [29]. All the inoculated plates were incubated anaerobically at 30°C for 48 h. The following two media were used to isolate lactic acid bacteria (LAB). MRS (MRS, Merck, Darmstadt, Germany) with 1.5% agar (M641, HiMedia, Mumbai, India) and pH adjusted to 4.2 (MRS 4.2) and incubated anaerobically at 30°C for 96 h was used as a selective medium for lactic acid bacteria. MRS is an inhibitory medium for *Carnobacterium*. Anaerobic MRS with Vancomycin and Bromocresol green (LAMVAB), incubated at 30°C for 96 h was used as an elective and selective medium for *Lactobacillus* spp. [30]. Anaerobic incubation of the three media was made in an anaerobic Gas-Pack system (LE002, HiMedia, Mumbai, India) with a mixture of 80% N_2, 10% H_2 and 10% CO_2. Colonies were selected either randomly, or in case of less than 10 colonies per each plate, all the samples were counted according to the method described by Thapa et al. [31]. Purity of the isolates was checked again by streaking them onto fresh agar plates of the isolation media, followed by microscopic examinations. Identified strains of lactobacilli were kept in MRS broth with 15% (v/v) glycerol at -20°C.

2.3. Characterization procedures for lactic acid bacteria

Eighty four strains were randomly selected for identification procedures based on the phenotypical characteristics. Cell morphology and motility of all isolates were observed using a phase contrast microscope (CH3-BH-PC, Olympus, Japan). Isolates were gram-stained and tested for catalase production test. Preliminary identification and grouping was based on the cell morphology and phenotypic properties such as CO_2 production from glucose, hydrolysis of arginine, growth at different temperatures (10, 15 and 45°C), and at different pH (3.9 and 9.6). As well as the ability to grow in different concentrations of NaCl (6.5% (w/v), 10% (w/v) and 18% (w/v)) in MRS broth was checked as well. The configuration of lactic acid produced from glucose was determined enzymatically using d-lactate and l-lactate dehydrogenase test kits (Roche Diagnostic, France). The presence of diaminopimelic acid (DAP) in the cell walls of LAB was determined using thin-chromatography on cellulose plates. Fermentation of carbohydrates was determined using API 50 CHL (API 50 CH is a standardized system, associating 50 biochemical tests for the study of carbohydrate metabolism in microorganisms. API 50 CH is used in conjunction with API 50 CHL Medium for the identification of *Lactobacillus* and related genera) strips according to the manufacturer's instructions (Biomerieux, Marcy l' Etoile, France). The APILAB PLUS database identification software (bioMe'rieux, France) was used to interpret the results. Identification was undertaken according to the method described by Kandler and Weiss [12] and Hammes and Vogel [32].

2.4. Statistical analysis

Statistical analysis using Student's t-test (SPSS, Version 11.0) was performed to find significant difference on lactobacilli count between LAMVAB and MRS 4.2. Pearson's correlation coefficient was used to investigate the correlation of lactobacilli count between LAMVAB and MRS 4.2 (SPSS Inc., Version 11.0, Chicago, USA). A significance level of $p < 0.05$ was used.

2.5. Screening of *Lactobacillus* strains for their inhibitory potential

In a first test series, the ability of each of the *Lactobacillus* isolates to exert an antibacterial effect against *Listeria monocytogenes* ATCC 19115 and *Salmonella* Typhimurium PTCC 1186 were examined by using three methods: the spot-on-lawn method, standardized agar disk diffusion method and the well diffusion method as described by Schillinger and Lucke [33], Benkerroum et al. [34] and Tagg & Mc Given [35]. Throughout, cell-free supernatants (CFS) of strains were obtained by centrifugation at 10,000 ×g for 20 min and then adjusted to pH 6.5 by applying NaOH (to exclude the effect of organic acid) before sterilization by filter (0.2 μm, Sigma, UK). Based on the screening tests, the inhibitory spectrum of potential bacteriocin-producing isolates was assessed against 42 indicator strains using a standardized agar disk diffusion test. The strains were kept frozen in 20% (v/v) glycerol at -20°C. For this purpose, an aliquot of 20 ml CFS was applied on disks (6 mm) and set on agar plates previously inoculated with each individual indicator strain suspension, which corresponded to a 10^5 CFU/ml. Plates were incubated 24 h at optimum temperatures of the test organism. Antimicrobial activity was detected as a translucent halo in the bacterial lawn surrounding the disks.

2.6. Characterization of the inhibitory effect

In order to determine the biological nature of the antimicrobial activity of bacteria, CFS (pH 6.0) of 24-h lactobacilli cultures of two selected isolates (*Lactobacillus casei* AP8 & *Lactobacillus plantarum* H5) incubated at 30°C, were tested for their sensitivity to the proteolytic enzymes. One ml of CFS was treated for 2 h with 1 mg ml^{-1} final concentration of the following enzymes: papain, trypsin, proteinase K, pronase E and α-amylase (Sigma, London). To clarify whether the antimicrobial activity detected derives from the production of hydrogen peroxide, 2600 IU/ml of catalase (Sigma, London) were added to 1 ml portions of extracellular extracts of LAB exhibiting antimicrobial activity and incubated for 24 h at ambient temperature. Chemicals were added to the CFS and the samples incubated for 5 h before being tested for antimicrobial activity. To determine the sensitivity of potential bacteriocin activities to the temperatures, samples of CFS were incubated under defined conditions. The effect of pH on bacteriocin activity was determined by adjusting the pH of the CFS (cell free supernatant (pH 6.5) of 24-h lactobacilli cultures incubated at 30°C) with diluted appropriate volumes of HCl and NaOH (Table 3). After incubating for 2 h, the pH of the samples was readjusted to 6.5 followed by sterilization (0.2 μm, Sigma, UK). In all cases, the remaining bacteriocin activity was assessed exemplarily by using strain *L. monocytogenes* ATCC 19115 as the indicator bacterium and by applying the agar disk diffusion plate bioassay. Untreated cell-free supernatants were used as controls and experiments were performed in duplicate.

2.7. Growth dynamics and antimicrobial compounds production

The time course of inhibitory substance production was performed by inoculating 10 mL of an overnight culture of selected *Lactobacillus* isolates into 100 mL of MRS broth followed by incubation at 30°C. Cells were subsequently removed by centrifugation at 10,000 ×g for 20 min. At appropriate intervals, changes in pH and optical density (600 nm) of the cultures were measured to monitor bacterial growth using a spectrophotometer (Hitachi U 1100, Tokyo, Japan). Antibacterial activity was evaluated every hour by using serial twofold dilutions of each culture used as a neutralized cell-free supernatant (CFS) tested against *L. monocytogenes* ATCC 19115 based on the agar disk diffusion plate bioassay. In a separate experiment, the inhibitory effect of CFSs of lactobacilli strains on target cells in liquid medium was also examined against *L. monocytogenes* ATCC 19115 as indicator strain. For this purpose, 20 mL of each filter-sterilized bacteriocin-containing cell-free supernatant were added to a 100 mL culture of the indicator organism at early exponential phase (4 h old). These experiments were also repeated with stationary-phase cells. The optical density at 600 nm and viable cell count were determined every hour during an observation period of 20 h. Indicator cells without CFSs were used as control.

2.8. Adsorption of bacteriocin to producer cells

Bacteriocin-producing cells were cultured for 18 h at 30 °C. The pH of the cultures was adjusted to 6.0 with 1 M NaOH to allow maximal adsorption of the bacteriocin to the producer cells, according to the method described by Yang et al. [36]. The cells were then harvested (10,000 ×g 20 min, 4 °C) and washed with sterile 0.1 M phosphate buffer (pH 6.5). The pellet was re-suspended in 10 ml of 100 mM NaCl (pH 2.0) and stirred slowly for 1 h at 4 °C. The suspension was then centrifuged (10,000 ×g 20 min, 4 °C), the CFS was neutralized to pH 7.0 with sterile 1 M NaOH followed by testing the bacteriocin activity as described above.

2.9. Partial purification and characterization of the bacteriocin

Bacteriocin producer strains were grown in MRS broth, and incubated without agitation for 18 h at 30°C. The cells were harvested (10,000 ×g, 20 min, 4 °C) and the bacteriocin precipitated from the CFS with 60% saturated ammonium sulphate [45]. The precipitate in the pellet and floating on the surface were collected and re-suspended in one-tenth volume 25 mM ammonium acetate buffer (pH 6.5). The sample was stored at -20 °C for one week and activity tests were performed as described above. For the determination of the molecular size of the bacteriocins, precipitated peptides re-suspended in 25 mM ammonium acetate buffer (pH 6.5) were separated by Tricine-SDS-PAGE, according to Schägger and Von Jagow [38]. Low molecular weight markers, ranging from 2.5 to 45 kDa (Pharmacia, Sweden) were used. One half of the gel containing the molecular marker was fixed for 20 min in 5% (v/v) formaldehyde, then rinsed with water and stained with Coomassie Brilliant Blue R250 (Bio-Rad) overnight. The other half of the gel (not stained and extensively pre-washed with sterile distilled water) was overlaid with a culture of 10^6 cfu/ml *L. monocytogenes* ATCC 19115 embedded in BHI agar. The position of the active bacteriocin was visualized by an inhibition zone around the active protein band [39].

3. Results

3.1. Isolation of lactobacilli

Intestinal content of 22 fish were analysed for the presence of lactobacilli. To determine the most appropriate medium for isolating lactobacilli from fish intestines, two media (MRS agar, LAMVAB) were used. LAMVAB was highly selective to quantify lactobacilli, as 99% of 143 randomly picked colonies and purified isolates were identified as *Lactobacillus* spp. and confirmed according to [12] (Table 1). Counts of intestinal lactobacilli for Persian sturgeon and beluga were detected at the range of approximately $10^{5.3}$ to $10^{6.4}$ cfu/g, respectively. The physiological and biochemical characterization of *Lactobacillus* isolates and the presumptive *Lactobacillus* species found in two fish species are shown in Table 2. From 84 isolates, 2 metabolic groups of *Lactobacillus* were recovered: facultative and obligate heterofermentatives. *L. sakei* and *L. plantarum* were the most often found isolates (Table 2). MRS 4.2 was suitable to quantify lactobacilli. As 30 randomly picked colonies on the highest dilution were identified as lactobacilli and coccoid forms were not found. Means of counts of 90 samples were not statistically different to LAMVAB counts in the Student's t-test (P=0.29) and were correlated with LAMVAB counts (r = 0.85; P<0.001). The correlation of counts on MRS 4.2 with those on LAMVAB and the absence of coccoids suggests that lactobacilli were the most important acidophilic lactic acid bacteria in the samples analysed. Facultative anaerobic flora recovered in CAB medium provided the highest counts in the samples analysed (Table 1).

Fish species	No.	CAB (cfu/g)	LAMVAB (cfu/g)	MRS 4.2 (cfu/g)
Acipenser persicus	12	7.84	5.32	4.85
Huso huso	10	8.21	6.45	5.64

CAB: Columbia blood agar; LAMVAB: *Lactobacillus* spp. Anaerobic MRS with Vancomycin and Bromocresol green; MRS 4.2: deMan, Rogosa and Sharp

Table 1. Average bacterial counts of intestinal bacteria (Log cfu/g of intestinal content) for Persian sturgeon and beluga in different media

3.2. Screening of *Lactobacilli* strains for antimicrobial activity and bacteriocin production

Eighty four lactobacilli strains previously isolated from two species of Sturgeon fish identified and their cell free supernatant extracts were assayed for antimicrobial activity and

Presumptive	L. sakei	L. plantarum	L. coryneformis	L. alimentarius	L. brevis	L. casei	L. oris
Lactobacillus species							
No. of isolates	30	18	12	10	7	5	2
Diaminopimelic acid	ND	+	ND	ND	ND	ND	ND
CO_2 from glucose	-	-	-	-	+	-	+
NH_3 from arginine	-	-	-	-	+	-	+
10°C	+	+	+	+	+	+	+
15°C	+	+	+	+	+	+	+
45°C	-	-	-	2	-	-	-
Glycerol	-	+	-	1	-	+	-
L-Arabinose	+	+	-	2	2	-	+
Ribose	+	-	-	+	+	+	+
D-Xylose	26	-	-	-	-	-	+
Galactose	29	-	-	-	-	+	-
Rhamnose	-	-	+	-	2	+	-
Inositol	-	+	-	-	-	+	+
Mannitol	-	+	5	-	+	+	-
Sorbitol	-	+	-	-	-	+	-
1-Methyl-D-mannoside	-	+	-	-	-	-	+
1-Methyl-D-glucoside	-	+	-	7	+	-	+
N-Acetyl glucosamine	28	+	+	+	+	+	+
Amygdaline	10	+	-	+	-	+	+
Arbutine	1	+	-	+	-	+	+
Esculine	+	+	+	+	1	+	+
Salicin	+	+	-	+	-	+	+
Cellobiose	27	+	-	+	-	+	+
Maltose	19	+	-	+	+	+	+
Lactose	26	+	-	+	-	+	+
Melibiose	+	+	+	2	+	-	+
Sucrose	+	+	+	8	+	+	+
Trehalose	+	+	-	+	-	+	-
Melezitose	-	+	-	-	+	+	+
D-Raffinose	29	-	-	2	+	-	-
Starch	-	-	-	-	-	+	-
Xylitol	-	+	3	-	-	-	+
2-Gentiobiose	+	+	-	+	-	+	+
D-Turanose	-	-	-	-	+	-	-
D-Tagatose	1	-	-	+	-	+	-
D-Arabitol	-	+	5	-	-	-	+
Gluconate	+	-	-	+	+	+	+
2-keto-gluconate	-	-	1	2	-	-	+
5-keto-gluconate	-	-	-	1	-	+	+
Lactic acid configuration	DL	DL	DL	DL	DL	DL	D

† +: Positive reaction of all the isolates. Numbers are the positive isolates. All isolates fermented D-Glucose, D-Fructose, D-Mannose, however they didn't ferment erythrol, D-Arabinose, L-Xylose, Adonitol, 2-Methyl-xyloside, L-Sorbose, Dulcitol, Inulin, Glycogen, D-Fucose, L-Fucose, L-Arabitol. ND: Not data

Table 2. Biochemical characteristics of *Lactobacillus* species isolated from the intestines of Persian sturgeon and beluga

bacteriocin production against *Listeria monocytogenes* ATCC 19115 and *Salmonella* Typhimurium PTCC 1186 by using spot-on-lawn method, standardized agar disk diffusion method and well diffusion method. In each instance, diameters of inhibition were quantified. Fifteen strains (18%) exhibited inhibitory activity against both indicator organisms. Consequently, all candidate isolates (Inhibition zone> 8mm) subjected to different tests such as growth at different temperatures, pH, salt content, antibiotic resistance, etc. Based on the result of aforementioned tests, two strains *Lactobacillus* casei AP8 and *Lactobacillus plantarum* H5, isolated from Persian sturgeon and beluga respectively, were chosen as active strains and were subjected to further examinations.

Presumptive Lactobacillus species	L. sakei	L. plantarum	L. coryneformis	L. alimentarius	L. brevis	L. casei	L. oris
Acipenser persicus	**	**	*	**	_	**	*
Huso huso	**	*	-	*	**	*	*

* = Presence of lactobacilli. ** = High number of lactobacilli presence

Table 3. *Lactobacillus* species isolated from the intestines of sturgeon fish

3.3. Inhibitory spectrum of bacteriocin

As the results in screening test showed that greater inhibition was observed by agar disk diffusion tests of cell-free supernatant extracts, so this method was selected as the best technique for examining the antibacterial activity of *L. casei* AP8 and *L. plantarum* H5 CFSs against forty two Gram-positive and Gram-negative bacteria. The CFS preparations from both strains showed a broad inhibitory spectrum against a wide range of LAB of different species and some food-borne pathogens and spoilage bacteria including *Listeria innocua, L. monocytogenes, Staphylococcus aureus, Aeromonas hydrophila, Aeromonas salmonicida, Bacillus cereus, Bacillus pumilus, Bacillus subtilis, Brochotrix thermosphacta*, Gram-negative *E. coli, Salmonella* and *Pseudomonas, Clostridium perfringens* and *Vibrio parahaemolyticus* (Table 4). Result showed that the Gram-positive bacteria tested were more sensitive to the bacteriocin produced by the isolates than Gram-negative bacteria. The largest spectrum of inhibition was showed by *L. casei* AP8 bacteriocin, which inhibited 33 out of 42 indicator strains.

3.4. Characterization of inhibitory effect

Table 5 and table 6 depict the stability of inhibitory substances at different physic-chemical conditions. To determine the biological nature of the antimicrobial activity of bacteria, CFSs were tested for their sensitivity to the proteolytic enzymes. Antimicrobial activities exhibited by *L. casei* AP8 and *L. plantarum* H5 were sensitive to proteolytic enzymes since proteolytic, but not lipolytic or glycolytic enzymes, completely inactivated the antimicrobial effect of both cell-free supernatants, confirming the proteinaceous nature of the inhibitors (Table 3). The effect of several chemicals on the antimicrobial activity was also evaluated. Interestingly, the cell-free extracts remained active after treatment with chemicals such as catalase, SDS, Triton X-100, Tween 20, Tween 80 and EDTA after 5 h of exposure (Table 2). Enhancing the antimicrobial activity in case of *L. casei* AP8 bacteriocin was observed after treating by EDTA and SDS against *L. monocytogenes* ATCC 19115. The stability study of

Indicator organism	Medium*	Temp. [°C]	Bac AP8	Bac H5
Gram Negative Group				
Aeromonas hydrophilus MJ 1120	BHI	37	++	0
Aeromonas hydrophilus MJ 1240	BHI	37	+++	+
Aeromonas salmonicida CC 1546	BHI	37	+	+
Aeromonas salmonicida RT 7895	BHI	37	++	+
Brochothrix thermosphacta RF 35	BHI	37	++	+
Escherichia coli ATCC 25922	BHI	37	++	0
Escherichia coli PTCC 1325	BHI	37	++	++
Photobacterium damselae ssp. Piscida	BHI	37	0	0
Pseudomonas aeruginosa PTCC 1310	BHI	37	++	+
Pseudomonas fluorescens HFC 1236	BHI	37	++	0
Salmonella enteritidis ATCC 13076	BHI	37	++	++
Salmonella spp SM 162	BHI	37	+++	++
Vibrio anguillarum MI12	BHI	37	++	+
Vibrio parahaemolyticus MI 23	BHI	37	+++	0
Vibrio parahaemolyticus MI 56	BHI	37	+++	+
Gram Positive Group				
Bacillus cereus ATCC 9634	BHI	37	+++	+++
Bacillus coagulans	BHI	37	+++	++
Bacillus licheniformis PTCC 1331	BHI	37	++	0
Bacillus subtilis ATCC 9372	BHI	37	+++	+
Clostridium perfringens ATCC 3624	RCM	37	++	+
Clostridium sporogenes PTCC 1265	RCM	37	++	+
Lactobacillus acidophilus ATCC 4356	MRS	30	++	+
Lactobacillus alimentarius AP 10	MRS	30	+	++
Lactobacillus brevis H56	MRS	30	++	++
Lactobacillus brevis AP 83	MRS	30	++	++
Lactobacillus casei PTCC 1608	MRS	30	0	++
Lactobacillus casei RN 78	MRS	30	0	0
Lactobacillus casei LB 10	MRS	30	0	+
Lactobacillus casei LB 46	MRS	30	0	+
Lactobacillus plantarum PTCC 1050	MRS	30	0	0
Lactobacillus plantarum AP 76	MRS	30	+	0
Lactobacillus plantarum H12	MRS	30	+	0
Lactobacillus sakei AP 43	MRS	30	0	+
Lactobacillus sakei	MRS	30	0	0
Lactococcous sp	MRS	30	+	0
Lactobacillus curvatus	MRS	30	0	+
Listeria innocua AN 15	BHI	37	++	++
Listeria monocytogenes ATCC 7644	BHI	37	+++	+++
Listeria monocytogenes PTCC 1163	BHI	37	++	++
Listeria monocytogenes PTCC1297	BHI	37	++	++
Staphylococcus aureus ATCC 25923	BHI	37	+++	+
Staphylococcus aureus PTCC 1112	BHI	37	+++	+

* BHI: brain hearth infusion, MRS: de Man-Rogosa-Sharpe agar and RCM: reinforced clostridial medium. 0 no zone of inhibition; +, 1 mm<zone<5 mm; ++, 5 mm<zone<8 mm; +++, zone>8 mm.; PTCC: Persian Type Culture Collection; ATCC: American Type Culture Collection.

Table 4. Antimicrobial activity of potential bacteriocin producing strain *L. casei* AP8 *and L. plantarum* H5 as examined with selected bacterial indicator strains.

inhibitory compounds of *L. casei* AP8 and *L. plantarum* H5 in different conditions indicated the high resistance of these agents. The antimicrobial compounds were able to resist most of these factors to which it was exposed even during prolong incubation period (Table 6). Cell free extracts prepared from both the isolates are found to be thermo-stable. When *L. casei* AP8 bacteriocin was heated at 40-100° C for 30 min, it retained inhibitory activity against *L. monocytogenes* ATCC 19115. However, a loss in activity in the ranges of 35% was observed when heated at 120°C for 15 min (Table 6). The Antilisterial activity of *L. plantarum* H5 bacteriocin was resistant to heat treatments of 40-100°C for 30 min and remained constant after heating at 121ºC for 15 min. Both investigated bacteriocins were most stable at 4°C and - 20°C and able to retain their antilisterial activity for 30 days without any decrease. *L. casei* AP8 bacteriocin was active in a wide range of pH, as full activity was retained at pH values between 3 and 10. *L. plantarum* H5 bacteriocin remained stable after incubation for 2 h at pH values between 2.0 - 12.0.

Treatment	Concentration	%Residual antimicrobial activity	
		L. casei AP8	*L. plantarum* H5
Enzymes			
Trypsin	1 mg/ml^{-1}	0	0
Papain	1 mg/ml^{-1}	100	100
Proteinase K	1 mg/ml^{-1}	0	0
Pronase E	1 mg/ml^{-1}	0	0
α- amylase	1 mg/ml^{-1}	100	100
Catalase	1mg/ml^{-1}	100	100
Organic solvents			
Butanol	10% [v/v]	100	100
Ethanol	10% [v/v]	100	100
Methanol	10% [v/v]	92	100
Ethyl ether	10% [v/v]	100	100
EDTA	5 mmol l^{-1}	100	83
Sodium deoxycholate	1mg ml^{-1}	100	100
Sulphobetaine 14	1mg ml^{-1}	92	100
SDS	1% [w/v]	100	100
Tween 20	1% [v/v]	100	100
Tween 80	1% [v/v]	92	100

Table 5. Effect of enzymes and chemicals on the antimicrobial activity of two selected strains *L. casei* AP8 and *L. plantarum* H5. For details see text

Treatment (Storage, Temperature and pH stability)	Residual antimicrobial activity	
	L. casei AP8	L. plantarum H5
4 ºC, -20ºC/ 30 d	+	+
40-100 ºC/30 min	+	+
121 ºC/10 min	+ [-35%]	+
121 ºC/15 min	-	+
pH= 2	-	+
pH= range 3-10	+	+
pH= 11	-	+
pH= 12	-	+

No inhibition= -; inhibition= +

Table 6. Effect of cold storage, different temperatures and pH on inhibitory activity against *Listeria monocytogenes* ATCC 19115. For details see text.

3.5. Growth and bacteriocin production

Figure 1 shows the growth and bacteriocin production curves of *L. casei* AP8 and *L. plantarum* H5 cultured at 30°C. For *Lactobacillus* casei AP 8 cell growth reached the stationary phase at 12 h of cultivation. Kinetics of bacteriocin production showed that its synthesis and/or secretion started at 4 h growth in the exponential phase of growth and maximum activity was observed at the early stationary phase of growth (1800 AU ml^{-1}) and had stabled for 6 h before the bacteriocin activity decreased (Figure 1). The pH values decreased from 6.5 to 3.7 at the end of incubation. For *L. plantarum* H5, bacteriocin activity was detectable in the culture supernatant after 5hr when an absorbance of 0.55 at 600 nm of the culture broth. Production of bacteriocin increased throughout logarithmic growth. In the stationary phase, *L. plantarum* H5 showed maximum bacteriocin activity (3400 AU/mL) and stabilized for 2 hr. But since then, bacteriocin activity declined gradually and stabilized at 1600 AU/ml during the following 4 h. In the stationary phase, extracellular pH was maintained, however, bactericidal activity decreased, excluding a possibility of lactic acid as a bactericidal mechanism.

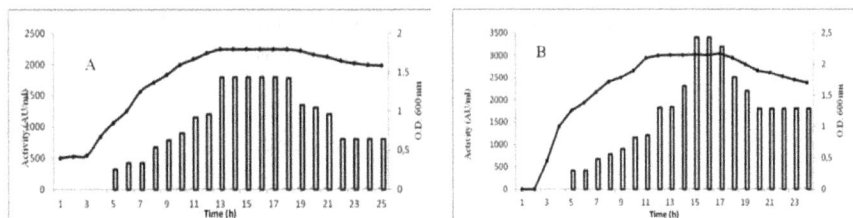

Figure 1. Antimicrobial activity [bars] against *L. monocytogenes* ATCC19115 of *L. casei* AP8 [A] and *L. plantarum* H5 [B] observed during growth in MRS medium [●] and expressed in AU/ml. Results are represent the mean of three independent experiments.

To investigate the reduction of viable cells of target organism in presence of inhibitory substances, twenty mL of each filter-sterilized bacteriocin-containing cell-free supernatant were added to 100 mL of *L. monocytogenes* ATCC 19115 (4 h old at 30°C). The optical density at 600 nm and viable cell count were determined every hour during 24 h. In the control samples inoculated with indicator strain the viable cell count reached to 10^{11}CFU/ml after 24 h incubation at 37°C. The inhibition kinetics using the bacteriocin AP8 (Figure 2) indicated a bactericidal mode of action against *L. monocytogenes*. Addition of the bacteriocin *L. casei* AP8 to early logarithmic-phase cells of indicator strain resulted in grows inhibition after 1h, followed by complete growth inhibition (slow decline) for the remaining time (20 h). In the case of *L. plantarum* H5 bacteriocin the inhibition kinetics showed a bacteriostatic mode of inhibition against indicator strain. Addition of bacteriocin H5 to culture of *L. monocytogenes* showed a growth inhibition after 1 h followed by slow growth. Experiment with stationary-phase cells did not showed any inhibition. No increase in the activity of bacteriocin AP8 and H5 were observed after treatment of the producer cells with 100 mmol/l NaCl at low pH, suggesting that these bacteriocins do not adhere to the surfaces of the producer cells.

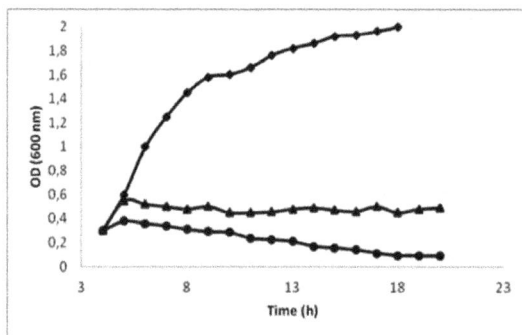

Figure 2. Antimicrobial effect of the CFS of *L. casei* AP8 [▲] and *L. plantarum* H5 [●] on the growth of *L. monocytogenes* ATCC 19115 at 30°C. Growth of *L. monocytogenes* ATCC 19115 without added bacteriocins [control, ◆].

3.6. Partial purification and molecular size of bacteriocins AP8 and H5

Ammonium sulfate precipitation method with 60% saturated ammonium sulphate is used for partial purification of both bacteriocins. Results showed an increase (10-15%) in the inhibitory activity of both bacteriocins against *L. monocytogenes* ATCC 19115 after precipitation. The SDS-PAGE analysis of the partially purified samples showed peptide bands for bacteriocins AP8 and H5 in size of approximately 5 and 3 kDa respectively (Figure 3).

4. Discussion

In this study, we isolated, quantified and characterized *Lactobacillus* species from two species of sturgeon fish inhabiting Caspian Sea to make a bank collection of strain for further research. These fishes are highly valuable species for fisheries and aquaculture in

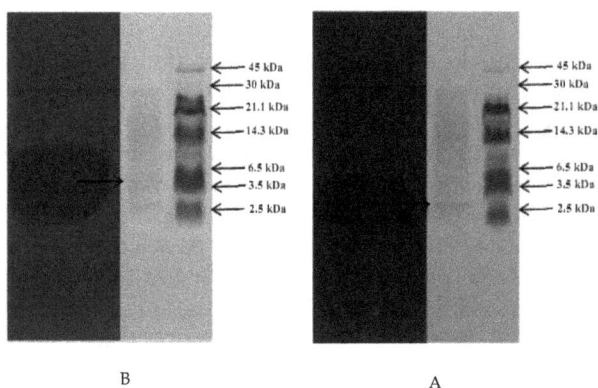

B   A

Figure 3. Tricine-SDS-PAGE gel of partially purified bacteriocins [precipitated by 60% saturated ammonium sulphate] *L. casei* AP8 [A] and *L. plantarum* H5 [B] along with the standard MW markers. The gel was overlaid with *L. monocytogenes* ATCC 19115 [approx. 10^6 CFU/ml], embedded in BHI agar, after incubation at 30 °C for 24.

Iran. Presumptive lactobacilli species found in this study were relatively similar to the species described by Bucio Galindo et al. [28]. These authors reported *L. alimentarius, L. coryneformis, L. casei, L. sakei, L. pentosus, L. plantarum, L. brevis and L. oris*, as lactobacilli presented in the intestinal content of studied fish. However, the fish species analysed in that study were different from the two species in this study which were collected from a lake environment. The biochemical characteristics used for identification of *Lactobacillus* may suggest some ideas in relation to the occurrence of the strains in nature. Most of *Lactobacillus* examined in this study (80%) had the capacity to ferment lactose and galactose. Generally, most lactobacilli are able to ferment lactose, by uptake of this disaccharide by a specific permease and splitting it by S-galactosidase for further phosphorylation of galactose and glucose [12]. Because, lactose is only present in milk and milk derivates, it is possible that these strains have evolved from environments related with mammals, as was suggested for other lactose positive *Lactobacillus* [40]. Lactose may be present or was present in the environment as a waste; resulting from livestock production, and disposal effluents from dairy factories. Another component, often fermented by the strains was the amino-sugar N-acetyl-glucosamine, a compound present in peptidoglycans, in blood, chitin and as one of the main constituents of mucus in the gastrointestinal tract [41]. The carbohydrate portion constitutes above 40% of the weight of the mucus [42] or higher values [41].

It could be shown that two strains, *Lactobacillus casei* AP8 and *Lactobacillus plantarum* H5 isolated from intestinal bacterial flora of beluga (*Huso huso*) and Persian sturgeon (*Acipenser persicus*) were able to produce antibacterial substances. According to the findings it was likely that the antibacterial effect was due to the formation of bacteriocin. Results from enzyme inactivation studies demonstrated that antimicrobial activity of isolates AP 8 and H5 was lost or unstable after treatment with all the proteolytic enzymes, confirming the protein status of metabolites and indicating the presence of bacteriocins. Furthermore, treatment with lipolytic or glycolytic enzymes did not affect the activity of antimicrobial compound produced by strain, suggesting that produced bacteriocins do not belong to the controversial group IV of the bacteriocins, which contain carbohydrates or lipids in the active molecule structure [45-47]. It is important to note that, their activities were not due to hydrogen peroxide or acidity, as antimicrobial activity was not lost after treatment with catalase. Both of the presumptive were considered to be heat stable. Although heat stability of antibacterial substances produced by *Lactobacillus* spp. has been well established [39,48,49,50-53] heat stability of *L. casei* AP8 121ºC for 10 min is novel. The result of pH stability were not coherent with previous report that had indicated the tolerance of bacteriocins to acidic pH rather than alkaline [36,54]. The loss of antimicrobial activity of AP8 bacteriocin at pH > 10 might be ascribed to proteolytic degradation, protein aggregation or instability of proteins at this extreme pH [39,48,55]. *L. casei* AP8 bacteriocin showed an increase in the inhibitory activity after treatments with SDS and EDTA, may be due to the ability of these compounds to break down the proteinaceous complex from its large form into smaller more active unite [21]. Similar to Lactocin RN78 and Plantaricin LC74, both bacteriocin *L. casei* AP8 and *L. plantarum* H5 were found to be stable after treatment with organic solvents like butanol, ethanol and methanol confirming their proteinaceous and soluble nature [18,21,56]. Pronounced inhibitory potential against various species of Gram-positive bacteria were shown, including pathogenic and spoilage microorganisms such as *A. hydrophila, A. salmonicida, C. perfringens, B. cereus* and *L. monocytogenes*. Observed effects were consistent with reports about bacteriocins produced by other strains of LAB [1,3,17,19,20,25,41,49,55,57 59,60]. Although bacteriocins from LAB usually are ineffective against Gram-negative bacteria and rather relate to a narrow antimicrobial spectrum [9,51,53], both presumptive bacteriocins AP8 and H5 showed broad antimicrobial activity against several genera of Gram-positive and Gram-negative bacteria. Even representatives of *Pseudomonas, Salmonella, E. coli, A. hydrophila, A. salmonicida and V. anguillarum* could be inhibited. Moreover a high level of inhibitory activity against *Listeria monocytogenes* was observed. Earlier studied have shown that several marine bacteria may produce inhibitory substances against bacterial pathogens in aquaculture systems [1,16,19]. Hence the use of such bacteria releasing antimicrobial substances in now gaining importance in fish farming as a natural alternative to administration of antibiotics [1,61-63]. In kinetic studies, both crude bacteriocins were continuously produced during logarithmic phase followed by optimal production during stationary growth phase, suggesting that these peptides may be secondary metabolites. Similar results were reported for some bacteriocins produced by some LAB isolates [5,64,65] and is contrary for other *Lactobacillus* species bacteriocins [1,16,25,55,60,66]. Bacteriocin H5 showed a decrease in activity towards the end of stationary growth may be due to proteolytic degration, protein aggregation, and feedback regulation as has been observed for Lactacin ST13BR, Lactacin B, Helveticin J and Enterocin1146 [53,55,67]. *L. casei* AP8 crude bacteriocin demonstrated a bactericidal mode of action, as the immediate decrease in the optical density of *L. monocytogenes* was observed in mix culture. In the case of H5 bacteriocin a bacteriostatic mode of action was observed. Crud H5 bacteriocin showed a growth inhibition, followed by decrease activity for remained time, suggesting that indicator organism became resistant to the bacteriocin or bacteriocin was destroyed by proteolytic enzymes [55]. Treating of bacteriocins AP 8 and H5 with NaCl at low pH did not result in increased levels of antilisterial activity, suggesting no adsorption of bacteriocins to their producer cells in agreement with result reported before for *Lactobacillus* strains bacteriocins [55,64,66].

More accurate techniques could be used to determine the molecular mass of molecules, yet the SDS-PAGE technique provides valuable information about the presence of the peptides [3]. In recent years, a large number of new bacteriocins produced by *L. plantarum* have been identified and characterized and the molecular masses of all the bacteriocins produced have been reported in the range of 3-10 kDa [5,39,55]. However, to our knowledge, there is no bacteriocin produced by any *L. casei* strain with a molecular mass of 5 kDa with similar characteristics to strain investigated in this study. Thus, it is possible that this bacteriocin may be a novel bacteriocin produced by *L. casei*. The physiochemical properties of bacteriocins from *L. casei* AP8 and *L. plantarum* H5 were similar to those of other bacteriocins of lactobacilli belonging to the group IIa lactic acid bacteria with respect to molecular weight, heat and pH stability and also sensivity to proteolytic enzymes [9,45,51]. Characteristics unifying all members of class IIa bacteriocins are 1) below 10kDa [1] their potent activity against *Listeria* spp., 2) their resistance to elevated temperatures and extreme pHs, and 3) their cystibiotic feature attributed to the presence of at least one disulfide bridge, which is crucial for antibacterial activity [15,45,51,55]. Class IIa bacteriocins were formerly considered as "narrow"-spectrum antibiotics, with antimicrobial activity directed against related strains. Recently, some class IIa bacteriocins, such as bacteriocin OR-7, enterocin E50-52, and enterocin E760, have been shown to be active against both Gram-negative and Gram-positive bacteria, including *Campylobacter jejuni*, *Yersinia* spp., *Salmonella* spp., *Escherichia coli* O157:H7, *Shigella dysenteriae*, *Staphylococcus aureus*, and *Listeria* spp. [15, 54, 45, 51, 3].

5. Conclusion

Bacteriocins AP8 and H5 showed a wide spectrum of antibacterial activity against seafood borne pathogens like *Listeria*, *Clostridium*, *Bacillus* spp, *S. aureus* and even Gram-negative pathogens like *Pseudomonas*, *Salmonella* and *E. coli*. Some of these foodborne pathogens can produce toxins resulting in human illness. In addition to the broad inhibition spectrum, theirs technological properties and especially cold, heat and storage stability, indicate that bacteriocins AP8 and H5 have potential for application not only as biopreservative agents to control pathogens in food products that are pasteurized and cook-chilled but also as bioprotect compounds at aquaculture. Accordingly *L. casei* may be of great interest as probiotics strains because of their ability to adhere to intestinal epithelial cells and being of human origin. Several authors have reported the production of bacteriocins by *L. casei* and *L. plantarum* strains from plant, dairy or meat origin. However, very few bacteriocins from *L. plantarum* have been reported to be isolated from fish and also based on our knowledge this is the first report of a *L. casei* bacteriocin isolated from fish.

Author details

Mahdi Ghanbari and Mansoureh Jami
University of Zabol, Faculty of Natural Resources, Department of Fishery; Zabol, Iran
BOKU -University of Natural Resources and Life Sciences, Department of Food Sciences and Technology, Institute of Food Sciences; Vienna, Austria

Masoud Rezaei
Tarbiat Modares University, Faculty of Marine Science, Department of Seafood Science and Technology, Noor, Mazandaran, Iran

Acknowledgement

The authors thank the personnel of the Shahid Rajaei Aquaculture Center for their assistance. We are also grateful to Professor Mehdi Soltani from Faculty of Veterinary Medicine, University of Tehran, Dr Ali Shahreki from Medicine Science University of Zahedan, Iran, Professor Masoud Rezaei from Tarbiat Modares University, Iran, Professor Wolfgang Kneifel and Professor Konrad Domig from University of natural Resources and Life Sciences, Vienna, Austria for their helpful opinions. Some part of this research was funded by Tarbiat Modares University, Tehran, Iran.

6. References

[1] Indira, K., S. Jayalakshmi, A. Gopalakrishnan, and M. Srinivasan. 2011. Biopreservative potential of marine *Lactobacillus* spp. African Journal of Biotechnology. 5[16]: 2287-2296

[2] Okafor, N. and B. C. Nzeako. 1985. Microbial flora of fresh and smoked fish from Nigerian fresh waters. Food Microbiology. 2: 71-75.

[3] Pinto, A. L., M. Fernandes, C. Pinto, H. Albano, F. Castilho, P. Teixeira, and P. A. Gibbs. 2009. Characterization of anti-Listeria bacteriocins isolated from shellfish: Potential antimicrobials to control non-fermented seafood. International Journal of Food Microbiology. 129: 50–58

[4] Ghanbari, M. 2007. The Use of Gram positive bacilli isolated from intestinal flora of fish to control growth of *Listeria monocytogenes* in cold smoked Roach. M.Sc Thesis. Trabiat Modares University. Iran.

[5] Todorov, S.D., P. Hob, M. Vaz-Velho, and L. M. T. Dicks. 2010. Characterization of bacteriocins produced by two strains of *Lactobacillus plantarum* isolated from Beloura and Chouriço, traditional pork products from Portugal. Meat Science. 84: 334–343

[6] Leroi, F., 2010. Occurrence and role of lactic acid bacteria in seafood products. Food Microbiology. 27, 1-12.

[7] Glvez, A., H. abriouel, R. L. Lopez and N. B. Omar. 2007. Bacteriocin-based strategies for food biopreservation. International Journal of Food Microbiology. 120: 51-70.

[8] Holzapfel, W. H., R. Geisen, and U. Schillinger. 1995. Biological preservation of foods with reference to protective cultures, bacteriocins and food grade enzymes. International Journal of Food Microbiology. 24: 343-362.

[9] Cleveland J., T. J., Montville, I. F. Nes, and M. L. Chikindas. 2001. Bacteriocins: safe natural antimicrobials for food preservation. International Journal of Food Microbiology. 71: 1-20.

[10] Kandler, O. and N. Weiss. 1986. In: Bergey's Manual of Systematic Bacteriology Vol. 2, Baltimore: Williams and Wilkins. 1209–1234.

[11] Maragkoudakis, P.A., G. Zoumpopoulou, C. Miaris, G. Kalantzopoulos, B. Pot and E. Tsakalidou. 2006. Probiotic potential of *Lactobacillus* strains isolated from dairy products. International Dairy Journal. 16: 189-199.

[12] Kandler, O and Weiss, N [1986]. Genus *Lactobacillus* beijerinck 1901, 212[AL]. In: Sneath, PHA; Mair, NS; Sharpe, ME and Holt, JG [Eds.], *Bergey's manual of systematic bacteriology*. [Eds.], Vol. 2, Baltimore: Williams and Wilkins. PP: 1209-1234.

[13] Havenaar, R; Ten Brink, B and Huis in't Veld, JHJ [1992]. Selection of strains for probiotic use. In: Fuller, R [Ed.], *Probiotics: the scientific basis*. [1st Edn.], London, Chapman and Hall. PP: 209-224.

[14] Walstra, P; Geurts, TJ; Noomen, A; Jellema, A and van Boekel, MAJS [1999]. *Dairy technology, principles of milk properties and processes*. Eds., New York, Marcel Dekker, Inc., P: 727.

[15] Belguesmia, Y., K. Naghmouchi, N. E. Chihib, and D. Drider. 2011. Class IIa bacteriocins: current knowledge and perspectives, p 171-195. *In* D. Drider and S. Rebuffat [ed.], Prokaryotic Antimicrobial Peptides: From Genes to Applications, DOI 10.1007/978-1-4419-7692-5_10, © Springer Science+Business Media, LLC 2011.

[16] Campos, C.A., O. Rodríguez, P. Calo-Mata, M. Prado, and J. Barros-Velásquez. 2006. Preliminary characterization of bacteriocins from *Lactococcus lactis, Enterococcus faecium* and *Enterococcus mundtii* strains isolated from turbot [*Psetta maxima*]. Food Research International 39: 356–364.

[17] Karska-Wysocki, B., M. Bazo and W. Smoragiewicz. 2010. Antibacterial activity of *Lactobacillus acidophilus* and *Lactobacillus casei* against methicillin-resistant *Staphylococcus aureus* [MRSA]. Microbiology Research 165: 674-686.

[18] Ghanbari, M., M. Rezaei, M. Soltani, G. R. Shahosseini., 2009. Production of bacteriocin by a novel Bacillus sp. strain RF 140, an intestinal bacterium of Caspian Frisian Roach [*Rutilus frisii kutum*]. Iranian Journal of Veterinary Research 10 [3]: pp. 267-272.

[19] Chahad, O. B., M. El Bour, P. Calo-Mata, A. Boudabous, J. Barros-Velazquez. 2011. Discovery of novel biopreservation agents with inhibitory effects on growth of food-borne pathogens and their application to seafood products. Research in Microbiology. doi:10.1016/j.resmic.2011.08.005

[20] Delves-Broughton, J., P. Blackburn, R. J. Evans, and J. Hugenholts. 1996. Application of the bacteriocin, nisin. Antonie van Leeuwenhock. 69: 193-202.

[21] Mojgani, N., and C. Amirinia. 2007. Kineitics of growth and bacteriocin production in *L. casei* RN 78 isolated from a dairy sample in IR Iran. International Journal of Dairy Science. 2[1]: 1-12

[22] Bucio, A., R. Hartemink, J. W. Schrama, J. Verreth, F. M. Rombouts. 2006. Presence of lactobacilli in the intestinal content of freshwater fish from a river and from a farm with a recirculation system. Food Microbiology. 23: 476–482.

[23] Ghanbari, M., M. Rezaei, R. M. Nazari, and M. Jami. 2009. Isolation and characterization of *Lactobacillus* species from intestinal contents of beluga [*Huso huso*] and Persian sturgeon [*Acipenser persicus*]. Iranian Journal of Veterinary Research 10 [2]: 152-157.

[24] Itoi, S., T. Abe, S. Washio, E. Ikuno, Kanomata, and H. Sugita. 2008. Isolation of halotolerant *Lactococcus lactis* subps. lactis from intestinal tract of coastal fish. International Journal of Food Microbiology. 121: 116121.

[25] Noonpakdee, W., P. Jumriangrit, K. Wittayakom, J. Zendo, J. Nakayama, K. sonomoto, and S. panyim. 2009. Two-peptide bacteriocin from *Lactobacillus plantarum* PMU 33 strainisolated from som-fak, a Thai low salt fermented fish product. Asia-Pacific Journal of Molecular Biology and Biotechnology. 17: 19-25

[26] Ringø, E., and F. J. Gatesoup. 1998. Lactic acid bacteria in fish: a review. Aquaculture 160: 177-203.

[27] Cai, YM; Suyanandana, P; Saman, P and Benno, Y [1999]. Classification and characterization of lactic acid bacteria isolated from the intestines of common carp and freshwater prawns. The Journal of General and Applied Microbiology 45: 177-184.

[28] Bucio Galindo, A; Hartemink, R; Schrama, JW; Verreth, JAJ and Rombouts, FM [2006]. Presence of lactobacilli in the intestinal content of freshwater fish from a river and from a farm with a recirculation system. Food Microbiology 23: 476-482.

[29] Hartemink, R and Rombouts, FM [1999]. Comparison of media for the detection of bifidobacteria, lactobacilli and total anaerobes from faecal samples. Journal of Microbiology Methods 36: 181-192.

[30] Hartemink, R; Domenech, VR and Rombouts, FM [1997]. LAMVAB - a new selective medium for the isolation of lactobacilli from faeces. Journal of Microbiology Methods 29: 77-84.

[31] Thapa, N; Pal, J and Tamang, JP [2006]. Phenotypic identification and technological properties of lactic acid bacteria isolated from traditionally processed fish products of the Eastern Himalayas. International Journal of Food Microbiology 107: 33-38.

[32] Hammes, WP and Vogel, RF [1995]. The genus *Lactobacillus*. In: Wood, BJB and Holzapfel, WH [Eds.], *The lactic acid bacteria, the genera of lactic acid bacteria*. [Eds.], Vol. 2, London, Blackie Academic and Professional. PP: 19-54.

[33] Schillinger, U. and F. Lucke, 1989. Antibactrial activity of *Lactobacillus sake* isolated from meat. Applied and Environmental Microbiology 55: 1901-1906.

[34] Benkerroum, N., Y. Ghouati, W. E. Sandine, and Tantaout-Elaraki. 1993. Methods to demonstrate the bactericidal activity of bacteriocins. Letters in Applied Microbiology 17: 78-81.

[35] Tagg J.R., and A.R. Mcgiven. 1971. Assay system for bacteriocins. Applied and Environmental Microbiology 21: 943-943.

[36]Yang, R., M.C. Johnson, and B. Ray, 1992. Novel method to extract large amounts of bacteriocin from lactic acid bacteria. Applied and Environmental Microbiology 58: 3355-3359.

[37] Sambrook, J.E., F. Eritsch, and J. Maniatis.1989. Molecular Cloning: A Laboratory Manual, 2nd ed. Cold Spring harbour Laboratory Press, Cold Spring Harbour, NY.

[38] Schägger, H., and G. Von Jagow. 1987. Tricine-sodium dodecyl sulphate-polyacrylamide gel electrophoresis for the separation of protein in the range from 1 to 100 kDa. Analytical Biochemistry 166: 368–379.

[39] Todorov, S.D., H. Nyati, M. Meincken, and L. M. T. Dicks. 2007. Partial characterization of bacteriocin AMA-K, produced by *Lactobacillus plantarum* AMA-K isolated from naturally fermented milk from Zimbabwe. Food Control 18: 656–664.

[40] 40. Garvie, EI [1984]. Taxonomy and identification of dairy bacteria. In: Davies, FL and Law, BA [Eds.], Advances in the microbiology and biochemistry of cheese and fermented milk, London: Elsevier Applied Science Publishers. PP: 35-65.

[41]Hicks, SJ; Theodoropoulos, G; Carrington, SD and Corfield, AP [2000]. The role of mucins in host-parasite interactions. Part I– Protozoan parasites. Parasitol. Today. 16: 476-481.

[42] Stephen, A.M. [1985]. Effect of food on the intestinal microflora. In: Food and the Gut [Hunter, J.O. and Alun Jones, V., Eds.]. Baillière Tindall. Sussex, England. pp .57-77

[43] Ouwehand, AC; Kirjavainen, PV; Grönlund, MM; Isolauri, E and Salminen, S [1999]. Adhesion of probiotic micro-organisms to intestinal mucus. International Dairy Journal 9: 623-630.

[44]Fuller, R [1989]. Probiotics in man and animals. Journal of Applied Bacteriology 66: 365-378.

[45] Cotter, P.D., C. Hill, and R. P. Ross. R.P. 2005. Bacteriocins: Developing innate immunity for food . Nature Reviews Microbiology 3: 777 – 788 .

[46] Lewus, C. B., S. Sun, and T. J. Montville. 1992. Production of an amylase-sensitive bacteriocin by an atypical *Leuconostoc paramesenteroides* strain. Applied and Environmental Microbiology 58:143-149.

[47] Keppler, K., R. Geisen and H. Holzapfelw. 1994. An amylase sensitive bacteriocin of *Leuconostoc carnosum.* Food Microbiology. 11, 39-45.

[48] Bhattacharya, S., and A. Das. 2010. Study of Physical and Cultural Parameters on the Bacteriocins Produced by Lactic Acid Bacteria Isolated from Traditional Indian Fermented Foods. American Journal of Food Technology *5: 111-120.*

[49] Chung, H. J., and A. E. Yousef. 2010. Synergistic effect of high pressure processing and *Lactobacillus casei* antimicrobial activity against pressure resistant *Listeria monocytogenes.* New Biotechnology 27 [4]: 403-408

[50] Hernandez, D., E. Cardell, and V. Zarate. 2005. Antimicrobial activity of lactic acid bacteria isolated from Tenerife cheese: initial characterization of plantaricin TF711, a bacteriocin-like substance produced by *Lactobacillus plantarum* TF711. Journal of Applied Microbiology 99: 77-84.

[51]Drider, D., G. Fimland, Y. Héchard, L. M. McMullen, and H. Prévost. 2006. The continuing story of Class IIa bacteriocins. Microbiology and Molecular Biology Reviews 70 [2]: 564–582.

[52] Ogunbanwo, S.T., A.I. Sanni, and A.A. Onilude. 2003. Characterization of bacteriocin produced by *Lactobacillus plantarum* F1 and *Lactobacillus brevis* OG1. African Journal of Biotechnology 2[8]: 219-227.

[53] Parente, E. and A. Ricciardi. 1999. Production, recovery and purification of bacteriocins from lactic acid bacteria. Applied Microbiology and Biotechnology 52: 628–638.

[54] Bhunia, A.K., M.C. Johnson, B. Ray, and N. Kalchayanand. 1991. Mode of action of pediocin AcH from *Pediococcus acidilactis* H on sensitive bacterial strains. Journal of Applied Bacteriology 70: 25-33.

[55] Todorov, S. D., and L. M. T. Dicks. 2004. Partial characterization of bacteriocins produced by four lactic acid bacteria isolated from regional South African barely beer. Annals of Microbiology 54[4]: 403-413.

[56] Rekhif, N., A. Atrih, M. Michel and G. Lefebvre. 1995. Activity of plantaricin SA6, a bacteriocin produced by *Lactobacillus plantarum* SA6 isolated from fermented sausages. Journal of Applied Bacteriology 78: 349–58.

[57] Cuozzo, S. A., F. Sesma, J. M. Palacios, A. P. de Ruiz Holgado, and R. R. Raya. 2000. Identification and nucleotide sequence of genes involved in the synthesis of lactocin 705, a two-peptide bacteriocin from *Lactobacillus casei* CRL 705. FEMS Microbiology Letters 185: 157-161

[58] Rammelsberg, M., E. Miiller, and F. Radler. 1990. Caseicin 80: purification and characterization of a new bacteriocin from *Lactobacillus casei*. Archives of Microbiology 154: 249 – 252

[59] Van Reenen, C. A., L. M. T. Dicks and M. L. Chikindas. 1998. Isolation, purification and partial characterization of plantaricin 423, a bacteriocin produced by *Lactobacillus plantarum* . Journal of Applied Microbiology 84: 1131–1137.

[60] Vignolo, G., S. Fadda, M. N. De Kairuz, A. A. P. de Ruiz Holgado, and G. Oliver. 1996. Control of *Listeria monocytogenes* in ground beef by Lactocin 705, a bacteriocin produced by *Lactobacillus casei* CRL 705. International Journal of Food Microbiology. 27: 397402.

[61] Venkat, H. K., N. P. Shau and K. J. Jain. 2004. Effect on feeding *Lactobacillus*-based probiotics on the gut microflora, growth and survival of postlarvae of *Macrobrachium rosenbergii* [de Man]. Aquaculture Research 35: 501-507.

[62] Verschuere, L., G. Rombaut, P. Sorgeloos and W. Verstraete. 2000. Probiotic bacteria as biological control agents in aquaculture. Microbiology and Molecular Biology Reviews. 64: 655-671.

[63] Vijayan, K. K., I. S. B. Singh, N. S. Jayaprakash, S. V. Alavandi, S. S. Pai, R. Preetha, J. J. S. Rajan and T. S. Santiago. 2006. A brackishwater isolate of *Pseudomonas* PS-102, a potential antagonistic bacterium against pathogenic vibrios in penaeid and non-penaeid rearing systems. Aquaculture 251: 192-200.

[64] Todorov, S., B. Onno, O. Sorokine, J. M. Chobert, I. Ivanova, and X. Dousset. 1999. Detection and characterization of a novel antibacterial substance produced by *Lactobacillus plantarum* ST 31 isolated from sourdough. International Journal of Food Microbiology. 48: 167-177

[65] Toméa, E.,V. L. Pereiraa, C. I. Lopesa, P. A. Gibbsc, and P. C. Teixeiraa. 2008. In vitro tests of suitability of bacteriocin-producing lactic acid bacteria, as potential biopreservation cultures in vacuum-packaged cold-smoked salmon. Food Control 19[5]: 535-543

[66] Karthikeyan, V., and S. W. Santosh. 2009. Isolation and partial characterization of bacteriocin produced from *Lactobacillus plantarum* . African Journal of Microbiology Research 3 [5]: 233-239

[67] Daba, H., S. Pandian, J. F. Gosselin, R. E. Simard, J. Huang, C. Lacroix. 1991. Detection and activity of bacteriocin produced by *Leuconostoc mesenteriodes*. Applied and Environmental Microbiology 57: 3450-3455.

[68] Stiles, M. E. and W. H. Holzaphel. 1997. Lactic acid bacteria of foods and their current taxonomy. International Journal of Food Microbiology. 36: 1 –29.

[69] Chung, H. J., and A. E. Yousef. 2009. Screening of Lactobacilli Derived from Fermented Foods and Partial Characterization of *Lactobacillus casei* OSY-LB6A for Its Antibacterial Activity against Foodborne Pathogens. Journal of Food Science and Nutrition *14: 162-167*

Chapter 4

Lactic Acid Bacteria and Their Bacteriocins: A Promising Approach to Seafood Biopreservation

Mahdi Ghanbari and Mansooreh Jami

Additional information is available at the end of the chapter

1. Introduction

The growing interest in a correct life style, including alimentation, and the parallel attention on food quality have contributed to orientate consumers towards fishery products which are considered safe, of high nutritional value and capable of influencing human health in a positive way [1]. The diverse nutrient composition of seafood makes it an ideal environment for the growth and propagation of spoilage micro-organisms and common food-borne pathogens [2]. It has been estimated that as much as 25% of all food produced is lost post-harvest owing to microbial activity [1,2]. It has been mentioned that as many as 30% of people in industrialized countries suffer from a food borne disease each year and in 2000 at least two million people died from diarrhoeal disease worldwide. It is clear that indigenous bacteria present in marine environment as well as the result of post contamination during process are responsible for many cases of illnesses [3,4]. In the last years, the traditional processes applied to seafood like salting, smoking and canning have decreased in favor of mild technologies involving lower salt content, lower cooking temperature and vacuum (VP) or modified atmosphere packing (MAP). The treatments are usually not sufficient to destroy microorganisms and in some cases psychrotolerant pathogenic and spoiling bacteria can develop during the extended shelf-life of these products [2,5]. As several of these products are eaten raw, it is therefore essential that adequate preservation technologies are applied to maintain its safety and quality. Among alternative food preservation technologies, particular attention has been paid to biopreservation to extent the shelf-life and to enhance the hygienic quality, minimizing the impact on the nutritional and organoleptic properties of perishable food products such as seafood [1,6]. Biological preservation refers to the use of a natural or controlled microflora and/or its antimicrobial metabolites to extend the shelf life and improve the safety of food. Lactic acid bacteria (LAB) are particularly interesting candidates for this technique [1,2,6,7]. Indeed, they are frequently naturally present in food products and are often strong competitors, by producing a wide range of antimicrobial metabolites such as organic acids, diacetyl, acetoin, hydrogen peroxide, reuterin, reutericyclin, antifungal peptides, and bacteriocins [8-10]. Hence, the last two decades have seen intensive investigation on LAB and their metabolites to discover new LAB strains that can be used in food preservation [1,7,11-13].

2. Bacterial hazards associated with fish and fish products

From the viewpoint of microbiology, fish and related products are a risky foodstuff group. Pathogenic bacteria associated with seafood can be categorized into three general groups [14]: 1) Bacteria (indigenous bacteria) that belong to the natural microflora of fish (*Clostridium botulinum*, pathogenic *Vibrio* spp., *Aeromonas hydrophila*); 2) Enteric bacteria (non-indigenous bacteria) that are present due to faecal contamination (*Salmonella* spp., *Shigella* spp., pathogenic *Escherichia coli, Staphylococcus aureus*); and 3) bacterial contamination during processing, storage, or preparation for consumption (*Bacillus cereus, Listeria monocytogenes, Staphylococcus aureus, Clostridium perfringens, Clostridium botulinum, Salmonella* spp.).

Vibrio parahaemolyticus has been isolated from sea and estuary waters on all continents with elevated sea water temperatures. *V. parahaemolyticus* is frequently isolated from fish, molluscs, and crustaceans throughout the year in tropical climates and during the summer months in cold or temperate climates [15]. Fish food associated with illnesses due to consumption of *V. parahaemolyticus* includes fish-balls, fried mackerel (*Scomber scombrus*), tuna (*Thunnus thynnus*), and sardines (*Sardina pilchardus*). These products include both raw and undercooked fish products and cooked products that have been substantially recontaminated [9,15]. The most affected by the pathogens are Japan, Taiwan, and other Asian coastal regions, though cases of disease have been described in many countries and on many continents [9,16]. Cases of diseases caused by *V. parahaemolyticus* are occasional in Europe. During 20 years, only two cases of gastroenteritis were recorded in Denmark. The interest in this organism has been widened by the finding that similar organisms, *V. alginolyticus* and group of F *Vibrio* sp. also cause serious disease in humans [17]. *V. cholerae* is often transmitted by water but fish or fish products that have been in contact with contaminated water or faeces from infected persons also frequently serve as a source of infection [1,9,19]. The organism would be killed by cooking and recent cases of cholera in South America have been associated with the uncooked fish marinade seviche (*Cilus gilberti*) [18].

E. coli is a classic example of enteric bacteria causing gastroenteritis. *E. coli* including other coliforms and bacteria as *Staphylococcus* spp. and sometimes enterococci are commonly used as indices of hazardous conditions during processing of fish. Such organisms should not be present on fresh-caught fish [9,20,21]. The contamination fish derived food with pathogenic *E. coli* probably occurs during handling of fish and during the production process [20,22]. An outbreak of diarrhoeal illness caused by ingestion of food contaminated with enterotoxigenic *E. coli* was described in Japan [23]. The illness was strongly associated with eating tuna paste. Brazilian authors [24] isolated 18 enterotoxigenic strains of *E. coli* (ETEC) from 3 of 24 samples of fresh fish originating from Brazilian markets; 13 of them produced a thermolabile enterotoxin. Infection with verocytotoxin _ producing strains of *E. coli* (VTEC) after ingestion of fish was recorded in Belgium [25]. An outbreak caused by salted salmon roe contaminated, probably during the production process, with enterohaemorrhagic *E. coli* (EHEC) O157 occurred in Japan in 1998 [22]. The roe was stored frozen for 9 months but it appears that O157 could survive freezing and a high concentration of NaCl and retained its pathogenicity for humans [26].

Aeromonas spp. has been recognized as potential foodborne pathogens for more than 20 years. Aeromonads are ubiquitous in fresh water, fish and shellfish and also in meats and fresh vegetables [27]. The epidemiological results so far are, however, very questionable. The organism is very frequently present in many food products, including raw vegetables, and very rarely has a case been reported. Up to 8.1% of cases of acute enteric diseases in 458 patients in Russia were caused by *Aeromonas* spp. [28]. In this study, *Aeromonas* spp. isolates with the same pathogenicity factors were isolated from river water in the Volga Delta, from fish, raw meat, and from patients with diarrhoea. Most *Aeromonas* spp. isolates are psychrotrophic and can grow at refrigerator temperatures [29]. This could increase the hazard of food contamination, particularly where there is a possibility of cross-contamination with ready-to-eat food products.

Salmonella has been isolated from fish and fishery product, though it is not psychrotrophic or indigenous to the aquatic environment [30]. The relationship between fish and *Salmonella* has been described by several scientists; some believe that fish are possible carriers of *Salmonella* which are harbored in their intestines for relatively short periods of time and some believe that fish get actively infected by *Salmonella* [31]. Most outbreaks of food poisoning associated with fish derive from the consumption of raw or insufficiently heat treated fish and cross-contamination during processing and the U.S. Food and Drug Administration's (FDA) data showed that *Salmonella* was the most common contaminant of fish and fishery products [31]. The highest *Salmonella* incidence in fishery products was determined in Central Pacific and African countries while it was lower in Europe and including Russia, and North America [32]. The most common serovar found in the world was *S.* sub Weltvreden [30, 31]. In seafood the commonest serotype encountered was *S.* sub Worthington followed by *S.* sub Weltevreden.

Enterotoxins produced by *Staphylococcus aureus* are another serious cause of gastroenteritis after consumption of fish and related products. In 3 of 10 samples of fresh fish, higher counts of *Staph. aureus* were detected than permitted by Brazilian legislation [20, 33]. In the southern area of Brazil, *Staph. aureus* was isolated from 20% of 175 examined samples of fresh fish and fish fillets (*Cynoscion leiarchus*). *Staph. aureus* has also been detected during the process of drying and subsequent smoking of eels in Alaska in 1993 [34]. During the process, *S. aureus* populations increased to more than 10^5 CFU g^{-1} of the analyzed sample, after 2 to 3 days of processing. Subsequent laboratory studies showed that a pellicle (a dried skin-like surface) formed rapidly on the strips when there was rapid air circulation in the smokehouse and that bacteria embedded in/under the pellicle were able to grow even when heavy smoke deposition occurred.

In ready-to-eat products, cooking, preservation ingredients, and storage atmosphere inhibit the Gram-negative organisms, resulting in a longer shelf life. Such conditions favor the growth of psychotropic pathogens such as *Listeria monocytogenes*, allowing them to grow to dangerous levels [9,35,36]. *L. monocytogenes* is a serious threat to consumer health and safety and has been implicated in several deadly outbreaks around the world [1,2]. This organism is halotolerant (up to 28% w/v for short periods), resistant to freezing temperatures, can grow and multiply during refrigeration, where other competing organisms cannot, and is able to survive at low water activity (aw) [9,14,37]. *L. monocytogenes* is widely distributed in the general environment including fresh water, coastal water and live fish from these areas. Contamination or recontamination of seafood may also take place during processing [37-39]. Moreover, *L. monocytogenes* is a psychrotrophic pathogen with the ability to grow from under 0 to 45°C [40]. This ability to grow at storage temperatures means that this bacterium is the main hazard in this kind of product. The pathogenic bacteria *L. monocytogenes* may grow on fresh seafood. *Listeria* has been found in farmed rainbow trout [41]. The outbreak of listeriosis related to vacuum packed gravad and cold-smoked fish was described in at least eight human cases for 11 months in Sweden [42]. Cold-smoked and gravad rainbow trout (*Oncorhynchus mykiss*) and salmon (*Salmo salar*) have been focused on during recent years as potential sources of infection with *L. monocytogenes* and there are several report on isolation of this food borne pathogen from fish-processing plants environments [14,37,39,43-45]. Seafood treatment is necessary to prevent food-borne illness. However, the pervasive nature of *L. monocytogenes* makes it difficult for processors to fully eliminate the organism from the environment.

Development of new-generation foods, which are mildly processed, contain few or no preservatives, are packaged in vacuum or modified atmospheres to ensure long shelf life and rely primarily on refrigeration for preservation, has raised concerns of potential increases in botulism risk caused by psychrotrophic nonproteolytic group II *Clostridium botulinum* [46]. An average of 450 outbreaks of foodborne botulism with 930 cases have been reported annually worldwide [47]. The main habitat of clostridia is the soil but they are also found in sewage, rivers, lakes, sea water, fresh meat, and fish [48,49]. Most critical are the hygienic conditions for handling the product after smoking. There is a risk of botulism due to the growth of *C. botulinum* type E in smoked fish. The bacterium becomes a hazard when processing practices are insufficient to eliminate botulinal spores from raw fish, particularly improper thermal processing [21]. The growth of *C. botulinum* and toxin production then depends on appropriate conditions in food before eating: the temperature, oxygen level, water activity, pH, the presence of preservatives, and competing microflora [21]. A problem with *C. botulinum* has been encountered with some traditional fermented fish products. These rely on a combination of salt and reduced pH for their safety. If the product has insufficient salt, or fails to achieve a rapid pH drop to below 4.5, *C. botulinum* can grow. There was no evidence that the fish had been mishandled, but a low salt environment in the viscera allowed the bacterium to multiply and to produce toxin. *C. perfringens*, an important cause of both food poisoning and non-food-borne diarrhoeas in humans, was found in a number of fish owing to contamination with sewage, which is the main source of this organism [21].

3. Biopreservation

Seafood products are known to be especially susceptible to both microbiological and biochemical spoilage pathways. The development of effective processing treatments to extend the shelf life of fresh fish products is a must [2]. Additionally, the consumers' demand for high-quality and minimally processed seafood has recently captivated great attention [5, 9]. However, an increase in foodborne illness outbreaks is concomitant with the increase in consumer demand for less processed foods [1]. These trends highlight the importance of studying new microbial stress factors to extend the shelf-life of foods. Until now, approaches to reduce the risk of outbreaks of food poisoning have relied on the search for addition of more efficient chemical preservatives or on the application of more drastic physical treatments such as heating, refrigeration, high hydrostatic pressure (HHP), ionising radiation, pulsed-light, ozone, ultrasound, etc [1,5,50]. In spite of some possible advantage, these types of treatments have many drawbacks and limitation in seafood products: the proven toxicity of many of the commonest chemical preservatives (e.g. nitrites) (3), the alteration of the organoleptic and nutritional properties of seafood by physical treatments due to their delicate nature (e.g. freezing damage, discolouration in case of HHP and ionising radiation) [50,51] and especially recent consumer trends in purchasing and consumption, with demands for healthy seafood products that have been subjected to less extreme treatments (less heat and chill damage), with lower levels of salts, fats, acids, and sugars and/or the complete or the partial removal of chemically synthesized additives [1,2,7]. To harmonize consumer demands with the necessary safety standards, traditional means of controlling microbial spoilage and safety hazards in seafood are being replaced by an alternative solution that is gaining more and more attention: "biopreservation technology" [2,9,13,52,53]. It consists in inoculating food with microorganisms, or their metabolites, selected for their antibacterial properties and may be an efficient way of extending shelf life and food safety through the inhibition of spoilage and pathogenic bacteria without altering the nutritional quality of raw materials and food products [54, 55].

Lactic acid bacteria (LAB) possess a major potential for use in biopreservation because they are safe to consume, and during storage they naturally dominate the microbiota of many foods. Certain LAB species and strains isolated from seafood have been shown to exert strong antagonistic activity against spoilage and pathogenic microorganisms such as *Listeria, Clostridium, Staphylococcus,* and *Bacillus* spp [56-58]. The antagonistic and inhibitory properties of LAB are due to the competition for nutrients and the production of one or more antimicrobially active metabolites such as organic acids (lactic and acetic acid), hydrogen peroxide, and antimicrobial peptides (bacteriocins) [10]. Certain LAB are able to grow at refrigeration temperatures and are tolerant to modified-atmosphere packaging, low pH, high salt concentrations, and the presence of certain additives such as lactic acid, acetic acid, and ethanol. Because of these benefits, LAB can be used as protective cultures to restrict the growth of undesired organisms such as certain spoilage and pathogenic bacteria, with the subsequent benefits in terms of food safety [9,10,58]. Moreover, these microorganisms may have additional functional properties and, in some circumstances, they can be beneficial for the consumers [6]. LAB represent the microbial group most commonly used as protective cultures, as they are present in all fermented foods and have a long history of safe use [8]. Safety for the consumers is an aspect of great importance, in particular for some seafood products which are not cooked before consumptions, but also for other types of foods.

4. The role of lactic acid bacteria in biopreservation technology

4.1. Characterization and classification

Lactic acid bacteria (LAB) encompass a heterogeneous group of microorganisms having as a common metabolic property the production of lactic acid as the majority end - product from the fermentation of carbohydrates [59]. LAB are Gram (+), usually nonmotile, non - sporulating, catalase - negative, acid - tolerant, facultative anaerobic organisms and have less than 55 mol% G+C content in their DNA [60-62]. Except for a few species, LAB members are nonpathogenic organisms with a reputed generally recognized as safe status (GRAS). Taxonomic revisions of these genera and the description of new genera mean that LAB could, in their broad physiological definition, comprise around 20 genera [10]. However, from a practical, food-technology point of view, the following genera are considered the principal LAB: *Aerococcus, Carnobacterium, Enterococcus, Lactobacillus, Lactococcus, Leuconostoc, Oenococcus, Pediococcus, Streptococcus, Tetragenococcus, Vagococcus,* and *Weissella* [61]. The classification of lactic acid bacteria into different genera is largely based on morphology, mode of glucose fermentation, growth at different temperatures, configuration of the lactic acid produced, ability to grow at high salt concentrations, and acid or alkaline tolerance [62, 63]. An important characteristic used in the differentiation of the LAB genera is the mode of glucose fermentation under standard conditions. In this regard, the accepted definition is that given by Hommes and Vogel [64]: obligately homofermentative LAB are able to ferment hexoses almost exclusively to lactic acid by the Embden–Meyerhof–Parnas (EMP) pathway while pentoses and gluconate are not fermented as they lack phosphoketolase; facultatively heterofermentative LAB degrade hexoses to lactic acid by the EMP pathway and are also able to degrade pentoses and often gluconate as they possess both aldolase and phosphoketolase; finally, obligately heterofermentative degrade hexoses by the phosphogluconate pathway producing lactate, ethanol or acetic acid and carbon dioxide; moreover, pentoses are fermented by this pathway [62]. Several strains of groups 1 and 2 and some of the hetero fermentative group 3 are either used in fermented foods, but group 3 are also commonly associated with food spoilage. (For a more detailed discussion concerning the metabolic pathways, see [59]).

4.2. Antimicrobial components from LAB

4.2.1. Bacteriocins

Bacteriocins are ribosomally synthesized peptides, that exert their antimicrobial activity against either strains of the same species as the bacteriocin producer (narrow range), or to more distantly related species (broad range) [1,2,7]. It has been estimated that between 30% and 99% of all bacteria and archaea produce bacteriocins; their production by LAB is very significant from the point of view of their potential applications in food systems and thus, unsurprisingly, these have been most extensively investigated [6,10,12,60,65,66]. It has been noted that the activity of bacteriocins is frequently directed against bacteria that are related to the bacteriocin - producing strain or against bacteria found in similar environments [67]. It has also been noted that some bacteriocins can also play a role in cell signaling. Microorganisms that produce bacteriocins also possess immunity mechanisms to confer self - protection, that is, to protect bacteriocin producers from committing "suicide" [10,68,69]. Besides concern about antibiotic resistance, increasing consumer awareness of potential health risks associated with chemical preservatives has increased interest in bacteriocins. Bacteriocins are naturally produced so they are more easily accepted by consumers [54]. Bacteriocins are usually classified combining various criteria. The main ones being the producer bacterial family, their molecular weight and finally their amino acid sequence homologies and/or gene cluster organization [59,70]. Based on a relatively recent approach [69,71,72] bacteriocins produced by LAB have been categorized into two major classes: the lanthionine - containing bacteriocins or lantibiotics (class I) and the largely unmodified linear peptide antimicrobials (class II).

4.2.2. Organic acid production

An important role of meat LAB starter cultures is the rapid production of organic acids; this inhibits the growth of unwanted flora and enhances product safety and shelf life. The types and levels of organic acids produced during the fermentation process depend on the LAB strains present, the culture composition, and the growth conditions [74]. Fermentation of the carbohydrates, glucose, glycogen, glucose-6-phosphate and small amounts of ribose, in meat and meat products, produces organic acids by glycolysis (Embden-Meyerhof Parnas pathway, EMP pathway) or the Hexose Monophosphate, HMP pathway. L (+) lactic acid is more inhibitory than its D (-) counterpart [68]. The antimicrobial effect of organic acids lies in the reduction of pH, and in the action of undissociated acid molecules [75]. It has been proposed that low external pH causes acidification of the cytoplasm. The lipophilic nature of the undissociated acid allows it to diffuse across the cell membrane collapsing the electrochemical proton gradient. Alternatively, cell membrane permeability may be affected, disrupting substrate transport systems [72]. The LAB in particular are able to reduce the pH to levels where putrefactive (e.g. clostridia and pseudomonads), pathogenic (e.g. *Salmonella* s and *Listeria* spp.) and toxinogenic bacteria (*Staphylococcus aureus. Bacillus cereus, Clostridium botulinum*) will be either inhibited or killed [7]. Also, the undissociated acid, on account of its fat solubility, will diffuse into the bacterial cell, thereby reducing the intracellular pH and slowing down metabolic activities, and in the case of Enterobacteriaceae such as *E. coli* inhibiting growth at around pH 5.1.

4.2.3. Other antimicrobials of LAB

Hydrogen peroxide is produced from lactate by LAB in the presence of oxygen as a result of the action of flavoprotein oxidases or nicotinamide adenine dinucleotide (NADH) peroxidise [76]. The antimicrobial effect of H_2O_2 may result from the oxidation of sulfhydryl groups causing denaturing of a number of enzymes, and from the peroxidation of membrane lipids thus increasing membrane permeability [8]. Most undesirable bacteria such as *Pseudomonas* spp. and *S. aureus* are many times sensitive to H_2O_2. Carbon dioxide (CO_2) is mainly produced by heterofermentative LAB. CO_2 plays a role in creating an anaerobic environment which inhibits enzymatic decarboxylations, and the accumulation of CO_2 in the membrane lipid bilayer may cause a dysfunction in permeability [8]. CO_2 can effectively inhibit the growth of many food spoilage microorganisms, especially Gram-negative psychrotrophic bacteria [77]. Diacetyl, an aroma component, is produced by strains within all genera of LAB by citrate fermentation. It is produced by heterofermentative lactic acid bacteria as a by-product along with lactate as the main product [8]. Diacetyl is a high value product and is extensively used in the dairy industry as a preferred flavour compound. Diacetyl also has antimicrobial properties. Diacetyl was found to be more active against gram-negative bacteria, yeasts, and molds than against gram-positive bacteria. Diacetyl is thought to react with the arginine-binding protein of gram-negative bacteria and thereby interfering with the utilization of this amino acid [78].

5. LAB in fish and fish products

LAB are not considered as genuine microflora of the aquatic environment, but certain genera, including *Carnobacterium, Lactobacillus, Enterococcus,* and *Lactococcus,* have been found in fresh and sea water fresh fish [61,63,79-83]. The number of lactobacilli in the gastro-intestinal tract of Arctic char was smaller in those reared in sea water than in fresh water, while the number of *Leuconostoc* and enterococci remained the same [84]. It is well documented that lactobacilli are part, not dominant, of the native intestinal microbiota of Arctic charr (*Salvelinus alpinus* L.), Atlantic cod, Atlantic salmon (*Salmo salar* L.), and brown trout (*Salmo trutta*) [82,85]. Several studies have shown the presence of other lactic acid bacteria, specially carnobacteria such as *Carnobacterium maltaromaticum* and *Carnobacterium divergence* within the intestinal content of salmonid species like Arctic charr (*Salvelinus alpinus*), Atlantic salmon (*Salmo salar*), rainbow trout (*Oncorhyncus mykiss*) [63,86-89], Atlantic cod [89], common wolffish (*Anarhichas lupus* L.) [85], brown trout [82] and also wild pike [63,82]. Bacteria of the genus *Enterococcus* have been isolated from the intestine of common carp (*Cyprinus carpio*) and brown trout [80,82].

LAB dominating in spoiled vacuum-packaged cold-smoked fish products include the genera of *Lactobacillus, Leuconostoc, Lactococcus* and *Carnobacterium* [9]. Magnússon & Traustadóttir

[91] reported the complete dominance of homofermentative lactobacilli in vacuum-packaged cold-smoked herring. In vacuum packaged cold-smoked salmon and herring, *Lactobacillus curvatus* has been found in majority together with lower numbers of *Lactobacillus sakei, Lactobacillus plantarum, Lactococcus* spp. and *Leuconostoc mesenteroides* [58,]. Paludan-Müller, Huss, & Gram [92] identified *Carnobacterium piscicola* as the dominant microorganism isolated from spoiled vacuum-packaged cold-smoked salmon. Leroi et al. [93] also isolated carnobacteria during the first stage of storage of vacuum-packaged cold-smoked salmon, whereas *Lactobacillus farciminis, Lactobacillus sakei,* and *Lactobacillus alimentarius* were isolated at advanced storage times. Other studies have also confirmed that most bacteria in vacuum-packaged "gravad" fish products stored at refrigeration temperatures are carnobacteria [94] and *L. sakei,* and to a lesser extent *Leuconostoc* spp., *L. curvatus,* and *Weissella viridescens* [95]. Gancel et al [90] have isolated 78 strains belonging to the genus *Lactobacillus* from fillets of vacuum packed smoked and salted herring *(Clupea harengus).* LAB has been found to occur in marinated herring, herring fillets and cured stockfish [58]. In marinated or dried fish, the lactic acid bacteria flora maybe quite diverse since the presence of *Lactobacilli* and *Pediococci* has been reported [90]. Thai fermented fishery products were screened for the presence of LAB by Ostergaard et al. [96]. LAB was found to occur in the low salted fermented products in the range of 10^7-10^9 cfu/g. The high salt product "hoi dorng" had a lower LAB count of 10^3-10^5 cfu/g. Olympia et al [97] have isolated 10^8 LAB/g from a Philippine low salt rice-fish product burong bangus. Several studies have been mentioned that some species of *Carnobacteriuim* such as *C. divergens* and *C. maltaromaticum* are present in seafood and are able to grow to high concentrations in different fresh and lightly preserved products such as modified atmosphere-packed (MAP) [98-100], chilled MAP [101,102], high-pressure processing treated seafood products [103] and vacuum-packed cold smoked or sugar-salted ('gravad') seafood [53,93,95]. These studies clearly highlight the ability of LAB fish isolates to grow on different harsh condition rather than other organisms. Obviously many investigation have been shown that carnobacteria are common in chilled fresh and lightly preserved seafood, but at higher storage temperatures (15–25°C) other species could be dominate the spoilage microbial community of seafood.

6. Application of LAB in seafood

Treating catfish fillets with of 0.50% sodium acetate, 0.25% potassium sorbate with 2.50% lactic acid culture completely inhibited growth of Gram negative bacteria, improved catfish odor and appearance during 13 days storage [110]. Einarsson & Lauzon [111] treated shrimps with various bacteriocins from lactic acid bacteria and reported shelf life extension except carnocin UI49. Total mesophilic and psychotropic bacteria and MRS counts of the samples treated with carnocin UI49 were not different than those of controls at 4.5°C. In a study with five strains of lactic acid bacteria (four *Lactobacillus* and one *Carnobacterium*) on fermented salmon fillets, *L. sake* LAD and *L. alimentarius* BJ33 was regarded as suitable starters for fermentation of salmon fillets [112] based on starter growth (increase of more than 1log in 3 days) and acidification of muscle (e.g. pH reduction of approximately 0.7

units in 5 days) as well as sensory evaluation. Kisla & Ünlütürk [113] studied the microbial shelf life of rainbow trout treated with nisin-containing aqueous solution of *Lactococcus lactis* subsp. lactis NCFB 497and lactic acid. They reported the dipping of rainbow trout fillets into a lactic culture did not prolonged the shelf life due to the low inoculum level and type of lactic culture used. Elotmani & Assobhei [114] evaluated the inhibition of the microbial flora of sardine by using nisin and a lactoperoxidase system (LP), observing the efficiency of the nisin–LP combination in inhibiting fish spoilage flora. In another study growth of *L. monocytogenes* was significantly inhibited ($P < 0.05$) by *L. sakei* Lb706 in rainbow trout fillets stored under vacuum at 4°C during 10 days of storage while bacteriocin negative Lb706-B did not affect the growth of *L. monocytogenes*. In the presence of the sakacin A-producing strain of *L. sakei* (Lb706), the growth of *L. monocytogenes* was significantly inhibited ($P < 0.05$) in the first 3 days of storage at 10°C, after which its count increased to 10^7 CFU g^{-1} [115]. Altieri et al. [106] succeeded in inhibiting *Pseudomonas* spp. and *P. phosphoreum* in VP fresh plaice fillets at low temperatures by using a *Bifidobacterium bifidum* starter, and extending the shelf-life, especially under MAP. Bifidobacteria combined with sodium acetate (SA) extended refrigerated shelf-life of catfish fillets at 4°C [116]. The application of two *Lactobacillus sakei* CECT 4808 and *L. curvatus* CECT 904T protective cultures on refrigerated vacuum-packed rainbow trout (*Oncorhynchus mykiss*) fillets resulted in extension of shelf-life by 5 days by significantly improved in the counts of all microbiological spoilage indicator organisms (Enterobacteriaceae, *Pseudomonas* spp., H2S-producing bacteria, yeasts and moulds) and also significantly improved in all examined chemical parameters and off-odour [117].

Under biopreservation, combined coating of *Lactobacillus casei* DSM 120011 and *Lactobacillus acidophilus* 1M in *Streptomces* sp. NIOF metabolites, played effective role in lowering the biochemical and microbiological changes, extended shelf-life and safety of stored fish under low temperature as reported by Daboor & Ibrahim [118]. Tahiri et al. [119] suggest that selection of protective strains to improve the sensory quality of seafood products should focus on specific spoilage microorganism's inhibition. This approach was chosen by Matamoros et al., [120] who have isolated seven strains from various marine products on the basis of their activity against many spoiling and pathogenic, Gram positive and Gram negative marine bacteria. Among strains, two *Le. gelidum*, and two *Lc. piscium* demonstrated promising effect in delaying the spoilage of tropical shrimp and of VP CSS. However, no correlation with the classical quality indices measured was evidenced. A recent study demonstrated that this protective effect could be due to the inhibition of *B. thermosphacta* identified as one of the major spoiler organisms in cooked shrimp stored under MAP [121]. The inoculation of Tilapia (*Oreochromis niloticus*) fillets with *Lactobacillus casei* DSM 120011 and *Lactobacillus acidophilus* 1M at 2% concentration decreased both total volatile basic nitrogen (TVB-N), trimethylamine nitrogen (TMA-N) and thiobarbituric acid (TBA) values and improved the biochemical quality criteria, microbial aspects and safety of frozen fish fillets during 45 and 90 days storage. [122].

For Shirazinejad et al. [123] 2.0% lactic acid combined with nisin indicated the highest reduction in population of *Pseudomonas* spp. and H2S producing bacteria during storage time of Chilled Shrimp. Fall et al. [121] evidenced the in situ inhibition of *B. thermosphacta*, a major spoiling bacterium, by *L. piscium* that could explain the protective effect observed in shrimp. Additionally, those strains also showed an inhibitory effect on *L. monocytogenes* [124] and *Staph. aureus*. Recently, Sudalayandi & Manja [109] succeeded to preserve fresh fish through controlling spoilage bacteria and amines of Indian mackerel fish chank for two days at 37°C by inoculating them with different strains of LAB such as *Pediococcus acidilactici, Pediococcus pentosaceous, Streptococcus thermophilus, Lactococcus lactis, Lactobacillus plantarum, Lactobacillus acidophilus* and *Lactobacillus helveticus*. Using bacteriocin-like metabolite producer and non-producer strains of *Pediococcus* spp. [125] only slightly improved sensory quality of Horse Mackerel during cold storage. It was concluded that *Pediococcus* strains used in this study were not proper for preserving horse mackerel fillets especially at low storage temperatures. EntP-producing enterococci isolated from farmed turbot, under a spray-dried format exhibited antilisterial, antistaphylococcal, and antibacilli activities in turbot fillets either vacuum-packaged or subjected to modified-atmosphere packaging [2].

LAB Protective cultures have not been applied in many other seafood products except for cold smoked salmon (CSS), as they are normally flora of such products at the end of storage, and *L. monocytogenes* control. The effectiveness of bacteriocins to control growth of *L. monocytogenes* in vacuum packed cold smoked salmon has also been demonstrated by several researchers. Among them, Sakacin P has been found to be very potent against *L. monocytogenes* and is one of the most extensively studied bacteriocins [126-131]. Leroi et al. [132] succeeded in increasing the sensory use-by-date of CSS slices by inoculating them with strains of *Carnobacterium* sp. However the results varied depending on the batch treated. Addition of nisin to CO_2 packed cold smoked salmon resulted in a 1 to 2 log_{10} reduction of *L. monocytogenes* [11]. Using a strain of *C. maltaromaticum*, Paludan-Müller et al. [92] only slightly extended the shelf-life of smoked salmon. Budu-Amoako et al. [133] tested nisin combined with heat as anti Listerial treatment in cold- packed lobster meat, finding decimal reductions of inoculated *L. monocytogenes* of 3 to 5 logs, whereas heat or nisin alone resulted in decimal reductions of 1 to 3 logs.

Duffes et al. [65] isolated *C. divergens* and *C. maltaromaticum* strains that exhibited listericidal activity in a model experiment with cold-smoked fish. They found that *C. piscicola* V1 inhibited *L. monocytogenes* by the in situ production of bacteriocins in vacuum-packed cold-smoked salmon stored at 4°C and 8°C. In contrast, another related species, namely, *C. divergens* V41 and its divercin V41, only exhibited a bacteriostatic effect on the target microorganism. Two strains of *C. maltaromaticum* isolated from CSS demonstrated their efficiency to limit the growth of *L. monocytogenes* in VP CSS during 31 days of storage at 5°C [134]. In a study using vacuum-packed cold smoked rainbow trout, the combination of nisin and sodium lactate injected into smoked fish decreased the count of *L. monocytogenes* from 3.3 to 1.8 log_{10} over 16 days of storage at 8°C [135]. Sakacin P was added to vacuum-packed cold smoked salmon, a lightly processed high-fat (15–20%) product, together with a sakacin P-producing *L. sakei* culture in order to study the effect on the growth of *L. monocytogenes*. In this product, the combination of purified sakacin P and a live culture was found to be bactericidal against *L. monocytogenes*. The addition of sakacin P alone inhibited the growth of *L. monocytogenes* on this product for about 1 week [126]. Silva et al. [136] used a bacteriocin-producing *Carnobacterium* strain under a spray-dried format. This strain survived the process and retained antilisterial ability, although it lost activity against other Gram-positive targets such as *Staph. aureus*. Some authors have evaluated the antimicrobial activity of nisin combined with other bacteriocins. Bouttefroy & Milliere [137] tested combinations of nisin and curvaticin 13 produced by *L. curvatus* SB13 for preventing the regrowth of bacteriocin-resistant cells of *L. monocytogenes*, finding that this combination induced a greater inhibitory effect than the use of a single bacteriocin. Aasen et al. [131] studied the interactions of the bacteriocins sakacin P and nisin with food constituents in cold-smoked salmon, chicken cold cuts, and raw chicken. They stated that owing to the amphiphilic nature of these peptides, they can be adsorbed to food macromolecules and undergo proteolytic degradation, which may limit their use as preservation agents. More than 80% of the added sakacin P and nisin were rapidly adsorbed by proteins in the food matrix that had not been heat-treated, less than 1% of the total activity remaining after 1 week in cold- smoked salmon. In heat-treated foods, they found that, bacteriocin activity was stable for more than 4 weeks. No important differences were observed between sakacin P and nisin, but less nisin was adsorbed by muscle proteins at low pH. The growth of *L. monocytogenes* was completely inhibited for at least 3 weeks in both chicken cold cuts and cold-smoked salmon by the addition of sakacin P (3.5 μ/g), despite proteolytic degradation in the salmon.

In the presence of the bacteriocinogenic strain *C. maltaromaticum* CS526 isolated from surimi, the population of *L. monocytogenes* in CSS decreased from 10^3 to 50 CFU g^{-1} after 7 days at 4°C [138]. This activity could be linked to the production of the bacteriocin piscicocin CS526, since a non-bacteriocin producing strain had a lower effect on the growth of the pathogenic bacteria [138, 139]. The growth of the protective *Carnobacterium* strains did not modify the sensory characteristic of the product. One of these strains showing the strongest inhibition activity produces a bacteriocin, named Carnobacteriocin B2 that was involved in the antilisterial activity [105]. Three strains of bacteriocin producing *Carnobacterium* have been tested with the agar diffusion test method against a wide collection of *L. monocytogenes* (51 strains) isolated from seafood. All of the *Listeria* strains were sensitive. The inhibition was confirmed in co-culture with a mix of *L. monocytogenes* strains in sterile CSS [140]. One of these strains, *C. divergens* V41 showed its ability to maintain *L. monocytogenes* at the initial inoculating level of 20 CFU g^{-1} during 28 days of storage at 4°C and 8°C. The effect of this strain on sensory characteristics and physico-chemical parameters revealed that it did not spoiled the product [56].

A bacteriocinogenic strain of *L. sakei* isolated from CSS allowed a 4 log reduction of *Listeria innocua* after 14 days of storage at 4°C. A reduction of 2 log units after 24 h at 5°C was also demonstrated with that strain in CSS juice towards *L. monocytogenes* [141]. Mix of bacteriocin-producing LAB like *L. casei, L. plantarum* and *C. maltaromaticum* were successfully used to limit the growth of *L. innocua* in CSS [142]. *C. maltaromaticum* had no effect on the inhibition of the Gram positive spoilage bacteria *B. thermosphacta* in cooked shrimps [143]. The anti-listerial activity of 3 LAB strains used individually or as co-cultures was assayed on cold-smoked salmon artificially contaminated with *L. innocua* and stored under vacuum at 4°C [142]. The association of *L. casei* T3 and *L. plantarum* PE2 was the most effective, probably due to a competition mechanism against the pathogen. In their study Tomé et al. [144] have also selected a strain of *Enterococcus faecium* among five bacteriocinogenic LAB strains for its ability to induce a decrease of the population of *L. innocua* inoculated in CSS. However in these studies the inhibition activities were not confirmed on *L. monocytogenes*. For Matamoros et al. [145] two LAB strains, *Lactococcus piscium* EU2241 and *Leuconostoc gelidum* EU2247 were efficient to limit the growth of both pathogenic bacteria *L. monocytogenes* and *S. aureus* in a challenge test in cooked shrimp stored under VP from 2 to 3 log CFU g^{-1} units after 4 weeks at 8°C followed by 1 week at 20°C. The strain of *Leuconostoc* produced a bacteriocin-like compound but its activity was slight lower than the *Lactococcus* strain that was non-bacteriocinogenic. In another study, the application of *C. divergens* M35 towards *L. monocytogenes* in CSS resulted in a maximal decrease of 3.1 log CFU g^{-1} of the pathogenic bacteria after 21 days of storage at 4°C whereas a non bacteriocinogenic strain had no effect [119].

7. Conclusion and future prospective

The presence of LAB in many processed seafood product is now well documented and the bio-protective potential of many strains and/or their bacteriocin has been highlighted in the last years. In situ production is readily cost-effective provided that the bacteriocin producers are technologically suitable. To date, only nisin and pediocin PA - 1 have been applied commercially in food applications where they are used to protect against spoilage and pathogenic organisms. However, other bacteriocins could be at least as effective for food processors as it is possible to apply them with hurdle approaches, particularly in light of consumer demands for minimally processed, safe, preservative - free foods. Control of pathogenic bacteria has widely focused on *L. monocytogenes* considered as the main risk in ready-to-eat seafood. However, in these minimally processed products, the new combination of hurdles can give selective advantages to enhance food safety and quality, particularly effective against other pathogenic bacteria like clostridia, vibrio or staphylococci. These goals can be facilitated through the incorporation of live bacteriocin - producing strain(s) or through the use of bacteriocins as concentrated preparations, either through direct addition to the seafood or in an immobilized form on packaging as well as in conjunction with other factors such as high pressure or pulse electric fields, to achieve more effective preservation of foods. The great results obtained with protective culture, bacteriocins for improving safety and quality of seafood products clearly indicate that the application of LAB protective culture and/or their bacteriocins in seafood product can suggest several important benefits; 1) extended shelf life of seafood during storage time, 2) decrease the risk for transmission of foodborne pathogens in lightly preserved seafood products, 3) ameliorate the economic losses due to seafood spoilage, 4) reduce the application of chemical preservatives and drastic physical treatments such as heating, refrigeration, etc. causing better preservation nutritional quality of food, 5) good option for industry due to cost effective way and finally 6) a good response to consumer demands for minimally processed, safe, preservative - free foods. At present the new techniques and disciplines emerging in the post – genomic era, such as genomics, proteomics, metabolomics, and system biology, open new avenues for interpretation of biological data. In combination with classical and molecular techniques, these new methods will be invaluable in the rational optimization of LAB function in order to obtain safer traditional and new seafood products.

Author details

Mahdi Ghanbari* and Mansooreh Jami
*University of Zabol, Faculty of Natural Resources, Department of Fishery; Zabol, Iran
BOKU — University of Natural Resources and Life Sciences, Department of Food Sciences and Technology, Institute of Food Sciences, Vienna, Austria*

8. References

[1] Cortesi ML, Panebianco A, Giuffrida A, Anastasio A (2009) Innovations in seafood preservation and storage. Veterinary Research Communications.; Supplement 1: p. S15-S23.

[2] Campos A, Castro P, Aubourg SP, Velázquez JB (2012) Use of Natural Preservatives in Seafood. In McElhatton A, Sobral. Novel Technologies in Food Science, Integrating Food Science and Engineering Knowledge Into the Food Chain.: © Springer Science+Business Media;. p. 325-360.

[3] Feldhusen F (2000) The role of seafood in bacterial foodborne diseases. Method. Microbiol. 2: 1651-1660.

[4] ICMSF (2011) Fish and Seafood Products. In International Commission on Microbiological Specifications for Foods (ICMSF). Microorganisms in Foods.: Springer Science+Business Media, LLC. p. 107-133.

[5] Alzamora S, Welti-Chanes J, Guerrero S (2012) Rational Use of Novel Technologies:A Comparative Analysis of the Performance of Several New Food Preservation Technologies for Microbial Inactivation. In McElhatton A, Sobral PJA(). Novel Technologies in Food Science, Integrating Food Science and Engineering Knowledge Into the Food Chain.: © Springer Science+Business Media, LLC.

[6] Soomro AH, Masud T, Anwaar K (2002) Role of lactic acid bacteria (LAB) in food preservation and human health—A review. Pak. J. Nut. 1: 20-24.

[7] Gálvez A, Abriouel H, López R, Omar N (2007) Bacteriocin-based strategies for food biopreservation. Int. J. Food Microbiol. 120: 51-70.

---

* Corresponding Author

[8] Holzapfel WH, Geisen R, Schillinger U (1995) A review paper: biological preservation of foods with reference to protective cultures, bacteriocins and food-grade enzymes. Int. J. Food Microbiol. 24: p. 343-362.

[9] Calo-Mata P, Arlindo S, Boehme K, Miguel T, Pascola A, Barros-Velazquez J (2008) Current Applications and Future Trends of Lactic Acid Bacteria and their Bacteriocins for the Biopreservation of Aquatic Food Products. Food Biopro. Tech.1: 43–63.

[10] Collins B, Cotter P, Hill , Paul Ross R (2010) Applications of Lactic Acid Bacteria - Produced Bacteriocins. In Mozzi F, Raya R, GM V. Biotechnology of Lactic Acid Bacteria Novel Applications.: Blackwell Publishing. 89-109.

[11] Nilsson L (1997) Control of *Listeria monocytogenes* in cold-smoked salmon by biopreservation: Danish Institute for Fisheries Research and The Royal Veterinary and Agricultural University of Copenhagen, Denmark, Ph. D Dissertation.

[12] Cleveland J, Montville T, Nes I, Chikindas M (2001) Bacteriocins: Safe, natural antimicrobials for food preservation. Int. J. Food Microbiol. 71: p. 1 – 20.

[13] Dortu C, Thonart P (2009) Bacteriocins from lactic acid bacteria: interest for food products biopreservation. Biotech. Agr. Society Environ. 13: p. 143-154.

[14] Beaufort A, Rudelle S, Gnanou-Besse N, Toquin MT, Kerouanton A, Bergis H (2007). Prevalence and growth of *Listeria monocytogenes* in naturally contaminated cold-smoked salmon. Lett. Appl. Microbiol. 44: 406-411.

[15] Baffone W, Pianei A, Bruscolini F, Barbieri E, Cierio B (2000) Occurrence and expression of virulence-related properties of Vibrio species isolated from widely consumed seafood products. Int. J. Food Microbiol. 54: 9-18.

[16] InternationalDiseaseSurveillanceCenter (IDSC). 1999. *Vibrio parahaemolyticus*, Japan 1996-1998, Infectious Agents Surveillance Report (IASR), 20:1-2.

[17] Joseph SW, Colwell RR, Kaper JB (1982) *Vibrio parahaemolyticus* and related halophilic Vibrios. Crit. Rev. Microbiol. 10(1): 77-124.

[18] Kam KM, Leung TH, Ho YY, Ho NK, Saw TA (1995) Outbreak of *Vibrio cholerae* 01 in Hong Kong related to contaminated fish tank water. Public Health. 109(5): p. 389-395.

[19] Colwell RR (1996) Global climate and infectious diseases: the cholera paradigm. Sci. 274: 2025-2031.

[20] Ayulo AM, Machado R, Scussel V (1994) Entero toxigenic *Escherichia coli* and Staphylococcus aureus in fish and seafood from the southern region of Brazil. Int. J. Food Microbiol. 24:171–178.

[21] Chattopadhyay P (2000) Fish – catching and handling. In Robinson RK. Encyclopedia of Food Microbiol. London: Academic Press; 153 p.

[22] Asai Y, Murase T, Osawa R, Okitsu T, Suzuki R, Sata S, Terajima J,Izumiya H, Watanabe H (1999) Isolation of Shiga toxin-producing *Escherichia coli* O157:H7 from processed salmon roe associated with the outbreaks in Japan, 1998, and a molecular typing of the isolates by pulsed-field gel electrophoresis. Kansenshogaku Zasshi. 73: 20-24.

[23] Mitsuda, T; Muto, T; Yamada, M; Kobayashi, N; Toba, M; Aihara, Y; Ito, A; Yokota, S (1998) Epidemiological study of a food-borne outbreak of enterotoxigenic *Escherichia coli* O25:NM by pulsed-field gel electrophoresis and randomly amplified polymorphic DNA analysis. J. Clin. Microbiol. 36: 652-656.

[24] Vieira RHSF, Rodrigues DP, Gocalves FA, Menezes FGR, Aragao JS, Sousa OV (2001) Microbicidal effect of medicinal plant extracts (*Psidium guajava* Linn.and *Carica papaya* Linn.) upon bacteria isolated from fish muscle and known to induce diarrhea in children. Rev. Inst. Med. trop. S. Paulo 43 (3): 145-148.

[25] Pierard D, Crowcroft N, de Bock S, Potters D, Crabbe G,VLF, Lauwers S (1999) A case-control study of sporadic infection with O157 and non-O157 verocytotoxin-producing *Escherichia coli*. Epid. Infec. 122: 359-365.

[26] Semanchek JJ, Golden DA (1998) Influence of growth temperature on inactivation and injury of *Escherichia coli* O157:H7 by heat, acid, and freezing. J. Food Prot. 61: 395-401

[27] Isonhood JH, Drake M (2002) *Aeromonas* species in foods. J. Food Prot. 65: 575–582.

[28] Pogorelova NP, Zhuravleva LA, Ibragimov FKH, Iushchenko GV (1995) Bacteria of the genus *Aeromonas* as the causative agents of saprophytic infection. Zh Mikrobiol Epidemiol Immunobiol. 4: 9-12.

[29] Fernandes CF, Flick GJ, Thomas TB (1998) Growth of inoculated psychrotrophic pathogens on refrigerated fillets of aquacultured rainbow trout and channel catfish. J. Food Prot. 61(3): 313-317.

[30] Novotny L, Halouzka R, Matlova L, Vavra O, Dvorska L, Bartos M (2010) Morphology and distribution of granulomatous inflammation in freshwater ornamental fish infected with mycobacteria. J. Fish Dis. 33: 947-955.

[31] Olgunoğlu IA (2012) *Salmonella* in Fish and Fishery Products. In Mahmoud BSM. *Salmonella* - A Dangerous Foodborne Pathogen.: InTech; 2012.

[32] Heinitz ML, Ruble RD, Wagner DE, Tatini SR (2000) Incidence of *Salmonella* in fish and seafood. J. Food Prot. 63(5): 579-592.

[33] Vieira RHSF, Rodrigues DP, Gocalves FA, Menezes FGR, Aragao JS, Sousa OV (2001) Microbicidal effect of medicinal plant extracts (*Psidium guajava* Linn. and *Carica papaya* Linn.) upon bacteria isolated from fish muscle and known to induce diarrhea in children. Revista do Instituto de Medicina Tropical de São Paulo. 43:145-148.

[34] Eklund MW, Peterson ME, Poysky FT, Paranjpye RN, Pelroy GA (2004) Control of bacterial pathogens during processing of cold-smoked and dried salmon strips. J. Food Prot. 67: 347-351.

[35] Francis GA, O'Beirne D (1998) Effects of the indigenous microflora of minimally processed lettuce on the survival and growth of L. monocytogenes. Int. J. Food Sci. Tech. 33: 477-488.

[36] Alves VF, De Martinis ECP, Destro MT, Vogel BF, Gram L (2005) Antilisteral activity of a *Carnobacterium piscicola* isolated from brazilian smoked fish (Surubim (*Pseudoplatystoma* sp.)) and its activity against a persistent strain of *Listeria monocytogenes* isolated from surubim. J. Food Prot.11: 2068-2077.

[37] Zunabovic M, Domig K, Kneifel W (2011)Practical relevance of methodologies for detecting and tracing of Listeria monocytogenes in ready-to-eat foods and manufacture environments - A review. LWT- Food Sci. Tech. 44(2): 351-362.

[38] Huss HH, Jørgensen LV, Vogel BF (2000) Control options for Listeria monocytogenes in seafoods. Int. J. Food Microbiol. 62(3): 267-74.

[39] Gudmundsdóttir S, Gudbjörnsdottir B, Lauzon H, Einarsson H, Kristinsson KG, Kristjansson M (2005) Tracing Listeria monocytogenes isolates from cold smoked salmon and its processing environment in Iceland using pulsed-field gel electrophoresis. Int. J. Food Microbiol. 101: 41-51.

[40] Bayles DO, Annous BA, Wilkinson BJ (1996) Cold stress proteins induced in Listeria monocytogenes in response to temperature downshock and growth at low temperatures. Appl. Environ. Microbiol. 62: 1116-1119.

[41] Miettinen H, Wirtanen G. Prevalence and location of Listeria monocytogenes in farmed rainbow trout (2005) Int. J. Food Microbiol. 104: 135-143.

[42] Tham W, Ericsson H, Loncarevic S, Unnerstad H, Danielsson-Tham ML (2000) Lessons from an outbreak of listeriosis related to vacuum-packed gravad and cold-smoked fish. Int. J. Food Microbiol. 62(3): 173-175.

[43] Fonnesbech Vogel B, Huss HH, Ojeniyi B, Ahrens P, Gram L (2001) Elucidation of Listeria monocytogenes contamination routes in cold-smoked salmon processing plants detected by DNA-based typing methods. Appl. Environ. Microbiol. 67(6): p. 2586-2595.

[44] Hoffman AD, Gall KL, Norton DM, Wiedmann M (2003) Listeria monocytogenes contamination patterns for the smoked fish processing environment and for raw fish. J. Food Prot. 66: p. 652-670.

[45] Thimothe J, Kerr Nightingale K, Gall K, Scott VN, Wiedmann M (2004) Tracking of Listeria monocytogenes in smoked fish processing plants. J. Food Prot. 67: 328–341.

[46] Peck MW(1997) Clostridium botulinum and the safety of refrigerated processed foods of extended durability. Trend.Food Sci. Tech. 8: 186-192.

[47] Hatheway CL (1995) Hath Botulism: the present status of the disease. Curr. Top. Microbiol. Imm. 195.

[48] Haagsma J. (1991) The distribution of Pathogenic anaerobic bacteria and the environment. Sci. Technic. Rev. Office Int. des. 10: 49-764.

[49] Sramova H, Benes C (1998) Occurrence of botulism in the Czech Republic (in Czech). Zpravy CEM (SZU Praha). 7: 395-397.

[50] Zhou GH, Xu XL, Liu Y (2010) Preservation technologies for fresh meat. Meat Sci. 86: 119-128.

[51] Devlieghere F, Vermeiren L, Debevere J (2004) New preservation technologies: Possibilities and limitations. Int. Dairy J.14: 273-285.

[52] Rodgers S (2001) Preserving non-fermented refrigerated foods with microbial cultures - a review. Trends. Food Sci. Tech. 12: 276-284.

[53] Pilet MF, Leroi F (2011)Applications of protective cultures , bacteriocins and bacteriophages in fresh seafood and seafood product. In Lacroix C. Protective cultures,

antimicrobial metabolites and bacteriophages for food and beverage biopreservation.: ©
2011 Woodhead Publishing Limited.

[54] Galvez A, Abriouel H, Benomar N, Lucas R (2010) Microbial antagonists to food-borne
pathogens and biocontrol. Cur. Opin. Biotech. 21: 142-148.

[55] Garcia P, Rodriguez L, Rodriguez A, Martinez B (2010) Food biopreservation:Promising
strategies using bacteriocins, bacteriophage and endolysins. Trends. Food Sci. Tech.
373-382.

[56] Brillet A, Pilet MF, Prévost H, Cardinal M, Leroi F (2005) Effect of inoculation of
inoculation of *Carnobacterium divergens* V41, a biopreservative strain against *Listeria
monocytogenes* risk, on the microbiological, and sensory quality of cold-smoked salmon.
Int. J. Food Microbiol. 104: 309-324.

[57] Pinto AL, Fernandes M, Pinto C, Albano H, Castilho F, Teixeira P, Gibbs PA (2009)
Characterization of anti- *Listeria* bacteriocins isolated from shellfish. Int. J.Microbiol.
129: 50-58.

[58] Leroi F (2010) Occurrence and role of lactic acid bacteria in seafood products. Food
Microbiol. 27: 698-709.

[59] Mozzi F, Raya RR, Vignolo GM, editors (2010) Biotechnology of Lactic Acid Bacteria:
Novel Applications: Blackwell Publishing.

[60] Stiles E (1996) Biopreservation by lactic acid bacteria. Antonie van Leeuwenhoek. 70:
331-345.

[61] Ghanbari M, Rezaei M, Jami M, Nazari M (2010) Isolation and characterization of
Lactobacillus species from intestinal content of Beluga(*Huso huso*) and persian sturgeon
(*Acipenser persicus*). Iran. J. Vet. Res. 10(2): 152-157.

[62] Mayo B, Aleksandrzak - Piekarczyk T, Fernández M, Kowalczyk M, Álvarez - Martín P,
Bardowski J (2010) Updates in the Metabolism of Lactic Acid Bacteria. In Mozzi F, Raya
RR, Vignolo GM, editors. Biotechnology of Lactic Acid Bacteria: Novel Applications.:
Blackwell Publishing.

[63] Ringo E, Gatesoupe F (1998) Lactic acid bacteria in fish: a review. Aquacult. 160: 177-
203.

[64] Hammes WP, Vogel RF (1995) The genus *Lactobacillus* Glasgow: Blackie Academic &
Professional.

[65] Duffes F, Corre C, Leroi F, Dousset X, Boyaval P (1999) Inhibition of *Listeria
monocytogenes* by in situ produced and semipurifi ed bacteriocins of Carnobacterium
spp. on vacuum-packed, refrigerated. J. Food Prot. 62: 394–1403.

[66] Campos C, Rodríguez O, Calo-Mata P, Prado M, Barros-Velazquez J (2006) Preliminary
characterizationof bacteriocins from *Lactococcus lactis* , *Enterococcus faecium* and
Enterococcus mundtii strains isolated from turbot (*Psetta maxima*). Food Res. Int. 39: 356-
64.

[67] Drider D, Fimland G, Hechard Y, McMullen L, Prevost H (2006) The continuing story of
class IIa bacteriocins. Microbiology and Molcular Biology Reviews. 70: 564 – 582.

[68] Ouwehand A, Vesterlund S (2004) Antimicrobial Components from Lactic Acid Bacteria. In Salminen S, Wright v, Ouwehand A. Lactic Acid Bacteria Microbiological and Functional Aspects.: Marcel Dekker, Inc.

[69] Cotter PD, Hill C, Ross RP (2005) Bacteriocins: Developing innate immunity for food . Nat. Rev. Microbiol 3, 777 – 788 .

[70] Nes I, Yoon S, Diep D (2007) Ribosomally Synthesiszed Antimicrobial Peptides (Bacteriocins) in Lactic Acid Bacteria: A Review. Food Sci. Biotech.. 16(5): 675-690.

[71] Cotter PD, Draper LA, Lawton EM, McAuliffe O, Hill C, Ross RP (2006) Overproduction of wild-type and bioengineered derivatives of the lantibiotic lacticin 3147 . Appl. Environ. Microbiol. 72, 4492 – 4496

[72] Gillor O, Etzion A, Riley M (2008) The dual role of bacteriocins as anti- and probiotics. Appl. Microbiol. Biotech.. 81: 591-606.

[73] Nes I (2011) History, Current Knowledge, and Future Directions on Bacteriocin Research in Lactic Acid Bacteria. In Drider D, RS, (eds.). Prokaryotic Antimicrobial Peptides: From Genes to Applications.: Springer Science+Business Media, LLC p. 3-12.

[74] Lindgren SE, Dobrogosz WJ (1990) Antagonistic activities of lactic acid bacteria in food and feed fermentations. FEMS Microbiol. Lett. 87(1-2): 149-164.

[75] Podolak PK, Zayas JF, Kastner CL, Fung DYC (1996) Inhibition of *Listeria monocytogenes* and *Escherichia coli* O157:H7 on beef by application of organic acids. J. Food Prot. 59: 370-373.

[76] Ammor MS, Mayo B (2007) Selection criteria for lactic acid bacteria to be used as functional cultures in dry sausage production: An update. Meat Sci. 76: 138–146.

[77] Devlieghere F, Debevre J (2000) Influence of dissolved carbon dioxide on the growth of spoilage bacteria. Lebensmittel- und Wissenschaft-Technologie. 33: 531-537.

[78] Lanciotti E, Santini C, Lupi E, Burrini D (2003) Actinomycetes, cyanobacteria and algae causing tastes and odours in water of the River Arno used for the water supply of Florence. J. Water Sup. Res. Tech. 52(7): 489–500.

[79] Kvasnikov EI, Kovalenko NK, Materinskaya LG (1997) Lactic acid bacteria of freshwater fish. Microbiol. 46: 619-624.

[80] Cai Y, Suyanandana P, Saman P (1999) Classification and characterization of lactic acid bacteria isolated from the intestines of common carp and freshwater prawns. The J. Gen. Appl. Microbiol. 45: 177-184.

[81] Huss HH, Jeppesen VF, Johansen C, Gram L (1995) Biopreservation of fish products a reviewof recent approaches and results. J. Aquat. Food Prod. Tech. 4: 5-26.

[82] González CJ, Encinas JP, García-López ML, Otero A (2000) Characterization and identification of lactic acid bacteria from freshwater fishes. Food Microbiol. 17: 383-391.

[83] Bucio A, Hartemink R, Schrama JW, Verreth J, Rombouts FM (2006) Presence of lactobacilli in the intestinal content of freshwater fish from a river and from a farm with a recirculation system. Food Microbiol. 23(5): 476-482.

[84] Ringo E, Strom E (1994) Microflora of Arctic char, *Salvelinus alpinus* (L.); gastrointestinal microflora of free-living fish, and effect of diet and salinity on the intestinal microflora. Aquacult. Fish. Manag. 25: 623-629.

[85] Ringo E (2004) Lactic acid bacteria in fish and fish farming. In Salminen S, Wright A, Ouwehand A, editors. Lactic acid bacteria : Microbiological and Functional Aspects. 3rd ed. New-York: CRC Press; 581-610.

[86] Ringø E, Strøm E, Tabachek JA (1995) Intestinal microflora of salmonids: a review. Aqua. Res. 26: 773–789.

[87] Ringø E, Olsen RE (1999) The effect of diet on aerobic bacterial flora associated with intestine of Arctic charr (*Salvelinus alpinus* L.). J. Appl. Microbiol. 86: 12–28.

[88] Spanggaard B, Huber I, Nielsen J, Nielsen T, Appel KF, Gram L (2000) The microflora of rainbow trout intestine. A comparison of traditional and molecular identification. Aquacult. 182: 1–15.

[89] Seppola M, Olsen RE, Sandaker E, Kanapathippillai P, Holzapfel W, Ringø E (2006) Random amplification of polymorphic DNA (RAPD) typing of carnobacteria isolated from hindgut chamber and large intestine of Atlantic cod (*Gadus morhua* L.). Sys. Appl. Microbiol. 29: p. 131–137.

[90] Gancel F, Dzierszinski F, Tailliez R (1997) Identification and characterization of *Lactobacillus* species isolated from fillets of vacuum-packed smocked and salted herring (*Clupea harengus*). J. appl. Microbiol. 82: 722-728.

[91] Magnússon H, Traustadóttir K (1982) The microbial flora of vacuum-packed smoked herring fillets. J. Food Tech. 17: 695–702.

[92] Paludan-Müller C, Dalgaard P, Huss H, Gram L (1998) Evaluation of the role of *Carnobacterium piscicola* in spoilage of vacuum and modified atmosphere-packed-smoked salmon stored at 5°C. Int. J. Food Microbiol. 39: 155-166.

[93] Leroi F, Joffraud JJ, Chevalier F, Cardinal M (1998) Study of the microbial ecology of cold smoked salmon during storage at 8°C. Int. J. Food Microbiol. 39: 111-121.

[94] Leisner J, Laursen B, Prevost H, Drider D, Dalgaard P (2007) Carnobacterium:positive and negative effects in the environment and in foods. FEMS Microbiol. Rev. 13: 592-613.

[95] Leisner JJ, Millan JC, Huss HH, Larsen LM (1994) Production of histamine and tyramine by lactic acid bacteria isolated from vacuum-packed sugar-salted fish. J. Appl. Bacter. 76: 417–423.

[96] Østergaard A, Ben Embarek PK, Wedel-Neergaard C, Huss HH, Gram L (1998) Characterization of anti-listerial lactic acid bacteria isolated from Thai fermented fish products. Food Microbiol. 15: 223–233.

[97] Olympia M, Ono H, Shinmyo A, Takano M (1992) Lactic acid bacteria in fermented fishery, burong bangus. J. Fer. Bioeng. 73(3): 193-197.

[98] Mauguin S, Novel G (1994) Characterization of lactic acid bacteria isolated from seafood. J. Appl. Bacter. 76 : 616–625.

[99] Emborg J, Laursen BG, Rathjen T, Dalgaard P (2002) Microbial spoilage and formation of biogenic amines in fresh and thawed modified atmosphere-packed salmon (Salmo salar) at 2 degrees C. J. Appl. Microbiol. 92(4). 790-799

[100] Franzetti L, Scarpellini M, Mora D, Galli A (2003) *Carnobacterium* spp. in seafood packaged in modified atmosphere. Annal. Microbiol. 53: 189-193.

[101] Emborg J, Laursen BG, Rathjen T, Dalgaard P (2002) Microbial spoilage and formation of biogenic amines in fresh and thawed modified atmosphere-packed salmon (*Salmo salar*) at 2°C. J. Appl. Microbiol. 92: 790-799.

[102] Dalgaard P, Madsen HL, Samieian N, Emborg J (2006) Biogenic amine formation and microbial spoilage in chilled garfish (*Belone belone*) effect of modified atmosphere packaging and previous frozen storage. J. Appl. Microbiol. 101: 80 - 95.

[103] Lakshmanan R, Dalgaard P (2004) Effect of high-pressure processing on *Listeria monocytogenes*, spoilage microflora and multiple compound quality indices in chilled cold-smoked salmon. J. Appl. Microbiol. 96: 398-408.

[104] Wessels S, Huss HH (1996) Suitability of *Lactococcus lactis* ATCC 11454 as a protective culture for lightly preserved fish products. Food Microbiol. 13: 323-332.

[105] Nilsson L, Ng YY, Christiansen JN, Jorgensen BL, Grotinum D, Gram L (2004) The contribution of bacteriocin to inhibition of *Listeria monocytogenes* by *Carnobacterium piscicola* strains in cold-smoked salmon systems. J. Appl. Microbiol. 96: 133-143.

[106] Altieri C, Speranza B, Del Nobile MA, Sinigaglia M (2005) Suitability of bifidobacteria bacteria and thymol as biopreservatives in extending the shelf life of fresh packed plaice fillets.. J. Appl. Microbiol. 99: 1294-1302.

[107] Yin LJ, Wu CW, Jiang ST (2007) Biopreservative effect of pediocin ACCEL on refrigerated seafood. Fish. Sci. 73: 907-912.

[108] Ringo E (2008) The ability of carnobacteria isolated from fish intestine to inhibit growth of fish pathogenic bacteria. Aqua. Res. 39: 171-180.

[109] Sudalayandi, K.M (2011) Efficacy of lactic acid bacteria in the reduction of trimethylamine-nitrogen and related spoilage derivatives of fresh Indian mackerel fish chunks. Afr. J. Biotech. 10: 42-47.

[110] Kim CR, Hearnsberger JO (1994) Gram negative bacteria inhibition by lactic acid culture and food preservatives on catfish fillets during refrigerated storage. J. Food Sci. 59: 513-516.

[111] Einarsson H, Lauzon HL (1995) Biopreservtaion of brined shrimp (*Pandalus borealis*) by bacteriocins from lactic acid bacteria. Appl. Environ. Microbiol. 61: 669-675.

[112] Morzel M, Fransen NG, Arendt EK (1997) Defined starter cultures for fermentation of salmon fillets. J. Food Sci. 62(6): 1214-1217.

[113] Kışla D, Ünlütürk A (2004) Microbial shelf life of rainbow trout fillets treated with lactic culture and lactic acid. Adv. Food Sci. 26: 17-20.

[114] Elotmani F, Assobhei O (2004) In vitro inhibition of microbial flora of fish by nisin and lactoperoxidase system. Lett. Appl. Microbiol. 38: 60–65.

[115] Aras Husar S, Kaban G, Husar O, Yanik T, Kaya M (2005) Effect of *Lactobacillus* sakei Lb706 on Behavior of *Listeria monocytogenes* in Vacuum-Packed Rainbow Trout Fillets. Tur. J. Vet. Anim. Sci. 29: 1039-1044.

[116] Kim Y, Ohta T, Takahashi T, Kushiro A, Nomoto K, Yokokura T, Okada N, Danbara H (2006). Probiotic *Lactobacillus* casei activates innate immunity via NF-κB and p38 MAP kinase signaling pathways. Microb. Infec.. 8: 994–1005.

[117] Katikou P, Ambrosiadis IGD, Koidis P, Georgakis SA 2(2007) Effect of *Lactobacillus* cultures on microbiological, chemical and odour changes during storage of rainbow trout fillets. J. Sci. Food Agri. 87: 477–484.

[118] Daboor SM, Ibrahim SM (2008) Biochemical and microbial aspects of tilapia (*Oreochromis niloticus* L.) biopreserved by Streptomces sp. metabolites. In 4th International Conference of Veterinary Research Division, National Research Center (NRC); Cairo, Egypt. p. 39-49.

[119] Tahiri M, Desbiens E, Kheadr C, Lacroix IF (2009) Comparison of different application strategies of divergicin M35 for inactivation of *Listeria monocytogenes* in cold-smoked wild salmon. Food Microbiol. 26: p. 783-793.

[120] Matamoros S, Pilet MF, Gigout F, Prévost H, Leroi F (2009) evaluation of seafood-borne psychrotrophic lactic acid bacteria as inhibitors of pathogenic and spoilage bacteria. Food Microbiol. 26: 638-644.

[121] Fall PA, Leroi F, Cardinal M, Chevalier F, Pilet MF (2010) Inhibition of *Brochothrix thermosphacta* and sensory improvement of tropical peeled cooked shrimp by *Lactococcus piscium* CNCM I-4031. Lett. Appl. Microbiol. 50: 357-361.

[122] Ibrahim SM, Salha GD (2009) Effect of antimicrobial metabolites produced by lactic acid bacteria on quality aspects of frozen Tilapia (*Oreochromis niloticus*) fillets. World Journal of Fish and Marine Sciences. 1: 40-45.

[123] Shirazinejad AR, Noryati I, Rosma A, Darah I (2010) Inhibitory Effect of Lactic Acid and Nisin on Bacterial Spoilage of Chilled Shrimp. World Acad. Sci. Eng. Tech.

[124] Fall PA, Leroi F, Chevalier F, C G, Pilet MF (2010) Protective effect of a non-bacteriocinogenic Lactococcus piscium CNCM I-4031 strain against *Listeria monocytogenes* in sterilised tropical cooked peeled shrimp. J. Aquat. Food Prod. Tech. 19: 84-92.

[125] Cosansu S, Mol S, Ucok Alakavuk D, Tosun ŞY (2011) Effects of *Pediococcus* spp. on the quality of vacuum-packed Horse Mackerel during Cold Storage. J. Agri. Sci. 17: 59-66.

[126] Katla T, Moretro T, Aasen IM, Holck A, Axelsson L, Naterstad K (2001) Inhibition of *Listeria monocytogenes* in cold smoked salmon by addition of sakacin P and/or live *Lactobacillus sakei* cultures. Food Microbiol. 18: 431-439.

[127] Blom H, Katla T, Hagen BF, Axelsson L (1997) A model assay to demonstrate how intrinsic factors affect diffusion of bacteriocins. Int. J. Food Microbiol. 38: 103-109.

[128] Brurberg MB, Nes IF, Eijsink VGH (1997) Pheromone-induced production of antimicrobial peptides in *Lactobacillus*. Mol. Microbiol. 26: 347-360..

[129] Eijsink VGH, Skeie M, Middelhoven H, Brurberg MB, Nes IF (1998) Comparative studies of pediocin-like bacteriocins.. Appl. Environ. Microbiol. 64: 3275-3281.

[130] Ganzle MG, Weber S, Hammes WP (1999) Effect of ecological factors on the inhibitory spectrum and activity of bacteriocins. Int. J. Food Microbiol. 46: 207 – 217.

[131] Aasen IM, Moretro T, Katla T, Axelsson L, Storro I (2000) Infuence of complex nutrients, temperature and pH on bacteriocin production by *Lactobacillus* sakei CCUG 42687. Appl. Microbiol. Biotech.. 53: 159-166.

[132] Leroi F, Arbey N, Joffraud J, Chevalier F (1996) 'Effect of inoculation with lactic acid bacteria on extending the shelf-life of vacuum-packed cold-smoked salmon. Int. J. Food Sci. Tech. 1996; 31: p. 497-504.

[133] Budu-Amoako B, Albert RF, Harris J, Delves-Broughton J (1999) Combined effect of nisin and moderate heat on destruction of *Listeria monocytogenes* in cold-pack lobster meat. J. Food Prot. 62: 46–50.

[134] Nilsson L, Gram L, Huss H (1999) Growth control of *Listeria monocytogenes* on cold smoked salmon using a competitive lactic acid bacteria flora. J. Food Prot. 62: 336-342.

[135] Nykanen A, Weckman K, Lapvetelainen A (2000) Synergistic inhibition of *Listeria monocytogenes* on cold-smoked rainbow trout by nisin and sodium lactate. Int. J. Food Microbiol. 61: 63-72.

[136] Silva J, Carvalho AS, Teixeira P, Gibbs PA (2002) Bacteriocin production by spray-dried lactic acid bacteria. Lett. Appl. Microbiol. 34(2): 77-81.

[137] Bouttefroy A, Millière JB (2000) Nisin-curvaticin 13 combinations for avoiding the regrowth of bacteriocin resistant cells of *Listeria monocytogenes* ATCC 15313. Int. J. Food Microbiol. 62: 65-75.

[138] Yamazaki K, Suzuky M, Kawai Y, Inoue N, Montville TJ (2003) Inhibition of *Listeria monocytogenes* in cold-smoked salmon by *Carnobacterium piscicola* CS526 isolated from frozen surimi. J. Food Prot. 66: 1420-1425.

[139] Yamazaki K, Suzuki M, Kawai Y,IN, Montville TJ (2005) Purification and characterization of a novel class IIa bacteriocin, piscicocin CS526, from surimi associated *Carnobacterium piscicola* CS526. Appl. Environ. Microbiol. 71: p. 554-557.

[140] Brillet A, Pilet MF, Prevost H, Bouttefroy A, Leroi F (2004). Biodiversity of *Listeria monocytogenes* sensitivity to bacteriocin-producing Carnobacterium strains and application in sterile cold-smoked salmon. J. Appl. Microbiol. 97: 1029-1037.

[141] Weiss A, Hammes WP (2006) Lactic acid bacteria as protective cultures against *Listeria* spp. on cold-smoked salmon. Eur. Food Res. Tech. 222: 343-346.

[142] Vescovo M, Scolari G, Zacconi C (2006) Inhibition of *Listeria innocua* growth by antimirobial-producing lactic acid cultures in vacuum-packed cold-smoked salmon. Food Microbiol. 23: 689-693.

[143] Laursen BG, Bay L, Cleenwerck I, Vancanneyt M, Swings J, Dalgaard P (2005). *Carnobacterium divergens* and *Carnobacterium maltaromicum* as spoilers or protective cultures in meat and seafood: phenotypic and genotypic characterisation. Sys. Appl. Microbiol. 28: 151-164.

[144] Tomé E, Pereira VL, Lopes CI, Gibbs PA, Teixeira PC (2008) In vitro tests of suitability of bacteriocin-producing lactic acid bacteria, as potential biopreservation cultures in vacuum-packaged cold-smoked salmon. Food Control. 19: 535-543.

[145] Matamoros S, Leroi F, Cardinal M, Gigout F, Kasbi Chadli F, Cornet J, Prevost F, Pilet M.F (2009). Psychrotrophic lactic acid bacteria used to improve the safety and quality of vacuum-packaged cooked and peeled tropical shrimp and cold smoked salmon. J. Food Prot. 72: 365-374.

New Fields of Application

Exploring Surface Display Technology for Enhancement of Delivering Viable Lactic Acid Bacteria to Gastrointestinal Tract

Shirin Tarahomjoo

Additional information is available at the end of the chapter

1. Introduction

Anchoring of proteins to the cell surface is a common theme in nature and the processes governed by different surface proteins are bases of many biological phenomena, such as cell-cell recognition, signal transduction, adherence, colonization, and immunoreactions (Westerlund & Korohonen, 1993). The utilization of cellular surface anchoring systems for the display of heterologous proteins on the surface of microbial cells has been developed into an active research area that holds a great promise for a variety of biotechnological applications including the production of whole cell biocatalysts, microbial adsorbents, live vaccines, antibody fragments, and screening of novel proteins (Hansson et al., 2001; Kondo& Ueda, 2004; Lee et al., 2003). Generally construction of these systems is accomplished by the expression of heterologous peptides or proteins as fusions with anchoring domains, which are able to attach to the cell surface (Fig 1.). Anchoring domains are usually cell surface proteins or their fragments. Depending on the characteristics of target and anchor proteins, N-terminal fusion, C-terminal fusion or sandwich fusion strategy can be considered (Lee et al., 2003).

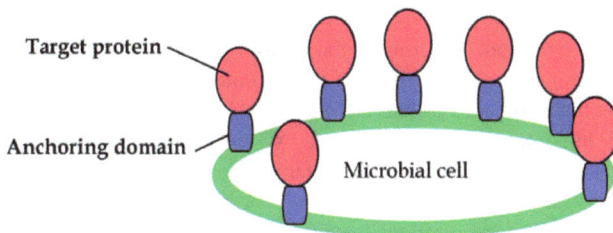

Figure 1. Schematic representation of a microbial surface display system

LAB are gram positive, non-spore forming, fastidious, acid tolerant, and strictly fermentative that secret lactic acid as the major end product of sugar fermentations (Axelsson, 1998). LAB are naturally present in media rich in organic nutrients such as food products and digestive tracts. They are a genetically diverse group of bacteria with GC contents varying from 34 to 53%, including rod shaped bacteria such as lactobacilli and also cocci such as lactococci, enterococci, pediococci, and leuconostoc (Stiles & Holzapfel, 1997). Since time immemorial, LAB have been used for the fermentation and preservation of food products, particularly dairy products, fermented meats, and vegetables. Consequently, several strains of LAB have a long record of safe association with humans and human foodstuffs (Mckay & Baldwin, 1990). The display of proteins on the surface of LAB cells can broaden or improve applications of these bacteria. In this chapter, we intend to describe cell surface anchoring domains used in LAB surface display systems. Then applications of surface engineered LAB are depicted and key factors affecting their performances are highlighted. Moreover, we explained comprehensively a novel application of the protein display in LAB, which is potentially useful for enhancement of the delivery of viable LAB to the gastrointestinal tract (GIT).

2. Anchoring domains in surface display systems of LAB

The cell surface anchoring domains utilized for the development of LAB surface display systems are associated with the cell surface either covalently or noncovalently. Depending on the applied anchoring domains, two modes of the protein display can be considered, including internal and external mode of the protein display. In the case of the internal mode of protein display, fusions of target proteins to anchoring domains are expressed in LAB cells, and therefore target proteins are displayed on the surface of expression hosts, whereas in the case of the external mode of protein display, expression and display hosts are different from each other. If the association of anchoring domains with the cell surface is of a non covalent nature, they can bind to cells when they are added from the outside. Therefore, fusions of target proteins to anchoring domains are produced in suitable expression hosts capable of their correct folding and modifications. The fusion proteins are then purified and incubated with desired display hosts in order to attach to the cell surface. As a result, the external mode of protein display can retain the nongenetically modified status of cells and is valuable for food and vaccine development.

The surface display systems based on the internal mode of protein display are often associated with the limitations in terms of the translocation of target proteins to the cell surface and the control of surface intensity of target proteins. The mislocalization of target proteins can affect their functions negatively(Dieye et al., 2003; Van Der Vaart et al., 1997; Wan et al., 2002). In contrast, the surface display systems based on the external mode of protein display can ensure the full exposure of target proteins outside of the cell wall and the surface intensity of target proteins can readily be adjusted by selecting appropriate display hosts and suitable concentrations for the fusion proteins in the incubation mixture. However, regarding noncovalent interaction of target proteins with the cell surface, the possibility of dissociation of target proteins from the cell surface should be considered.

2.1. Covalent anchors

2.1.1. LPXTG motif containing anchors

The most widely used surface display systems of LAB are based on cell wall anchored proteins that contain an LPXTG motif. These proteins are first synthesized as a preprotein containing an N-terminal signal peptide and a C-terminal cell wall anchor domain. The anchor domain starts at its C-terminus with a short tail of positively charged residues (five to seven amino acids) that remain in the cytoplasm. Upstream of the cytoplasmic domain, a stretch of approximately 30 hydrophobic amino acids is preceded by the highly conserved pentapeptide LPXTG. The charged tail and hydrophobic domain are thought to function as a temporary stop to position the LPXTG motif for proteolytic cleavage. Correct positioning results in cleavage between the threonine and glycine residues followed by amid linkage of the threonine residue to the peptide crossbridge in the peptidoglycan of cell wall, by the action of a sortase (Navarre & Schneewind, 1994). The amino acid composition of the peptide crossbridge varies among the different LAB species and is flexible with respect to the sorting reaction (Strauss et al., 1998; Ton-That et al., 1998). The anchor domain is preceded by a wall associated region of about 50 to 125 residues and is characterized by a high percentage of proline/glycine and/or threonine/serine residues (Fischetti et al., 1990). In surface display genetic constructs, secretion signal peptides are fused to the N-termini of target proteins to transport them to the outside of cell and the LPXTG motif containing anchors are fused to the C-termini of target proteins in order to attach them to the cell surface. The cell surface associated proteinases of *Lactococcus lactis* and *Lactobacilli casei* (PrtPs), M6 protein of *Streptococcus pyogenes*, and Protein A of *Staphylococcus aureus* (SpA) are among LPXTG type anchors, which have mainly been used for the construction of surface display systems in LAB (Maassen et al., 1999; Norton et al., 1996; Piard et al., 1997; Pouwels et al., 1996; Steidler et al., 1998).

Slight deviations from the LPXTG consensus sequence is found in some LAB. Recently Kleerebezem (2003) found that sortase substrates of *Lactobacillus plantarum* contain LPQTXE motifs instead of the LPXTG motifs. In addition, in the carboxy end of cell surface proteinase of *Lactobacillus delbrueckii* (PrtB), a degenerated LPKKT motif is surrounded by two imperfect repeats of 59 residues, which are lysine rich. However, downstream of the LPKKT motif, there is no hydrophobic domain and no charged tail at the extreme C-terminus. It was shown that the C-terminal region of PrtB was able to attach to the cell wall of *L. lactis* and the capacity of attachment was drastically reduced by absence of the duplicated sequences. The high content of total positive charges in the anchoring region of PrtB suggests interactions of the anchor with the negatively charged teichoic acids of the cell wall. The mechanism of PrtB attachment to the cell wall probably implicates electrostatic forces (Germond et al., 2003).

2.1.2. Lipoprotein anchors

Lipoproteins are lipid modified proteins produced as secretory precursors with a signal peptide linked to their amino termini. The C-terminal region of their signal peptide contains a well conserved lipobox motif of four amino acid residues and invariably, the last residue is cysteine. The covalent binding of lipoproteins to the cell membrane is generally achieved via diacylglyceryl modification of the indispensible cysteine residue in the lipobox by a diacylglyceryl transferase. Lipidation of the cysteine residue is a perquisite for cleavage of the signal peptide by a lipoprotein specific signal peptidase (SPase II)(Yamaguchi et al., 1988; Venema et al., 2003). The lipoprotein anchors should be fused at their C-termini to N-termini of target proteins in order to display them on the cell surface. Poquet (1998) identified four lactococcal lipoprotein anchors using nuclease of *S. aureus* as an export specific reporter enzyme. The nuclease activity was shown to require an extracellular location in *L. Lactis* demonstrating its suitability to report the protein export. The enzyme activity was detected in a plate test by the presence of pink halos. Fusions of the lipoprotein anchors to the nuclease expressed in *L. Lactis* were associated with the cell fraction and the recombinant lactococcal cells showed strong nuclease activities indicating the cell surface anchoring function of the lipoproteins. For one of the anchors (NlpI), the surface location of the fusion protein was also confirmed by proteinase K treatment of *L. lactis* cells (Poquet et al., 1998). Basic membrane protein A (BmpA) of *L. Lactis* is a putative lipoprotein that has been used for the protein display on the surface of lactococci (Berlec et al., 2011).

2.1.3. Transmembrane anchors

The strategy to insert target amino acid sequences in the exterior loop between transmembrane spanning domains (TMSs) can limit the insert size in order not to disturb the membrane protein topology. Therefore, a fusion approach is often preferred in which a target protein is simply linked at its N-terminus to one or more TMSs of a cytoplasmic membrane protein. *L. Lactis* bacteriocin transport accessory protein (LcnD) and *Bacillus subtilis* poly-γ-glutamate synthetase A protein (PgsA) are transmembrane proteins, which were fused to the N-termini of target proteins in order to display them on the cell surface of *L. Lactis* and *L. casei*, respectively (Franke et al., 1996; Narita et al., 2006). In addition, in the same random procedure as described above for the lipoprotein anchors Poquet (1998) identified seven lactococcal gene fragments encoding TMSs that function as membrane anchors in *L. Lactis*.

2.2. Noncovalent anchors

2.2.1. S- layer protein anchors

Some LAB strains possess a surface layer (S-layer) of proteins as the outermost structure of the cell envelope. These S-layers are composed of regularly arranged subunits of a single protein (SLP) and may constitute up to 20% of the total cell protein content. S-layers self assemble in entropy driven process during which multiple noncovalent interactions between individual SLPs and the underlying cell surface take place. These two types of interactions in SLPs can be assigned to two separate domains including the self assembly domain and the cell wall binding domain. These domains have been characterized in SLPs of *Lactobacillus acidophilus* ATCC 4356 (Sᴀ), *Lactobacillus crispatus* JCM 5810 (CbsA), and

Lactobacillus brevis ATCC 8287 (SlpA). The C-terminal regions of S$_A$ and CbsA showed the cell surface anchoring function and the N-terminal regions were involved in the self assembly process. In contrast, the self assembly domain of SlpA was located in the C-terminal region and its cell wall binding domain was found in the N-terminal region. The (lipo)teichoic acids were identified as the cell wall ligands of S$_A$ and CbsA. Moreover, the specific cell wall component that interacts with SlpA was shown to be the neutral polysaccharide moiety of the cell wall (Antikainen et al., 2002; Avall-Jaaskelainen et al., 2008; Smit et al., 2002). Avall-Jaaskelainen (2002) decribed the construction of recombinant *L. brevis* strains expressing poliovirus epitope VP1 of 10 amino acid residues inserted in the *slpA* gene. Insertion sites in the *slpA* gene were selected on the basis of the hydrophilicity profile of the SlpA protein. The four most hydrophilic parts of the SlpA protein were selected for testing because it was expected that parts of them were likely to be sites where the epitope would be accessible to the cell surface. One of the insertion sites was at the N-terminus of SlpA and the others were at its C-terminus. Only one site at the C-terminus showed strong colour response in whole cell enzyme linked immunosorbent assay (ELISA) using anti epitope antibody demonstrating that the epitope was accessible on the surface of the recombinant *L. brevis*. In another study, the C-terminal region of SLP of *L. crispatus* K2-4-3 (LcsB) isolated from the chicken intestine was used for the construction of surface display systems. Green fluorescent protein (GFP) was fused to the N-terminus of LcsB. The fusion protein (GFP-LcsB) was expressed in *Escherichia coli*. It was then purified and mixed with various LAB. The binding of the fusion protein to LAB cells was viewed by the fluorescence microscopy. GFP-LcsB was associated with the cell surface of various LAB including *L. delbrueckii*, *L. brevis*, *Lactobacillus helveticus*, *Lactobacillus johnsonii*, *L. crispatus*, *Streptococcus thermophilus*, *L. lactis* and *Lactobacillus salivarius*. GFP alone did not bind to the cells. These results indicated that binding of GFP to the surface of LAB cells is directed by LcsB. However, the fusion protein could not attach to the cell surface of *L. casei*. The reason for this observation requires further studies to elucidate the target ligand of LcsB on the cell surface of LAB (Hu et al., 2011).

2.2.2. Lysin motif containing anchors

The lysine motif (LysM) was first discovered in the lysozyme of *Bacillus* phage ø29 as a C-terminal repeat composed of 44 amino acids separated by 7 amino acids. The cell wall attachment of several bacterial proteins in both gram positive and gram negative organisms occurs through LysMs, often repeated several times in the protein sequence. Many LysM containing proteins are cell wall hydrolases (Buist et al., 2008). The C-terminal region of peptidoglycan hydrolase (AcmA) of *L. lactis* MG 1363 (CpH) contains three 44 amino acid residue lysMs separated by stretches of 21 to 31 amino acids rich in the serine, threonine, and asparagine residues. CpH is able to bind to the cell surface of lactococci and several strains of lactobacilli. Moreover, CpH is able to bind both to the cell surface of LAB treated with sodium dodecyl sulfate (SDS) to remove cell wall associated proteins and LAB treated with trichloroacetic acid (TCA) to remove carbohydrates and (lipo) teichoic acids. These findings suggest that peptidoglycan is the binding ligand of the CpH domain (Buist et al.,

1995). The C-terminus of endolysin Lyb5 of *Lactobacillus fermentum* bacteriophage øPYB5 (Ly5C) contains three LysMs. Each of LysMs is composed of 41 amino acids and they are separated by intervening sequences varying in length and composition. Ly5C fused to GFP was expressed in *E. coli*. After mixing the fusion protein with various cells *in vitro*, GFP was successfully displayed on the surfaces of *L. lactis, L. casei, L. brevis, L. plantarum, L. fermentum, L. delbrueckii, L. helveticus,* and *S. thermophilus* cells. Increases in the fluorescence intensities of TCA treated *L. lactis* and *L. casei* cells compared to those of nontreated cells showed that the cell wall peptidoglycan was the cell surface binding target of Ly5C. Concentration of sodium chloride and pH influenced the binding capacity of the fusion protein, and optimal conditions of these factors were determined empirically in order to obtain high fluorescence intensities of *L. lactis* and *L. casei* cells (Hu et al., 2010). N-terminus of putative muropeptidase (MurO) of *L. plantarum* also contains two LysMs composed of about 43 amino acids separated by 22 amino acid residue sequences. The LysM domain fused to GFP was expressed in *E. coli* and it was able to bind to the cell surface of *L. plantarum* after being mixed with the cells (Xu et al., 2011). Examination of supernatant fractions from broth cultures of *L. fermentum*, revealed the presence of a 27-kDa protein termed Sep. The N-terminus of Sep contains a LysM. Sep fused N-terminally to a six histidine epitope was expressed in *L. fermentum, Lactobacillus rhamnosus,* and *L. lactis*. The protein was found associated with the surface of the expression hosts. However, it was largely present in the supernatant of the cell cultures (Turner et al., 2004).

2.2.3. WxL anchors

The C-terminal cell wall binding domain designated WxL was first identified in proteins of *Lactobacillus* and other LAB based on *in silico* analysis (Kleerebezem et al., 2010). WxL domain contains a WxL motif followed by a proximal well conserved YXXX(L/I/V)TWXLXXXP motif. This domain was found in gene clusters that also encode additional extracellular proteins with C-terminal membrane anchors and LPxTG motif containing anchors, suggesting that they form an extracellular protein complex (Siezen et al., 2006). The C-terminal WxL domains identified in two proteins of *Enterococcus faecalis* were fused at their N-termini to an export reporter enzyme (nuclease of *S. aureus)* and a secretion signal peptide. The fusion proteins expressed in *E. faecalis* were detected in both cell wall and supernatant fractions of the recombinant enterococci. Removal of the WxL domains from the fusion proteins nearly eliminated them in the cell wall. Treatment of the cell wall fractions with SDS disrupted binding of the fusion proteins to these fractions. These results indicated that the fusion proteins had noncovalent interactions with the cell wall of *E. faecalis*. The fusion proteins were able to attach to the cell surface of *E. faecalis* and *L. johnsonii* when they were added exogenously (Brinster et al., 2007).

2.2.4. Other anchors

Basic surface protein A (BspA) is a surface located protein of *L. fermentum* BR11. Sequence comparisons have been shown that BspA is a member of family III of the solute binding

Exploring Surface Display Technology for Enhancement of Delivering Viable Lactic Acid Bacteria to
Gastrointestinal Tract

101

proteins. Most solute binding proteins are lipoproteins. However, BspA is not a lipoprotein and is attached to the cell envelope by electrostatic interactions. It has been used as a fusion partner to direct proteins to the cell surface of *L. fermentum* BR11. In these constructs, BspA was fused at its C-terminus to target proteins and the fusion proteins were expressed in *L. fermentum* BR11 (Turner & Giffard, 1999). The C-terminal region of cell associated dextransucrase of *Leuconostoc mesentroides* IBT-PQ (DsrP) contains five repeats of 65 amino acid residues. The domain expressed in *E. coli* was able to bind to the cell surface of *L. mesentroides* IBT-PQ cells after being mixed with the cells (Olvera et al., 2007). The carboxy end of PrtP of *Lactobacillus acidophilus* was used for the protein display on *L. acidophilus* using the internal mode. The association of this domain with the cell surface was mediated by electrostatic interactions (Kajikawa et al., 2011).

3. Applications of surface engineering of LAB

Research in the field of surface engineering of LAB has mainly been focused on the construction of vaccine delivery vehicles but other interesting applications have also been reported. In this section, we will describe different areas of biotechnology in which surface display of heterologous proteins on LAB have been investigated.

3.1. Development of vaccine delivery vehicles

Vaccination represents one of the most effective public health strategies to combat infectious diseases (Mielcarek et al., 2001). One of the technologies being developed for vaccine production is the use of bacteria as live vectors for the delivery of recombinant vaccine antigens to the immune system. Such vaccines have the potential for the production of protective antigens *in vivo* and are inexpensive to manufacture (Moore et al., 2001).

Most infections affect or initiate infectious processes at mucosal surfaces and mucosal local immune responses can block pathogens at the portal of entry. Live bacterial vaccines can induce mucosal, as well as systemic, immune responses when delivered via mucosal routes, such as oral or intranasal administration (Mielcarek et al., 2001). The mucosal, needle free, administration of vaccines can significantly decrease the need for syringes with their inherent added cost and risk of disease transmission, and it can increase compliance, and consequently the coverage of vaccination programs (Giudice & Campbell, 2006). The first live recombinant bacterial vectors developed were derived from attenuated pathogenic microorganisms. In addition to the difficulties often encountered in the construction of stable attenuated mutants of pathogenic organisms, attenuated pathogens may retain a residual virulence level that renders them unsuitable for the vaccination of partially immunocompetent individuals such as infants, the elderly or immunocompromised patients (Curtiss, 2002). These problems can be addressed by the application of nonpathogenic food grade LAB as antigen delivery vehicles. LAB therefore represent attractive alternatives as antigen carriers and their use has mainly been focused on the construction of mucosal vaccines. The cellular location of antigens can influence the elicited immunological responses. Cell surface anchored antigens are better recognized by the immune system than those produced intracellularly. Furthermore, intracellular production of antigens may limit their *in vivo* release. The vaccines constructed by the cell surface anchoring of antigens are of particulate nature. In contrast to most soluble antigens, which are ignored by the immune systems, particles are recognized as foreign and as danger eliciting effective immune responses (Storni et al., 2005). Tetanus toxin fragment C (TTFC) is an immunogen protective against tetanus. In a pioneer study by Norton (1996) three recombinant strains of *L. lactis* expressing TTFC in three cellular locations, intracellular, secreted or cell surface anchored via lactococcal PrtP were constructed. The recombinant lactococcal cells were used to immunize mice, which were then challenged by the subcutaneous inoculation of tetanus toxin. When compared in terms of the dose of expressed TTFC required to elicit protection against the lethal challenge, the cell surface displayed form of TTFC was significantly (10-20 fold) more immunogenic than the alternative forms of the protein. The result of this study indicated the advantage of antigen display on the cell surface for the construction of the lactococcal vaccines. In addition to TTFC, several other antigens were displayed on the surface of LAB and the protection studies were carried out to evaluate the efficiency of these vaccines (Bermudez-Humaran et al., 2005; Hou et al., 2007; Kajikawa et al., 2007; Lee & Faubert, 2006; Lee et al., 2006; Li etal., 2010; Lindholm et al., 2004; Liu et al., 2009; Medina et al., 2008; Poo et al., 2006; Tang & Li, 2009; Wei et al., 2010; Xin et al., 2003).

F18 fimbrial *E. coli* strains are associated with porcine postweaning diarrhea and pig edema disease. Adherence of F18 fimbrial *E. coli* to porcine intestinal epithelial cells is mediated by the FedF adhesin of F18 fimbriae. For the development of a mucosal vaccine against porcine postweaning diarrhea and edema disease, different expression cassettes for the display of FedF on the cell surface of *L. lactis* were constructed. Preliminary attempts to express the entire FedF protein as a fusion protein in *L. lactis* resulted in inefficient secretion and degradation of the adhesin. Therefore, only those regions of FedF required for binding specificity to porcine intestinal epithelial cells, were used in the construction of cell surface display systems. Initially, recombinant *L. lactis* clones secreting the partially overlapping receptor binding domains of FedF (42 and 62 amino acid residues) were prepared using two different signal peptides. Substantially higher levels of the fusion proteins (four- to six-fold) were secreted by the clones possessing *L. brevis* SlpA signal peptide than by those possessing *L. lactis* Usp45 signal peptide. In order to enhance the secretion of the fusion proteins, a synthetic sequence encoding the propeptide LEISSTCDA was inserted between the signal sequences and the receptor binding domains of FedF. For the construction of surface display systems, the secreted proteins were anchored to the cell wall of *L. lactis* via the CpH protein or the lactococcal PrtP protein. Three groups of expression vectors with *prtP* spacer sequences of 0.6, 0.8 and 1.5 kb were also designed. The spacers inserted between the receptor binding domains and the anchors. Whole cell ELISA for the detection of cell surface exposure of the FedF receptor binding regions showed that the CpH anchor performed significantly better than the PrtP anchor, particularly in a *L. lactis* mutant devoid of the extracellular housekeeping protease, HtrA. Among the cell surface display systems possessing the CpH anchor, only those with the longest PrtP spacer resulted in efficient binding of the recombinant *L. lactis* cells to porcine intestinal epithelial cells (Lindholm et al., 2004).

In another study, pneumococcal surface antigen A (PsaA), a conserved membrane anchored virulence factor, was expressed in different strains of LAB and it was associated with the surface of LAB cells. *L. plantarum* and *L. helveticus* were found to be more effective at inducing mucosal and systemic anti-PsaA immune responses than *L. casei* following intranasal vaccination of mice. Because all three *Lactobacillus* strains expressed almost the same amount of PsaA and were also recovered from mice nasal mucosa in the same period (3 days), the observed differences among their respective antibody responses may reflect their different intrinsic adjuvant properties. PsaA expressed by *L. lactis* at 2×10^{-8} ng/ colony forming unit (CFU), which was about 10% of PsaA amount produced by the *Lactobacillus* strains. The recombinant *L. lactis* remained in the nasal mucosa only 1 day after the inoculation. Therefore, the inability of *L. lactis* expressing PsaA to significantly induce serum IgG or secreted IgA in mice can be explained by the low level of antigen production in this bacterium compared with the *Lactobacillus* strains and also its shorter persistence in the nasal mucosa. Intranasal inoculation of the mice with *L. lactis* expressing PsaA did not exert any effect on *Streptococcus pneumoniae* recovery from the nasal mucosa upon colonization challenge, in comparison with inoculation of saline or the control *L. lactis* carrying the expression vector devoid of the antigen gene. On the other hand, all the recombinant *Lactobacillus* strains showed a significant reduction of *S. pneumoniae* colonization when compared with the saline group ($10^{0.6}$-$10^{1.35}$ CFU). However, only *L. helveticus* expressing PsaA showed a significant reduction of *S. pneumoniae* colonization in relation to control *L. helveticus* (10 CFU) (Oliviera et al., 2006). Among LAB, *L. lactis* and *Lactobacillus* strains have mostly been used for the construction of LAB vaccines. Selection of LAB strains for use as antigen carriers depends on their persistence in the host, capacity to express foreign antigens, and intrinsic adjuvanticity. *L. lactis* does not colonize the internal cavities of man or animals. Therefore, the use of lactobacilli, which are able to colonize the cavities such as the GIT transiently seems more advantages than that of *L. lactis* for developing LAB vaccines because the longer persistence of lactobacilli in the host body may enhance immunological responses. On the other hand, the progress in the genetics of lactobacilli is more recent than that of lactococci. Furthermore, the availability of a commercial powerful gene expression system for *L. lactis*, nisin inducible gene expression system, urged many researchers to establish LAB vaccines based on lactococci. It has been reported that several LAB strains particularly strains from the genera *Lactobacillus* are able to act as adjuvants. This aspect should be considered when selecting a vaccine strain as it is a natural way to potentiate the immune reaction against heterologous antigens produced by recombinant LAB. It might be speculated that a high level of the antigen expression will not be necessary when using immunostimulatory LAB strains. However, studies have not yet been reported for comparison of the adjuvanticity of *L. lactis* and different lactobacilli. *L. casei* and *L. plantarum* are among *Lactobacillus* strains, which can colonize the GIT of human and mice and they show immunostimulatory properties (Pouwels et al., 1998; Wells et al., 1996).

It has been reported that the flagellin of *Salmonella* has significant vaccine potential because it is the only surface antigen of *Salmonella* detected to have a mitogenic stimulatory effect on lymphocytes (Toyota-Hantani et al., 2008). Bacterial flagellins can also induce innate immune responses through their interaction with Toll-like receptor 5 (TLR5) (Ramos et al., 2004). Recombinant *L. casei* cells expressing the flagellin of *Salmonella* Enteritidis (LCF) on their cell surface via *L. casei* PrtP anchor were constructed. Intragastric immunization of mice with the recombinant lactobacilli resulted in a significant level of protective immunity against an oral challenge with *S*. Enteritidis. There was no significant difference in the level of protection after immunization with the recombinant lactobacilli compared with the free flagellin isolated from *S. Enteritidis*, although the amount of flagellin carried by LCF was less than that of the free flagellin. The immunization of mice with the recombinant lactobacilli did not result in antigen-specific antibody responses in either feces or sera but did induce the release of interferon (IFN)-γ on restimulation of primed lymphocytes *ex vivo*. These results suggested that the protective efficacy provided by flagellin expressing *L. casei* was mainly attributable to cell mediated immune responses. When the levels of IFN-γ produced by primed and flagellin restimulated lymphocytes were compared between the recombinant *L. casei* cells expressing flagellin on their cell surface, and a mixture of the purified flagellin and normal *L. casei*, the results indicated that the *Lactobacillus* strain showed adjuvanticity only when the flagellin was expressed on the cell surface (Kajikawa et al., 2007).

Two recombinant *L. acidophilus* strains displaying flagellin of *Salmonella typhimurium* on the cell surface were constructed using different anchor motifs. In one construct, the flagellin gene was fused at its carboxy end to the C-terminal region of PrtP of *L. acidophilus*. In other construct, the flagellin gene was fused in the same way to the anchor region of mucus binding protein (Mub) of *L. acidophilus* containing an LPXTG motif. The density of the flagellin fused protein at the cell surface of *L. acidophilus* displaying the flagellin by the PrtP protein (FliC-PrtP) was higher than that of *L. acidophilus* displaying the antigen by the Mub anchor (FliC-Mub). Both of the recombinant lactobacilli showed TLR5 stimulating activity, which indicated that the surface associated flagellin was recognized by TLR5. The magnitude of the TLR5 stimulating activity of *L. acidophilus* cells expressing FliC-PrtP was higher than that of *L. acidophilus* cells expressing FliC-Mub and this result showed that the magnitude of the TLR5 stimulating activity was dependent on the quantity of surface located flagellin. Moreover, the two recombinant lactobacilli exhibited dissimilar maturation and cytokine production by human myeloid dendritic cells (Kajikawa et al., 2011).

Human papillomavirus type 16 (HPV-16) has been associated with more than 50% of HPV related cervical cancer (CxCa) (Krinbauer et al., 1992). HPV-16 E7 oncoprotein is constitutively expressed in CxCa cells during malignant progression of HPV-16 induced cervical lesions and is therefore considered as an effective target for the cancer immunotherapy. *L. lactis* was engineered to express HPV-16 E7 on the cell surface (LL-E7) using the M6 anchor and its coadministration with another lactococci secreting IL-12 (LL-IL-12) was investigated for the immunization and immunotherapy of HPV-related CxCa. IL-12 is a heterodimeric cytokine that induces Th1 responses, enhances cytotoxic T-lymphocyte (CTL) maturation, promotes natural killer (NK) cell activity and induces IFN-γ production. Mice were vaccinated intranasally with LL-E7 and were then challenged by injection of tissue culture number 1 (TC-1) tumor cells. Thirty five percent of the vaccinated mice remained tumor free. Coadministration of LL-IL-12 with LL-E7 resulted in higher antitumor activities as half of the inoculated mice remained tumor free and the tumor median size in the remaining tumor bearing animals was less than that in LL-E7 immunized mice. Antitumoral activity elicited by covaccination with LL-E7 and LL-IL-12 appeared to be long lasting as when the tumor free animals were rechallenged 3 months later with TC-1 cells, they remained tumor free for up to 6 months. To investigate the therapeutic effects of the coadministration of LL-E7 and LL-IL-12, mice were challenged with the TC-1 tumor cell line prior to the initiation of immunotherapy. Once 100% of the mice had palpable tumor, the immunotherapy was started. Only LL-E7/LL-IL-12 treatment resulted in total tumor regression in 35% of the immunized animals. Moreover, the tumor median size in the remaining tumor bearing mice was lower than that measured in mice treated with LL-E7. In contrast, no tumor regression was observed in mice treated with LL-E7 alone. Mice immunized with LL-E7/LL-IL-12 also exhibited both systemic and mucosal humoral responses, which were induced at higher levels than those in mice vaccinated with LL-E7 (Bermudez-Humaran et al., 2005).

Boiling of *L. lactis* cells in TCA followed by washing and neutralization resulted in nonviable spherical peptidoglycan microparticles , which are deprived of surface proteins and their intracellular content is largely degraded. The proteins IgA1 protease (IgA1p), putative proteinase maturation protein A (PpmA) and streptococcal lipoprotein A (SlrA) were bound to the surface of the lactococcal particles after recombinant production of the antigens as hybrids with the CpH domain. TCA removes (lipo) teichoic acids from the lactococcal cell wall that results in enhancement of the binding capacity for CpH fusions. Mice immunized intranasally with the monovalent lactococcal particle based vaccines were not protected against an intranasal pnemococcal challenge. However, intranasal immunization with a trivalent vaccine containing PpmA, SlrA and IgA1p bound to the lactococcal particles by the CpH domain showed protection against fatal pneumococcal pneumonia in mice (Audouy et al., 2007).

3.2. Development of whole cell biocatalysts

Cellulosome is a multi enzyme complex in which various cellulytic enzymes assemble into a macromolecular structure by their attachment to a nonenzymatic central scaffold protein for the efficient degradation of cellulose (Bayer et al., 2004; Demain et al., 2005). Cellulosome of the gram positive thermophile *Clostridium thermocellum* is anchored to the surface of cells, resulting in one of the most efficient bacterial systems for the cellulose hydrolysis. All of the cellulosomal enzymes of *C. thermocellum* contain a twice repeated sequence, usually at their C-termini, called type I dockerin domain. These dockerin domains are considered to bind to the hydrophobic domains termed cohesins, which are repeated nine times in the central scaffold protein (CipA) of the bacterium. CipA also contains a cellulose binding module (CBM3a), allowing the different cellulases to act in synergy on the crystalline substrate, as well as a type 2 dockern domain, which binds the cell surface anchor proteins, ensuring the cellulosome's attachment to the cell surface. Association of the cellulosome with the cell surface yields formation of cellulose-enzyme-microbe ternary complexes and results in enhanced activity and synergy (Begum et al., 1996). The assembly of recombinant cellulosome inspired complexes on the cell surface of surrogate hosts such as LAB is highly desirable. LAB can produce commodity chemicals such as lactic acid and bioactive compounds (De Vuyst & Leory, 2007; Hofvendahl & Han-Hagerald, 2007; Siragusa et al., 2007). The economics of these processes would be greatly improved if LAB could utilize cellulosic substrates, which are cheap and abundant. While most of LAB can not assimilate cellulose, by the display of cellulosome on the surface of these bacteria, the hydrolysis of cellulosic substrates and the fermentation of hydrolysis products to desirable compounds can be carried out in a single step process, which has economical advantages. As a key step in the development of these recombinant LAB, fragments of CipA were functionally expressed on the cell surface of *L. lactis*. The fragments engineered to contain a single cohesin module, two cohesin modules, one cohesin and CBM3a, or only CBM3a. Cell toxicity from over expression of the proteins was circumvented by use of the nisin A (nisA) inducible promoter. Incorporation of the C-terminal anchor motif of the streptococcal M6 protein in the expression cassette resulted in the successful surface display of the fragments. All of the constructs containing cohesin modules were able to bind to an engineered hybrid reporter enzyme, *E. coli* ß-glucuronidase fused to the type I dockerin domain of a cellusomal enzyme. These results demonstrated that the cohesins were displayed on the cell surface of *L. lactis* cells in the functional form. In addition, the cell surface complex formation was dependent on the presence of both cohesin and dockerin modules (Wieczorek & Martin, 2010).

The process for the conversion of starch to lactic acid by LAB includes the enzymatic hydrolysis of starch followed by the fermentation of resultant oligosaccharides to lactic acid by LAB. These steps can be carried out simultaneously by α-amylase displaying LAB. Therefore, using these whole cell biocatalysts can result in economical benefits for the conversion of starch to lactic acid. The PgsA anchor protein was fused to the N-terminus of α-amylase of *Streptococcus bovis* 148 (AmY). The resulting fusion protein expressed in *L. casei* and it was associated with the membrane and cell wall fractions. However, the status of exposure of AmY outside of the cell wall was not clarified. The constructed whole cell biocatalyst was able to convert starch to lactic acid. Because the lactic acid concentration increased as the total sugar concentration decreased, it was concluded that the lactic acid was produced by simultaneous saccharification and fermentation of starch. The yield of lactic acid was improved by repeated utilizations of the recombinant *L. casei* cells (Narita et al., 2006).

The C-terminal region of peptidoglycan hydrolase (AcmA) of *L. lactis* IL 1403 (CphI) is a homolog of CpH of *L. lactis* MG 1363 and it contains three LysMs. CphI can bind to the surface of various LAB (Tarahomjoo et al., 2008a). We studied the capability of CphI for the development of α-amylase displaying LAB using the external mode of protein display. These whole cell biocatalysts are expected to be effective for the direct fermentation of starch to lactic acid. Starch is a large substrate that is not capable of penetrating the cell wall. For this reason, in order to achieve its efficient hydrolysis, the enzyme must be exposed on the outside of the cell wall such that it is accessible to starch. The display systems based on the internal mode of protein anchoring often have limited ability for the translocation of target proteins to the cell surface. In contrast, the display systems based on the external mode of protein anchoring can display the enzyme completely outside of the cell wall. Therefore, a whole cell biocatalyst constructed using the externally added cell surface adhesive α-amylase is considered as a suitable selection for our purpose. Moreover, the cell surface adhesive α-amylase can readily be recovered together with LAB cells at the end of starch conversion process for reuse. CphI fused to AmY either at its C-terminus (CphI-AmY) or at its N-terminus (AmY-CphI) was expressed in *E. coli*. Both of the fusion proteins were able to bind to the cell surface of *L. lactis* ATCC 19435. Therefore, CphI is considered as a bidirectional anchor protein. However, the number of bound molecules per cell in the case of CphI-AmY was 3 times greater than that in the case of AmY-CphI. The change in the fusion direction may cause conformational alterations in the fusion protein leading to a better accessibility of CphI for the cell surface binding and an increase in the number of bound molecules. Moreover, the specific activity for starch digestion of CphI-AmY was 11 fold higher than that of AmY-CphI. The starch binding domain of AmY is located at its C-terminus. As a result, the fusion of CphI to the N-terminus of AmY may help improve the adsorption of starch onto the enzyme and enhance starch degradation, resulting in a higher specific activity for starch digestion. In addition to *L. lactis* ATCC 19435, *L. plantarum* NRRL B531, *L. lactis* IL1403, *L. casei* NRRL B441, *L. delbrueckii* ATCC 9649 and *L. casei* NRRL B445 were examined in terms of the binding of CphI-AmY. Of the LAB tested, *L. lactis* ATCC 19435 showed the highest binding capability for CphI-AmY, up to 6×10^4 molecules per cell. The binding of CphI-AmY to *L. delbrueckii* ATCC 9649 cells was very stable and its dissociation rate constant at 37°C was 7×10^{-6} s^{-1} (the half life of binding ($t_{1/2}$) was 28 h). The binding of this protein to cells of *L. lactis* ATCC 19435 was also stable, with a dissociation rate constant of 5×10^{-5} s^{-1} at 30°C ($t_{1/2}$=4 h). Lactate production by lactic acid bacteria is maximal during the exponential growth phase. Therefore, for a successful application of the constructed whole cell biocatalysts in lactate production, suitable fermentation conditions should be specified to adjust the duration of the exponential growth phase with respect to the dissociation rate of the protein. These half lives are long enough for lactic acid fermentation when the inoculm size is adequet and/or suitable growth conditions with high specific growth rates are used (Tarahomjoo et al., 2008a). A CphI mutant devoid of its N-glycosylation sites was expressed extracellularly in *Pichia pastoris*. This domain was able to bind to the cell surface of *L. casei*. However, its dissociation rate constant from the cell surface was 3.5 fold lower than that of CphI. These results indicated that the protein engineering approaches can be useful for increasing the binding stability of noncovalent anchors (Tarahomjoo et al. 2008b).

3.3. Attachment of bacteria to host tissues

Display of adhesins capable of binding to a host tissue on the surface of LAB can provide them with a specific adhesion capability, which can be beneficial when LAB are used as mucosal delivery vehicles of bioactive compounds. Because the displayed adhesins can increase the persistence of bacteria in the host tissue and as a result, the desired effects of the delivered bioactive compounds can be enhanced. The N-terminal region of SlpA was recently shown to mediate adhesion to human intestinal cell lines *in vitro*. The SlpA adhesion mediating domain fused to the N-terminus of lactococcal CpH protein was expressed in *L. lactis*. To increase the surface accessibility of the hybrid protein, a part of PrtP of *L. lactis* subsp. *cremoris* Wg2 was used as a spacer protein, which could extend the SlpA receptor binding region out of the cell surface. The spacer was inserted between the receptor binding domain and the anchor protein. *In vitro* adhesion assay with the human intestinal epithelial cell line Intestine 407 indicated that the recombinant lactococcal cells had gained an ability to adhere to Intestine 407 cells significantly greater than that of parental nonrecombinant *L. lactis* cells (Avall-Jaaskelainen et al., 2003).

It has been reported that *Mycobacterium bovis* bacillus Calmette-Guerin (BCG) exerts antitumor effects against superficial bladder cancer. These antitumor effects are due to the activation of immune responses, which are mediated by the attachment of BCG to the bladder wall through fibronectin. In addition, *L. casei* strain Shirota showed an antitumor activity when it was administered to the mouse model of superficial bladder cancer. However, the *L. casei* cells exhibits no significant binding to fibronectin. Therefore, higher antitumor activity could be expected for the *L. casei* cells genetically engineered to acquire binding capacity to fibronectin. For this purpose, fibronectin binding domain (FbD) of *S. pyogenes* ATCC 21059 fused to the C-terminal region of PrtP 763 of *L. lactis* NCDO 763 was expressed in *L. casei* strain Shirota. The recombinant *L. casei* cells were able to bind to fibronectin. Furthurmore, FbD expressed on the *L. casei* cell surface promoted the adherence to murine fibroblast STO cells, which secret fibronectin in large amounts (Kushiro et al., 2001).

3.4. Enhancement of delivering viable bacteria to gastrointestinal tract

Probiotics are live microbial food supplements, which benefit the health of consumers by improving their intestinal microbial balance (Fuller, 1989). Since the viability and activity of a probiotic is essential at the site of action, it must survive the harsh environment of the upper GIT, and it must be able to function in the gut environment (Collins et al., 1998). Most commonly used probiotics are lactobacilli and bifidobacteria (Daly & Davis, 1998). However, several studies indicate that most of these bacteria may not be able to withstand the harsh acidity of the upper GIT (Conway et al., 1987; Lankaputhra & Shah, 1995). In a study by Wang (1999), the enhancement of survival of bifidobacteria grown in the presence of starch granules and mixed with them was reported. However, the exact mechanism underlying the protective effect of starch was not clarified. The adhesion of bacteria to starch was considered as a possible explanation for these observations. The results of their study suggested that starch granules can be used to protect living microbes from environmental stress factors.

Microencapsulation is an approach that has been proposed to protect probiotics from environmental stresses. It segregates cells from adverse environments; thus, it minimizes cell injury (Anal & Singh, 2007). Myllaerinen (1999) has recently developed a microencapsulation technology that involves entrapping bacteria in the hollow cores of partially hydrolyzed starch granules, which are then encapsulated in an outer coating of amylose. The aim of this technology is to protect the probiotic bacteria from adverse environmental conditions during processing, in products during storage, and during passage through the GIT, and it is based on the finding that starch granules can be used to protect living microbes from adverse environments.

We therefore aimed to investigate whether the conferment of starch adhesion ability to cells and using this characteristic to encapsulate the cells between starch granules can enhance their viability in simulated gastric conditions. However, using genetic engineering techniques to confer starch binding ability to probiotics is not favorable because of consumers' concerns about genetically modified food ingredients. CphI of *Lactococcus lactis* IL1403 is an efficient anchoring domain for the display of heterologous proteins on LAB cells, which can bind to the cell surface when added from the outside (Tarahomjoo et al., 2008a). This domain can be used to confer starch binding ability to LAB without making any genetic modifications in them. We therefore studied the capability of the CphI anchor to direct the display of a starch binding domain (SbD) on the surface of *L. casei* cells and the aggregation of cells with starch was examined as an alternative technique of microencapsulation. This is the only available report demonstrating the potential applicability of the cell surface display technology for increasing the delivery of viable microorganisms to the GIT.

3.4.1. Materials and methods

3.4.1.1. Bacterial strains and growth conditions

E. coli XL1-Blue was used for the construction of vectors and the expression of heterologous proteins. It was grown in Luria-Bertani (LB) liquid medium or on LB agar plates at 37°C. *L. casei* subsp. *casei* NRRL B-441 (Agriculture Research Service Culture Collection, Peoria, IL, USA) was used for the binding assay, and it was grown at 37°C in MRS broth (Difco Laboratories, Detroit, MI, USA).

3.4.1.2. DNA manipulation

The gene encoding the linker region and the first nine amino acid residues of SbD of the AmY was prepared by PCR from pQE31amyA (Shigechi et al., 2004) with 5'-aaggatccgggccaagctagccaagcagctc-3' and 5'-gcgccaattatctgggttttgg-3' as forward and reverse primers, respectively. The amplified fragment was digested with *Bam*HI and *Bst*XI and inserted at the same restriction sites into pQCA (Tarahomjoo et al., 2008a). The obtained plasmid was designated as pQCLS, in which the gene encoding CphI was fused at its C-terminus to the gene encoding the linker region and SbD of AmY. The correctness of the construct was confirmed by restriction digestion and sequencing.

3.4.1.3. Expression studies

E. coli cells harboring the desired plasmids were grown overnight at 37°C in LB broth supplemented with 100 µg/ml ampicillin and 15 µg/ml tetracycline. The cells were then harvested by centrifugation and transferred to fresh LB broth containing the antibiotics mentioned above, and incubated at 37°C until the OD_{600} reached 0.5. Isopropyl β-D-thiogalactoside (IPTG) was added to a final concentration of 1 mM to induce the expression of the target protein. At the same time, ampicillin was added to a final concentration of 400 µg/ml for plasmid maintenance. After further incubation for 4 h, the cells were collected and the expression was studied by resolving the whole-cell extracts by 12.5% sodium dodecyl sulfate polyacrylamide gel electrophoresis (SDS-PAGE) and staining the gel using Coomasie Brilliant Blue R250 (CBB).

3.4.1.4. Purification of fusion protein

Proteins were purified under native conditions by metal affinity chromatography, utilizing the interaction between the histidine tag and a nickel chelate column (Ni-NTA superflow column [1.5 ml]; Qiagen GmbH, Hilden, Germany). The induced cells from a 100-ml culture were harvested by centrifugation and were resuspended in binding buffer (50 mM NaH_2PO_4 (pH 8), 300 mM NaCl, 10 mM imidazole). Lysozyme was added to a final concentration of 1 mg/ml and the cell suspension was incubated for 1 h on ice. The cells were disrupted by sonication and the clear supernatant obtained by centrifugation was applied to the Ni-NTA column equilibrated with the binding buffer. The column was washed three times with the same buffer containing 20 mM imidazole and the bound proteins were eluted with the elution buffer, which was the same as the binding buffer except that it contained 250 mM imidazole. The buffer of the eluent was then exchanged to 20 mM Tris-Cl buffer (pH 8.0) by ultrafiltration. The protein preparation was applied to an anion exchange column (SuperQ-5PW, , Tosoh, Tokyo, Japan) equilibrated with 20 mM Tris-Cl buffer (pH 8.0). The absorbed proteins were eluted by a linear NaCl gradient (0–1M). Protein elution was monitored using a UV detector, and the desired fraction was collected and desalted by ultrafiltration. Purified proteins were subjected to 12.5% SDS-PAGE and the bands were visualized by CBB staining. Gels were scanned using a GT-F600 scanner (Epson, Suwa, Japan) and densitometrical analysis was performed with Scion image software (Scion, MD, USA) to quantify the proteins.

3.4.1.5. Cell surface binding assay

L. casei cells were grown as mentioned above until an OD_{660} of 1 was achieved. The cells from a 1.5 ml culture were dispersed in 0.15 ml of de-Man Rogosa Sharp (MRS) medium containing the purified fusion protein at 0.12 mg/ml and incubated at 30ºC for 2 h with gentle shaking. After washing the cells twice with 0.1 M phosphate buffer (PB) (pH 7.0), the cell pellets were resuspended in 2×SDS-PAGE loading buffer containing 20% (w/v) glycerol, 125 mM Tris-HCl (pH 6.8), 4% SDS, 5% (v/v) β-mercaptoethanol, and 0.01% bromophenol blue, and boiled for 5 min. Binding of the protein to the cells was studied by 12.5% SDS-PAGE followed by CBB staining and the amount of the fusion protein bound to the cells was determined by densitometrical analysis of CBB stained gels as mentioned above.

3.4.1.6. Starch binding assay

The fusion protein (0.06 mg/ml in PB) was mixed with an equal volume of a suspension of starch granules (Corn starch, Sigma-Aldrich Tokyo, Japan) (10 mg/ml) in the same buffer and incubated at 37°C for 2 h with gentle shaking. After centrifugation, the supernatant was examined for the presence of unbound proteins by 12.5% SDS-PAGE and CBB staining.

3.4.1.7. Aggregation of bacteria with starch and microencapsulation

After performing the cell surface binding assay as described above, the cells were washed and resuspended in PB to a final density of 10^9 cells/ml. The suspension of the starch granules in PB was mixed with an equal volume of the cell suspension for 30 min. The mixture was then allowed to stand at room temperature for 1 h. The formation of aggregates was studied both visually and with phase contrast microscopy. To determine the percentage of cells adhering to starch, a 0.5 ml sample was taken from below the liquid surface after the sedimentation. Optical density at 540 nm was measured and compared with those of controls including bacteria without starch and starch without bacteria to calculate the starch adhesion percentage as described by Crittenden (2001). For coating of the aggregates with amylose, a 1% solution of amylose in water (amylose from potato, Sigma –Aldrich, Tokyo, Japan) was prepared by heating it to a temperature of 170°C in a pressure heater (Taiatsu Techno, Tokyo, Japan), which was then cooled down to about 37°C. The aggregates were mixed gently with 0.5 ml of the amylose solution and the coating was allowed to form overnight at 4°C.

3.4.1.8. Survival of cells in simulated gastric juice

Simulated gastric juice (SGJ) was prepared as described by Lian (2003), which was a pepsin solution (3 g/l) in saline (0.5% NaCl). The SGJ was prepared freshly and its pH was adjusted to 2.0 or 3.0 with 5 M HCl. The cells (5×10^7 cfu) were mixed with 1 ml of the filter sterilized SGJ and incubated at 37°C. At specified time intervals, the gastric juice was removed following centrifugation and the cells were washed once with PB following with two washes with saline. Amylose coated cells were then resuspended in PB containing 30 U/ml α-amylase (Megazyme, Bray, Ireland) and incubated at 40°C for 20 min to aid the release of cells from the encapsulating materials. Viable bacteria were enumerated on MRS-agar after incubation for 24 h at 37°C and survival percentage was determined by dividing the final viable population (cfu/ml) with the initial viable population (cfu/ml) of *L. casei* cells exposed to the SGJ.

3.4.2. Results

3.4.2.1. Expression and purification of fusion protein

To investigate the capability of CphI for the construction of a cell surface adhesive SbD, this domain was fused at its C-terminus to the linker region and SbD of AmY (Fig. 2). The fusion protein (CphI-SbD) was expressed intracellularly in *E. coli* using the T5 promoter at 0.35 g/l. The molecular size was 56 kDa as expected and 75% of the protein was present in the soluble form. When the protein was purified under native conditions by the histidine tag affinity chromatography, two additional protein bands corresponding to 73 and 71 kDa were present in the protein preparation (Fig. 3, lane 1). After incubation of the protein preparation with corn starch, no band for CphI-SbD was detected in the supernatant indicating its adsorption to the starch granules (Fig. 3, lane 2). CphI-SbD was further purified using an anion exchange chromatography (Fig. 3, lane 4). The result of starch binding assay showed that the purified fusion protein was in the active form and it was able to adhere to the starch granules (data not shown).

Figure 2. Structure of expression cassette

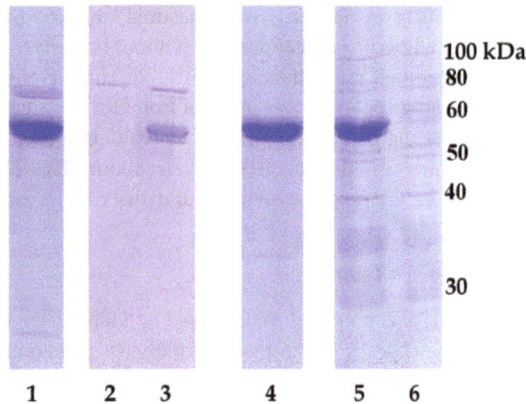

Figure 3. Purification of CphI- SbD and its binding to corn starch and *L. casei* cells. Lane 1, nickle chelate column purified protein preparation; lane 2, supernatant after starch binding assay; lane 3, control without starch; lane 4, ion exchange chromatography purified CphI-SbD; lane 5, cells bound to CphI-SbD; lane 6, cells only.

3.4.2.2. Binding of CphI-SbD to L. casei cells

L. casei cells were incubated with the purified fusion protein and studied in terms of the binding of the fusion protein by SDS-PAGE. As shown in Fig. 3 (lane 5), the fusion protein was associated with the cells. The result of the densitometrical analysis showed that 6×10^4 molecules of CphI-SbD were bound to each cell of *L. casei*.

3.4.2.3. Aggregation of bacteria with starch

For aggregation of bacteria with starch, an optimal ratio between bacteria and starch must be determined. Therefore, we examined the dependence of aggregate formation on the final starch concentration in the mixture at a constant cellular density. Free cells mixed with starch, and starch without cells were used as controls. For each concentration of starch, we compared the volume of sediment formed in the sample containing the mixture of bacteria bound to CphI-SbD and starch with those of the controls visually. The result is shown in Table 1. The volumes of the sediments formed in the controls after 1 h standing at room temperature, were almost the same at all the starch concentrations tested. When the starch concentration was 5 mg/ml, the volume of formed sediment in the sample containing the mixture of bacteria bound to CphI-SbD and starch was markedly larger than those of the controls. The adhesion percentage of *L. casei* cells to the starch granules under these conditions was determined by cosedimentation assay, which measures the reduction in optical density in bacterial suspensions after the addition of starch (Crittenden et al., 2001), and it was 32% for the bacteria bound to CphI-SbD and 4% for the free cells.

Starch concentration (mg/ml)	BPS[1]	BS[2]	S[3]
1	+	+	+
2	++	+	+
5	+++	+	+
10	+	+	+

[1]: Mixture of *L. casei* cells displaying CphI-SbD and starch
[2]: Mixture of *L. casei* cells and starch
[3]: Starch only

Table 1. Comparison of sediment formation at different starch concentrations.

Mixing of the bacteria bound to CphI-SbD with starch at a final concentration of 5 mg/ml resulted in the crosslinking of starch granules and the formed aggregates were observed by phase contrast microscopy (Fig. 4). In contrast, when starch was mixed with free cells or when no bacteria were present in the mixture, the starch granules were found to be separated from each other and no aggregates were observed by phase contrast microscopy (data not shown).

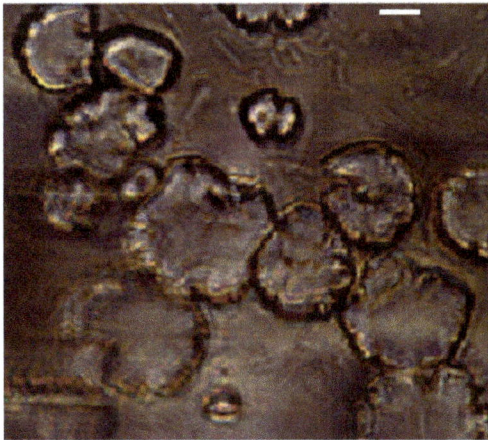

Figure 4. Aggregation of *L. casei* cells displaying CphI-SbD with starch granules (scale bar=5 μm)

3.4.2.4. Survival of encapsulated cells under simulated gastric conditions

When *L. casei* free cells were exposed to the SGJ at pH 3.0 and 2.0 for 1 h, the survival percentages of the cells were 0.074% and 0.002%, respectively. However, the survival percentages of amylose coated bacterial aggregates after 1 h incubation in the SGJ were 63.9% and 6.03% at pH 3.0 and 2.0, respectively (Fig. 5).

We studied the effects of starch, amylose coating and CphI-SbD on cell survival in the SGJ (pH 3.0, 1 h) (Table 2). It was observed that when free cells were mixed with the starch granules, the survival percentage was 3.1%, and the survival percentage of amylose coated free cells in the SGJ was 7.2%. The survival of the CphI-SbD displaying bacteria (0.093%) was not significantly different from that of free cells (0.074%) and the survival of amylose coated fusion protein displaying bacteria was comparable to that of amylose coated free cells. However, when fusion protein displaying bacteria were aggregated with the starch granules, the survival percentage was 7.7% higher than that of free cells mixed with the starch granules, and when the aggregates were coated with amylose, the survival percentage was 27.6% higher than that of the amylose coated mixture of the starch granules and free cells.

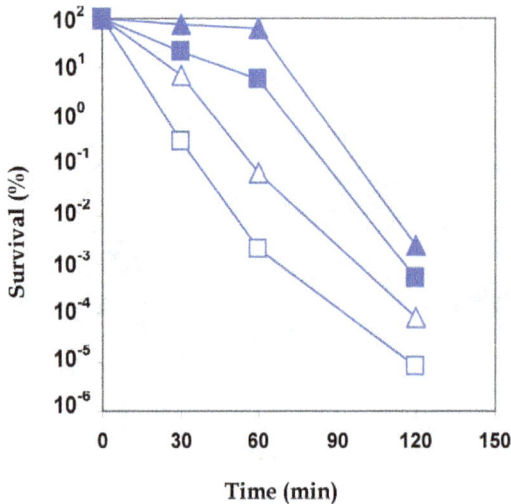

Figure 5. Time course of survival of *L. casei* cells under simulated gastric conditions. Open and filled triangles are free cells and amylose coated bacterial aggregates, respectively at pH 3.0; Open and filled rectangles, those at pH 2.0. The Data represent the means of two independent experiments.

Exploring Surface Display Technology for Enhancement of Delivering Viable Lactic Acid Bacteria to
Gastrointestinal Tract

115

Bacteria	CphI-SbD	Starch	Amylose	Survival (%)
+	–	–	–	0.074 ± 0.012
+	+	–	–	0.093 ± 0.018
+	–	+	–	3.1 ± 0.4
+	+	+	–	10.8 ± 2.0
+	–	–	+	7.2 ± 1.7
+	+	–	+	6.6 ± 2.5
+	–	+	+	36.3 ± 5.4
+	+	+	+	63.9 ± 2.7

Table 2. Effects of protective systems components on cell survival. Data represent means± standard deviations of three independent experiments.

3.4.3. Discussion

The objective of this study was to evaluate the possibility for enhancement of the delivery of viable microorganisms to the GIT through the conferment of starch binding ability to them. In this way, the bacteria are entrapped between starch granules to use the protective effect of starch for maintaining the cell viability under adverse conditions. The surface display technology based on the external mode was used to provide starch adhesion ability for the cells while retaining their nongenetically modified status because consumption of genetically modified microorganisms is not favorable for consumers. It was observed that CphI-SbD was able to bind both to the cell surface of *L. casei* and to the starch granules. Therefore, the fusion protein was able to mediate the adhesion of cells to the starch granules. We examined the aggregation of cells with starch as an alternative protective strategy, which entraps bacteria between starch granules. Compared with the previous method of entrapping bacteria within the porous starch granules prepared by an enzymatic digestion (Myllaerinen et al., 2001), our technique, is much simpler and faster. Moreover, starch granules can be used in their intact forms without any modifications.

When the aggregates of CphI-SbD displaying bacteria with starch were coated with amylose and exposed to the SGJ, there were significant increases in the survival percentage (63.9% at pH 3.0, and 6.03% at pH 2.0) with respect to those of free cells (0.074% at pH 3.0, and 0.002% at pH 2.0). The pHs of the SGJ measured for all the combinations of bacteria, fusion protein, starch and amylose coating (Table 2), in addition to that of the SGJ without any of these components both before and after the incubation at 37°C, were in the range of 3.00- 3.04. Therefore, the observed increases in the survival were not due to the modifications of the pH of SGJ.

Analysis of the effects of fusion protein, starch and amylose coating on the cell survival showed that the binding of fusion protein to *L. casei* cells did not have a significant effect on the cell survival (Table 2). When free cells were mixed with the starch granules, their survival percentage was 3.03% higher than that of free cells, which indicated the protective effect of starch on the cell survival. It was observed that, the entrapment of bacteria between the starch granules with the aid of CphI-SbD (the aggregation of bacteria with the starch granules) enhanced the protective effect of starch, and the survival percentage was increased to 10.8%. The effect of the amylose coating on the survival of CphI-SbD displaying bacteria was comparable to that of the free cells (6.6 and 7.2% respectively), and the observed difference was not statistically significant. Incorporation of the fusion protein in the protective system composed of the starch and amylose resulted in a 27.6% increase in the cell survival percentage, which showed that the simultaneous application of two protective strategies (the aggregation of bacteria with starch and the amylose coating) resulted in the highest cell survival percentage (63.9%). In conclusion, in this study we showed the potential usefulness of the cell surface display technology for protection of cells under adverse gastric conditions.

4. Conclusion

Surface display is an attractive technology that can be used to confer new functions to LAB. The effectiveness of these systems depends on the appropriate selection of several factors including the anchoring domains, secretion signals, and host strains. Moreover, a proper strategy for the fusion of anchoring domains to target proteins should be determined to protect the functionality of target proteins. So far, a limited number of surface display systems have been developed. The characterization of anchoring, secretion, and regulatory signals from genome sequences can expand the surface display systems. The low transformation efficiency of LAB is a major obstacle for the construction of surface display systems and the establishment of efficient transformation protocols is therefore necessary.

Author details

Shirin Tarahomjoo

Department of Biotechnology, Razi Vaccine and Serum Research Institute, Karaj, Iran
Food Industries and Biotechnology Research Center, Amirkabir University of Technology, Tehran, Iran

5. References

Anal, A. K., and Singh, H. (2007). Recent advances in microencapsulation of probiotics for industrial applications and targetd delivery. *Trends Food Sci Technol* 18, 240-251 (2007).

Antikainen, J., Anton, L., Sillanpää, J., Korhonen, T. K. (2002). Domains in the S-layer protein CbsA of *Lactobacillus crispatus* involved in adherence to collagens, laminin and lipoteichoic acids and in self-assembly. *Mol Microbiol* 46, 381-394.

Audouy, S. A. L., van Slem, S., van Roosmalen, M. L., Post, E., Kanninga, R., Neef, J., Estevao, S., Nieuwenhuis,, E. S., Adrian, P. V., Leenhouts, K. And Hermans, P. W. M. (2007). Development of lactococcal GEM based pneumococcal vaccines. *Vaccine* 25, 2497-2506.

Avall-Jääskeläinen, S., Kyla-Nikkila, K., Kahala, M., Miikkulainen, T. and Palav, A. (2002). Surface display of foreign epitopes on the *Lactobacillus brevis* S-layer. *Appl Environ Microbiol* 68, 5943-5951.

Avall-Jääskeläinen, S., Lindholm, A. And Palva, A. (2003). Surface display of the receptor binding region of the Lactobacillus brevis S-layer protein in *Lactococcus lactis* provides nonadhesive lactococci with the ability to adhere to intestinal epithelial cells. *Appl Environ Microbiol* 69, 2230-2236.

Avall-Jääskeläinen, S., Hynönen, U., Ilk, N., Pum, D., Sleytr, U. B. and Palva, A. (2008). Identification and characterization of domains responsible for self-assembly and cell wall binding of the surface layer protein of *Lactobacillus brevis* ATCC 8287. *BMC Microbiol* 8, 165-181.

Axelsson L. (1998). *Lactic acid bacteria: classification and physiology. In Lactic acid bacteria: Microbiology and functional aspects*, Salminen S. and Wright A. W. Marcel Dekker Inc., New York.

Bayer, E. A., Belaich, J. P., Shoham, Y. and Lamed, R. (2004). The cellulosomes: multienzyme machines for degradation of plant cell wall polysaccharides. *Ann Rev Microbiol* 58, 521-554.

Begume, P. and Lemaire, M. (1996). The cellulosome: an extracellular multiprotein complex specialized in cellulose degradation. *Critic Rev Biochem Mol Bio* 31, 201-236.

Berlec, A., Zadravec, P., Jevnikar, Z. and Strukelj, B. (2011). Identification of candidate carrier proteins for surface display on *Lactococcus lactis* by theoretical and experimental analysis of the surface proteom. *Appl Environ Microbiol* 77, 1292-1300.

Bermudez-Humaran, L. G., Cortes-Perez, N. G., Lefever, F., Guimaraes, V., Rabot, S., Alcocer-Gonnzalez, J. M., Gratadoux, J-J., Rodriguez-Padilla, C., Tamez-Guerra, R. S., Corthier, G., Gruss, A. and Langella, P. (2005) A novel mucosal vaccine based on live lactococci expressing E7 antigen and IL-12 induces systemic and mucosal immune responses and protects mice against human papillomavirus type 16-induced tumors. *J Immunol* 175, 7297-7302.

Brinster, S., Furlan, S., Serror, P. (2007). C-terminal WXL domain mediates cell wall binding in *Enterococcus faecalis* and other gram positive bacteria. *J Bacterio* 189, 1244-1253.

Buist, G., Kok, J., Leenhouts, K. J., Babrowska, M., Venema, G., and Haandrikman, A. J. (1995). Molecular cloning and nucleotide sequence of the gene encoding the major peptidoglycan hydrolase of *Lactococcus lactis*, a muramidase needed for cell separation. *J Bacteriol* 177, 1554-1563.

Buist, G., Steen, A., Kok, J. and Kuipers, O. P. (2008). LysM, a widely distributed protein motif for binding to (peptide)glycans. *Mol Microbiol* 68, 838-847.

Collins J.K., Thornton, and G., Sullivan, G. O. (1998). Selection of probiotic strains for human applications. *Int Dairy J* 8, 487-490.

Conway, P. L., Gorbasch, S. L., and Goldin, B. R. (1987). Survival of lactic acid bacteria in the human stomach and adhesion to intestinal cells. *J dairy Sci* 70, 1-12.

Crittenden, R., Laitila, A., Forssell, P., Matto, J., Saarela, M., Mattilla-Sandholm, T., and Myllarinen, P. (2001). Adhesion of bifidobacteria to granular starch and its implications in probiotic technologies. *Appl Environ Microbiol* 67, 3469-3475.

Curtiss, R. (2002) Bacterial infectious disease control by vaccine development. *J Clin Invest* 110, 1061-1066.

Daly, C., and Davis, R. (1998). The biotechnology of lactic acid bacteria with emphasis on applications in food safety and human health. *Agric Food Sci* 7, 219-250 (1998).

Demain, A. L., Newcomb, M. and Wu, J. H. D. (2005). Cellulase, clostridia and ethanol. *Microbiol Mol Biol Rev* 69, 124-154.

De Vuyst, L. and Leory, F. (2007). Bacteriocins from lactic acid bacteria: production, purification, and food applications. *J Mol Microbiol Biotechnol* 13, 194-199.

Dieye, Y., Hoekman, A. J. W., Clier, F., Juillard, V., Boot, H. J. and Piard, J-C. (2003). Ability of *Lactococcus lactis* to export viral capsid antiges: a crucial step for development of live vaccines. *Appl Environ Microbiol* 69, 7281-7288.

Fischetti, V. A., Pancholi, V. and Schneewind, O. (1990). Conservation of a hexapeptide sequence in the anchr region of surface proteins from gram positive cocci. *Mol Microbiol* 4, 1603-1605.

Franke , C. M., Leenhouts, K. J., Haandrikman, A. J., Kok, J., Venema, G. and Venema, K. (1996). Topology of LcnD, a protein implicated in the transport of bacteriocins from *Lactococcus lactis*. *J Bacteriol* 178, 1766-1769.

Fuller, R. (1989). Probiotics in man and animals. *J Appl Bacteriol* 66, 365-378.

Germond, J-E., Delley, M., Gilbert, C. and Atlan, D. (2003). Determination of the domain of the *Lactobacillus delbrueckii* subsp. *bulgaricus* cell surface proteinase PrtB involved in attachment to the cell wall after heterologous expression of the *prtB* gene in *Lactococcus lactis*. *Appl Environ Microbiol* 69, 3377-3384.

Giudice, E. L. and Campbell, J. D. (2006) Needle free vaccine delivery. *Adv Drug Deliv Rev* 58, 68-89.

Hansson, M., Samuelson, P., Gunneriusson, E. and Stahl, S. (2001). Surface display on gram positive bacteria. Comb. *Chem High throuput Screen* 4, 171-184.

Hofvendahl, K. and Han-Hagerdal, B. (2000). Factors affecting the fermentative lactic acid production from renewable resources. *Enz Microbial Technol* 26, 87-107.

Hou, X-L., Yu, L-Y., Liu, J. and Wang, G-H. (2007) Surface-displayed porcine epidemic diarrhea viral (PEDV) antigens on lactic acid bacteria. *Vaccine* 26, 24-31.

Hu, S., Kong, J., Kong, W., Guo, T. and Ji, M. (2010). Characterization of a novel LysM domain from *Lactobacillus fermentum* bacteriophage endolysin and its use as an anchor to display heterologous proteins on the surfaces of Lactic acid bacteria. *Appl Environ Microbiol* 76, 2410-2418.

Hu, S., Kong, J., Sun, Z., Han, L., Kong, W. and Yang, P. (2011). Heterologous protein display on the cell surface of lactic acid bacteria is mediated by the s-layer protein. *Microbial Cell Factories* 10, 86.

Kajikawa, A., Satoh, E., Leer, R. J., Yamamoto, S. and Igimi, S. (2007) Intragastric immunization with recombinant *Lactobacillus casei* expressing flagellar antigen confers antibody-independent protective immunity against *Salmonella enterica* serovar *Enteritidis*. *Vaccine* 25, 3599-3605.

Kajikawa, A., Nordone, S. K., Zhang, L., Stoeker, L. L., LaVoy, A. S., Klaenhammer, T. R. and Dean, G. A. (2011). Dissimilar properties of two recombinant *Lactobacillus acidophilus* strains displaying *Salmonella* FliC with different anchoring motifs. Appl Environ Microbiol 77, 6587-6596.

Kleerebezem, M., Boekhorest, J., van Kranenburg, R., Molenaar, D., Kuipers, O. P., Leer, R., Tarchini, R., Peters, S. A., SAndbrink, H. M., Fiers, M. W., Stiekema, W., Lankhorst, R. M., Bron, P. A., Hoffer, S. M., Groot, M. N., Kerkhoven, R., de Vries, M., Ursing, B., de Vos, W. M. and Siezen, R. J. (2003). Complete genome sequence of *Lactobacillus plantarum* WCFS1. *Proc Natl Acad Sci USA* 100, 1990-1995.

Kleerebezem, M., Hols, P., Bernard, E., Rolain, T., Zhou, M., Siezen, R. J. and Bron, A. (2010). The extracellular biology of the lactobacilli. FEMS Microbiol Rev 34, 199-230.

Kondo, A. and Ueda, M. (2004). Yeast cell-surface display- applications of molecular display. *Appl Microbiol Biotechnol* 64, 28-40.

Krinbauer, R., Booy, F., Cheng, N., Lowy, D. R. and Schiller, J. T. (1992) Papillomavirus L1 major capsid protein self assembles into virus-like particles that are highly immunogenic. *Proc Natl Acad Sci USA* 89, 12180-12184.

Kushiro, A., Takahashi, T., Asahara, T., Tsuji, H., Nomoto, K. and Morotomi, M. (2001). *Lactobacillus casei* acquires the binding activity to fibronectin by the expression of the fibronectin binding domain of *Streptococcus pyogenes* on the cell surface. *J Mol Microbiol Biotechnol* 3, 563-571.

Lankaputhra, W. E. V., and Shah, N. P. (1995). Survival of *Lactobacillus acidophilus* and *Bifidobacterium* spp. in the presence of acid and bile salts. *Cultured Dairy Products J* 30, 2-7.

Lee, J-S., Poo, H., Han, D. P., Hong, S-P., Kim, K., Cho, M. W., Kim, E., Sung, M-H. and Kim, C-J. (2006) Mucosal immunization with surface displayed severe acute respiratory syndrome coronavirus spike protein on *Lactobacillus casei* induces neutralization antibodies in mice. *J Virol* 80, 4079-4087.

Lee, P. and Faubert, G. M. (2006) Expression of the *Giardia lamblia* cyst wall protein 2 in *Lactococcus lactis*. *Microbiol* 152, 1981-1990.

Lee, S. Y., Choi, J. H. and Xu, Z. (2003). Microbial cell-surface display. *Trends Biotechnol* 21, 45-52.

Li, Y-J., Ma, G-P., Li, G-W., Qiao, X-Y., Ge, J-W., Tang, L-J., Liu, M. and Liu, L-W. (2010) Oral vaccination with the porcine rotavirus VP4 outer capsid protein expressed by *Lactococcus lactis* induces specific antibody production. *J Biomed Biotech* 2010, 1-9.

Lian W-H., Hsiao H-C., and Chou C-C. (2003). Viability of microencapsulated bifidobacteria in simulated gastric juice and bile solution. *Int J Food Microbiol* 86, 293-301.

Lindholm, A., Smeds, A. and Palva, A. (2004) Receptor binding domain of *Escherichia coli* F18 fimbrial adhesin FedF can be both efficiently secreted and surface displayed in a functional form in *Lactococcus lactis*. *Appl Environ Microbial* 70, 2061-2071.

Liu, J-K., Hou, X-L., Wei, C-H., Yu, L-Y., He, X-J., Wang, G-H., Lee, J-S. and Kim, C-J. (2009) Induction of immune responses in mice after oral immunization with recombinant *Lactobacillus casei* strains expressing enterotoxigenic *Escherichia coli* F41 fimbrial protein. *Appl Environ Microbiol* 75, 4491-4497.

Maassen, C. B. M., Laman, J. D., Heijne den bak-Glashouwer M. J., Tielen, F. J., van Holten-Neelen, J. C. P. A., Hoogteijling, L., Antonissen, C., Leer, R. J., Pouwels, P. H., Boersma,

W. J. A. and Shaw, D. M. (1999). Instruments for oral disease intervention strategies: recombinant *Lactobacillus casei* expressing tetanus toxin fragment C for vaccination or myelin proteins for oral tolerance induction in multiple sclerosis. *Vaccine* 17, 2117-2128.

Mckay, L. L. and Baldwin, K. A. (1990). Applications for biotechnology: present and future improvements in lactic acid bacteria. *FEMS Microbiol Lett* 87, 3-14.

Medina, M., Villena, J., Vintini, E., Hebert, E. M., Raya, R. and Alvarez, S. (2008) Nasal immunization with *Lactococcus lactis* expressing the pnemococcal protective protein A induces protective immunity in mice. *Infect Immun* 76, 2696-2705.

Mielcarek, N., Alonso, S. and Locht, C. (2001) Nasal vaccination using live bacterial vectors. Adv Drug Deliv Rev 51, 55-69.

Moore, R. J., Stewart, D. J., Lund, K. and Hodgson, L. M. (2001) Vaccination against ovine footrot using a live bacterial vector to deliver basic protease antigen. FEMS Microbiol Lett 194, 193-196.

Myllaerinen, P., Forssell, P., Von, W. A., Alander, M., Mattila-sandholm, T. and Poutanen, K. (2001) Starch capsules containing microorganisms and/or polypeptides or proteins and a process for producing them, In: *European patentEP1063976*,   Available   from: <http://www.freepatentsonline.com/EP1063976A1.html>.

Narita, J., Okano, K., Kitao, T., Ishida, S., Sewaki, T., Sung, M-H., Fukuda, H. and Kondo, A. (2006). Display of α-amylase on the surface of *Lactobacillus casei* cells by use of the PgsA anchor protein, and production of lactic acid from starch. 2006. *Appl Environ Microbiol* 72, 269-275.

Navarre, W. W. and Schneewind, O. (1994). Proteolytic cleavage and cell wall anchoring at the LPXTG motif of surface proteins in gram positive bacteria. *Mol Microbiol* 14, 115-121.

Norton, P. M., Brown, H. W. G., Wells, J. M., Macpherson, A. M., Wilson, P. W. and Le Page, R. W. F. (1996). Factors affecting the immunogenicity of tetanus toxin fragment C expressed in *Lactococcus lactis*. *FEMS Immunol Med Microbiol* 14, 167-177.

Oliviera, M. L. S., Areas, A. P. M., Campos, I. B., Monedro, V., Perez-Martinez, G., Miyaji, E. N., Leite, L. C. C., Aires, K. A. and Ho, P. L. (2006) Induction of systemic and mucosal immune response and decrease in *Streptococcus pneumoniae* colonization by nasal inoculation of mice with recombinant lactic acid bacteria expressing pneumococcal surface antigen A. *Microbes Infect* 8, 1016-1024.

Olvera, C., Fernandez-Vazquez, J. L., Ledezma-Candanoza, L. and Lopez-Munguia, A. (2007). Role of the C-terminal region of dextransucrase from *Leuconostoc mesentroides* IBT-PQ in cell anchoring. Microbiol 153, 3994-4002.

Piard, J. C., Hautefort, I., Fischetti, V. A., Ehrlich, S. D., Fons, M. and Gruss A. (1997a). Cell wall anchoring of *Streptococcus pyogenes* M6 protein in various lactic acid bacteria. *J Bacteriol* 179, 3068-3072.

Poo, H., Pyo, H-M., Lee, T-Y., Yoon, S-W., Lee, J-S., Kim, C-J., Sung, M-H. and Lee, S-H. (2006) Oral administration of human papilomavirus type 16 E7 displayed on *Lactobacillus casei* induces E7-specific antitumor effects in C57/BL6 mice. *Int J Cancer* 119, 1702-1709.

Poquet, I., Dusko Ehrlich, S. and Gruss, A. (1998). An export specific reporter designed for gram positive bacteria: application to *Lactococcus lactis*. J Bacteriol 180, 1904-1912.

Pouwels, P. H., Leer, R. J., and Boersma, W. J. A. (1996). The potential of Lactobacillus as a carrier for oral immunization: Development and preliminary characterization of vector systems for targeted delivery of antigens. *J Biotechnol* 44, 183-192.

Pouwels, P. H., Leer, R. J., Shaw, M., den Bak-Glashouwer, M-J. H., Tielen, F. D., Smit, E., Martinez, B., Jore, J. and Conway, P. L. (1998) Lactic acid bacteria as antigen delivery vehicles for oral immunization purposes. *Int J Food Microbiol* 41, 155-167.

Ramos, H. C., Rumbo, M. and Sirad, J-C. (2004) Bacterial flagellins: mediators of pathogenicity and host immune responses in mucosa. *Trends Microbiol* 12, 509-517.

Shigechi, H., Koh, J., Fujita, Y., Matsumoto, T., Bito, Y., Ueda, M., Satoh, E., Fukuda, H., and Kondo, A. (2004). Direct production of ethanol from raw corn starch via fermentation by use of a novel surface-engineered yeast strain codisplaying glucoamylase and α-amylase. *Appl Environ Microbiol* 70, 5037-5040.

Siezen, R., Boekhorest, J., Muscariello, L., Molenaar, D., Renckens, B., Kleerebezem, M. (2006). *Lactobacillus plantarum* gene clusters encoding putative cell surface protein complexes for carbohydrate utilization are conserved in specific gram positive bacteria. *BMC Genomics* 7, 126.

Siragusa, S., De Angelis, M., Di Cagno, R., Rizzello, G., Coda, R. and Gobbetti, M. (2007). Synthesis of γ-aminobutyric acid by lactic acid bacteria isolated from a variety of Italian cheeses. *Appl Environ Microbiol* 73, 7283-7290.

Smit, E., Jager, D., Martinez, B., Tielen. F. J. and Pouwels, P.H. (2002). Structural and functional analysis of the S-layer protein crystallisation domain of *Lactobacillus acidophilus* ATCC 4356: evidence for protein-protein interaction of two subdomains. *J Mol Biol* 324,953-964.

Steidler, L., Viaene, J., Fiers, W. and Remaut, E. (1998). Functional display of a heterologous protein on the surface of *Lactococcus lactis* by mean of cell wall anchor of *Staphylococcus aureus* protein A. *Appl Environ Microbiol* 64, 342-345.

Stiles, M. E., and Holzapfel, W. H. (1997). Lactic acid bacteria of foods and their current taxonomy. *Int J Food Microbiol* 36, 1-29.

Storni, T., Kundig, T. M., Senti, G., and Johansen, P. (2005). Immunity in response to particulate antigen delivery systems. Adv Drug Deliv Rev 57, 333-355.

Tang, L. and Li, Y. (2009) Oral immunization of mice with recombinant *Lactococcus lactis* expressing porcine transmissible gastroenteritis virus spike glycoprotein. *Virus Genes* 39, 238-245.

Tarahomjoo, S., Katakura, Y., Satoh, E. and Shioya, S. (2008a). Bidirectional cell-surface anchoring function of the C-terminal repeat region of peptidoglycan hydrolase of *Lactococcus lactis* IL1403. *J Biosci Bioeng* 105, 116-121.

Tarahomjoo, S., Katakura, Y., and Shioya, S. (2008b). Expression of C-terminal repeat region of peptidoglycan hydrolase of *Lactococcus lactis* IL1403 in methylotrophic yeast *Pichia pastoris*. J Biosci Bioeng 105, 134-139.

Toyota-Hantani, Y., Inoue, M., Ekawa, T., Ohta, H., Igimi, S. and Baba, E. (2008) importance of the major FliC antigenic site of *Salmonella enteritidis* as a subunit vaccine antigen. *Vaccine* 26, 4135-4137.

Turner, M. S. and Giffard, P. M. (1999). Expression of *Chlamydia psittaci* and human immunodeficiency virus derived antigens on the cell surface of *Lactobacillus fermentum* BR11 as fusions to BspA. *Infect Immun* 67, 5486-5489.

Turner, M. S., Hafner, L. M., Walsh, T. and Giffard, P. (2004). Identification and characterization of the novel LysM domain containing surface protein Sep from *Lactobacillus fermentum* BR11 and its use as a peptide fusion partner in *Lactobacillus* and *Lactococcus*. *Appl Environ Microbiol* 70, 3673-3680.

Van Der Vaart, J. M., Biesebeke, R., Chapman, J. W., Toschka, H. Y., Klis, F. M., and Verrips, T. (1997). Comparison of cell wall proteins of *Saccharomyces cervisiae* as anchors for cell surface expression of heterologous proteins. *Appl Environ Microbiol*, 63, 615-620.

Venema, R., Tjalsma, H., van Dijl, J. M., de Jong, A., Leenhouts, K., Buist, G. and Venema, G. (2003). Active lipoprotein precursors in the gram positive eubacterium *Lactococcus lactis*. *J Biol Chem* 278, 14739-14746.

Wan, H-M., Chang, B-Y., and Lin, S-C. (2002). Anchorage of cyclodextrin glucanotransferase on the outer membrane of *Escherichia coli*. *Biotechnol Bioeng* 79, 457-464.

Wang, X., Brown, I. L., Evans, A. J., and Conway, P. L. (1999). The protective effect of high amylose maize (amylomaize) starch granules on the survival of *Bifidobacterium* spp. in the mouse intestinal tract. *J Appl Microbiol* 87, 631-639.

Wei, C-H., Liu, J-K., Hou, X-L., Yu, L-Y., Lee, J-S. and Kim, C-J. (2010) Immunogenicity and protective efficacy of orally or intranasally administered recombinant *Lactobacillus casei* expressing ETEC K99. *Vaccine* 28, 4113-4118.

Wells, J. M., Robinson, K., Chamberlain, L. M., Schofield, K. M. and Le Page, R. W. F. (1996) Lactic acid bacteria as vaccine delivery vehicles. *Antonie van Leeuwen* 70, 317-330.

Westerlund, B., and Korhonen, T.K. (1993). Bacterial proteins binding to the mammalian extracellular matrix. *Mol Microbiol* 9, 687-694.

Wieczorek, A. and Martin, V. J. J. (2010). Engineering the cell surface display of cohesions for assembly of cellulosome inspired enzyme complexes on *Lactococcus lactis*. *Microbial Cell Fact* 9, 69.

Xin, K-Q., Hoshino, Y., Toda, Y., Igimi, S., Kojima, Y., Jounai, N., Ohba, K., Kushiro, A., Kiwaki, M., Hamajima, K., Klinman, D. and Okuda, K. (2003) Immunogenicity and protective efficacy of orally administered recombinant *Lactococcus lactis* expressing surface-bound HIV Env. *Blood* 102, 223-228.

Xu, W., Huang, M., Zhang, Y., Yi, X., Dong, W., Gao, X. and Jia, C. (2011). Novel surface display system for heterogonous proteins on *Lactobacillus plantarum*. *Lett Appl Microbiol* 53, 641-648.

Yamaguchi, K., Fujio, Y. and Inouye, M. (1988). A single amino acid determinant of the membrane localization of lipoproteins in *E. coli*. *Cell* 53, 423-432.

Chapter 6

Lactose and β-Glucosides Metabolism and Its Regulation in *Lactococcus lactis*: A Review

Tamara Aleksandrzak-Piekarczyk

Additional information is available at the end of the chapter

1. Introduction

1.1. Lactic acid bacteria

Lactic acid bacteria (LAB) are a group of Gram-positive, non-sporulating, low-GC-content bacteria that comprise 11 bacterial genera, such as *Lactococcus, Lactobacillus, Leuconostoc, Streptococcus* and others (Stiles & Holzapfel, 1997). LAB have a generally regarded as safe (GRAS) Food and Drug Administration (FDA) status, and some strains of different LAB species exhibit also probiotic properties (Gilliland, 1989). They are ubiquitous in many nutrient rich environments, such as milk, meat and plant material, and some of them are permanent residents of mainly mammalian intestinal tracts, while others are able to colonize them temporarily. Due to their ability to produce lactic acid as an end product of sugar fermentation, they are industrially important and are used as starter cultures in various food-fermentation processes. The importance of LAB for humans can be appreciated from the estimated 8.5 billion kg of fermented milk produced annually in Europe, leading to human consumption of 8.5×10^{20} LAB (Franz et al., 2010).

Understanding the mechanisms involved in carbohydrate metabolism and its regulation in LAB is essential for improving the industrial properties of these microorganisms. There are several ways to improve the metabolic potential of LAB cells, of which metabolic engineering offers a very efficient and effective tool.

1.2. Genus *Lactococcus*

Lactococci are homofermentative, mesophilic LAB that basically inhabit two natural environments, milk and plants, of which plants seem to constitute the primary niche. Occasionally, there have been reports that *L. lactis* was also isolated from soil, effluent water, the skin of cattle (Klijn et al., 1995), insects (leafhoppers, termites) (Bauer et al., 2000;

Latorre-Guzman et al., 1977; Schultz & Breznak, 1978) and fish (Itoi et al., 2008, 2009; Pérez et al., 2011). Adaptation of lactococcal strains from plants to the dairy environment has caused the loss of some functions, resulting in smaller chromosomes and acquisition of genes (often plasmidic) important for growth in milk (Kelly et al., 2010).

Since *Lactococcus lactis* was first described in 1919 (Orla-Jensen, 1919), its taxonomy has changed repeatedly and still is confusing in some aspects. This group of bacteria, previously designated lactic streptococci, was placed in the new *Lactococcus* taxon in 1985 (Schleifer *et al.*, 1985). The current taxonomy of *L. lactis* is based on phenotype and includes four subspecies (*lactis, cremoris, hordniae,* and the newly identified subsp. *tructae*) and one biovar (subsp. *lactis* biovar diacetylactis) (Schleifer et al., 1985; van Hylckama Vlieg et al., 2006; Pérez et al., 2011; Rademaker et al., 2007). Among them, only *L. lactis* subsp. *hordniae* and subsp. *tructae* have never been isolated from dairy products. The *lactis* and *cremoris* phenotypes are distinguished on the basis of several basic criteria, such as: arginine and maltose utilization, decarboxylation of glutamate to γ-aminobutyric acid (GABA), and 40°C, 4% NaCl and pH 9.2 tolerance. *L. lactis* subsp. *cremoris* strains are reported to be negative for all of these features (Nomura et al., 1999; Schleifer et al., 1985). Moreover, the biovar diacetylactis strains are able to metabolize citrate, which is converted to diacetyl, an important aroma compound. Additionally, numerous genetic studies (DNA–DNA hybridization, 16S rRNA and gene sequence analysis) of *L. lactis* isolates of dairy and plant origin have revealed the existence among them of two main genotypes that have also been called *L. lactis* subsp. *lactis* (*lactis* genotype) and *L. lactis* subsp. *cremoris* (*cremoris* genotype). Furthermore, it has been demonstrated that the genotype and phenotype do not always correspond within one isolate, thus introducing a degree of disorder into the taxonomy of this species (Tailliez et al., 1998). It has been observed that within the group of *cremoris* genotype, strains with both *lactis* (MG1363) and *cremoris* (SK11) phenotypes may occur, and, likewise, within the group of *lactis* genotype there are ones with *lactis* (KF147) as well as biovar diacetylactis (IL594) phenotypes (Bayjanov et al., 2009; Kelly et al., 2010; Nomura et al., 2002; Rademaker et al., 2007; Tanigawa et al., 2010). Hence, the *L. lactis* has an atypical taxonomic structure with two phenotypically distinct groups, such as *L. lactis* subsp. *lactis* and *L. lactis* subsp. *cremoris*, which may belong to two distinct genotype groups. As a result, in order to sufficiently describe the individual strains, it is necessary to specify both the genotype (*cremoris* or *lactis*) and the phenotype (*cremoris, diacetylactis,* or *lactis*).

Strains belonging to *L. lactis* subsp. *lactis* and *L. lactis* subsp. *cremoris* together with a diverse assortment of other LAB are widely used as dairy starters for the production of a vast range of fermented dairy products, including various types of cheeses, sour cream, buttermilk and butter (Daly, 1983; Davidson et al., 1996). In the dairy industry, the *lactis* subspecies are better for making soft cheeses and the *cremoris* subspecies for the hard ones. Overall, it is generally accepted that the *L. lactis* subsp. *cremoris* strains make better quality products than *L. lactis* subsp. *lactis* because of their important contribution to flavour development via their unique metabolic mechanisms (Salama et al., 1991; Sandine, 1988).

During growth in milk, the primary function of *L. lactis* is rapid conversion of lactose to lactic acid, which provides preservation of the fermented product by preventing growth of pathogenic and spoilage bacteria, it supports curd formation, and creates optimal conditions for ripening. Further, due to their proteolytic activity and amino acid conversion, lactococci contribute to the final texture (moisture, softness) and flavour of dairy products (Smit et al., 2005). Many of lactococcal functions vital for successful fermentations are borne on plasmids, which are a common feature in lactococci, even in strains isolated from non-dairy sources (Davidson et al., 1996). For example, specific plasmid-borne genes encode proteins involved in lactose transport and metabolism and in hydrolysis and utilization of casein (Davidson, et al., 1996; McKay, 1983). Hence, there is considerable selective pressure on dairy strains to retain these plasmids, since plasmid-cured derivatives grow poorly in milk. Since plasmids are mobile elements, they can be readily exchanged among different strains (via conjugal transfer) (Gasson, 1990).

Due to its industrial importance *L. lactis* has become the best studied LAB, and although most studies have been performed on a small number of laboratory strains of dairy origin, it is regarded as a model organism for this bacterial group. A number of genome sequences of *L. lactis* strains are available, including strains from *L. lactis* subsp. *lactis*, such as IL1403, KF147 and CV56, as well as strains from *L. lactis* subsp. *cremoris*, such as MG1363, A76, NZ9000 and SK11 (according to http://www.ncbi.nlm.nih.gov/genome/). Among them, *L. lactis* subsp. *lactis* IL1403 (Chopin et al., 1984) and *L. lactis* subsp. *cremoris* MG1363 (Gasson, 1983) are the most important laboratory strains, and they can be distinguished by differences in specific DNA sequences, including those encoding 16S rRNA (Godon et al., 1992), and by their genome organization (Le Bourgeois et al., 1995). These two strains are plasmid-cured derivatives of the dairy starter strains IL594 (IL1403) and NCDO 712 (MG1363) respectively, and due to their industrial importance, their metabolism, physiology and genetics have been extensively studied over the past years. Both belong to *L. lactis* subsp. *lactis* phenotypically, but the parent strain of IL1403 has a citrate permease plasmid (Górecki et al., 2011) and is able to metabolize citrate, placing it with *L. lactis* subsp. *lactis* biovar diacetylactis, whereas MG1363 has a *lactis* phenotype and a *cremoris* genotype (Kelly et al., 2010). Despite their physiological and 16S rRNA gene sequence similarities, they share only about 85% chromosomal sequence identity, which is comparable to the genetic distance between *Escherichia coli* and *Salmonella typhimurium* (McClelland et al., 2001; Salama et al., 1991; Wegmann et al., 2007). A derivative of MG1363 was created by the integration of the *nisRK* genes (involving the "NICE" system for nisin-controlled protein overexpression) into the *pepN* gene, yielding *L. lactis* NZ9000 (Kuipers et al., 1998).

2. Lactose metabolism

Most microorganisms have adapted to growth in milk habitat due to acquisition of the ability to the use its most abundant sugar, lactose, as a carbon source. This disaccharide consists of a galactose moiety linked at its C_1 via a β-galactosidic bond to the C_4 of glucose. Because of the efficiency and economic importance of its fermentation, a large number of studies have focused on the utilization of lactose by LAB.

Uptake of lactose into a bacterial cell can be mediated by several pathways, such as the lactose-specific phosphotransferase system (*lac*-PTS), ABC protein-dependent systems and secondary system transporters like lactose-galactose antiporters and lactose-H⁺ symport systems (de Vos & Vaughan, 1994). While ABC protein-dependent lactose transport has been demonstrated only in non-LAB, Gram-negative *Agrobacterium radiobacter* (Williams et al., 1992), the *lac*-PTS as well as secondary lactose transport systems have been described for many LAB species.

2.1. Lactose-specific phosphotransferase systems (*lac*-PTS)

Although LAB used as starter cultures may also convert pyruvate to a variety of end products, these pathways are not expressed during lactose fermentation, which is homolactic in most strains (Cocaign-Bousquet et al., 2002; Neves et al., 2005). Since the primary function of LAB in dairy fermentations is the conversion of lactose to lactic acid, the industrial strains are primarily selected on the basis of their ability for its rapid, homolactic fermentation (de Vos & Simons, 1988).

Starter lactococcal strains transport lactose exclusively by the most abundant in LAB uptake system for various sugars - the phosphoenolpyruvate-dependent phosphotransferase system (PEP-PTS). The *lac*-PTS has a very high affinity for this sugar and is bioenergetically the most efficient system since one lactose molecule is translocated and phosphorylated in a single step, at the expense of a single ATP equivalent. Concomitantly with transport, PTS catalyzes the phosphorylation of the incoming sugar. Phosphoenolpyruvate is the first phosphoryl donor, which phosphorylates Enzyme I (EI), and then the phosphoryl group is transferred in sequence to HPr, EIIA, EIIB, and finally, via transmembrane porter (EIIC), to the transported sugar (Lorca et al., 2010). After translocation via *lac*-PTS, lactose is hydrolyzed by P-β-galactosidase to glucose and galactose-6-P. While glucose enters the Embden-Meyerhof-Parnas glycolytic pathway through phosphorylation by glucokinase, galactose-6-P, before it also enters the glycolytic pathway, is further metabolized via the D-tagatose-6-P (Tag-6P) pathway. This involves three enzymes: (i) galactose-6-P isomerase (LacAB); (ii) tagatose-6-P kinase (LacC); and (iii) tagatose-1,6-diphosphate aldolase (LacD). The resulting triosephosphates (glyceraldehydes-3-P and dihydroxyacetone-P) are further metabolized via glycolysis. The operons engaged in this rapid, homolactic lactose fermentation are usually plasmid-located (*lac*-plasmids) and, in addition to the genes for the *lac*-PTS proteins and P-β-galactosidase, contain genes coding for the enzymes of the Tag-6P pathway. Their transcription is regulated by various repressors, with tagatose-6-P being the molecular inducer in *L. lactis* (van Rooijen et al., 1991).

It is believed that plasmid-encoded ability for rapid lactose fermentation characteristic for dairy strains was recently acquired by wild-type plant strains, as a result of their adaptation to milk-environment (Kelly et al., 2010).

2.2. Lactose permease-β-galactosidase systems

Another strategy developed by LAB for lactose metabolism depends on its uptake via secondary transport systems. These systems transport lactose in an unphosphorylated form via specific permeases belonging to the LacS subfamily (TC No. 2.A.2.2.3) of the 2.A.2 glycoside-pentoside-hexuronide (GPH) family (Saier, 2000). Carriers of the LacS subgroup are chimeric in nature: at their carboxy terminal end they contain an approximately 160 amino acid hydrophilic extension homologous to the EIIA domains of PTS. Thus, lactose transport is controlled by HPr-dependent phosphorylation (Gunnewijk et al., 1999; Gunnewijk & Poolman, 2000a; Gunnewijk & Poolman, 2000b). Due to this additional domain these lactose permeases are larger than the other carriers from the GPH family, which are generally about 500 amino acids in length. Depending on the organism, LacS can mediate lactose transport coupled to proton symport or by antiport with galactose. Following its import, lactose is hydrolyzed by β-galactosidase (David et al., 1992; Vaughan et al., 1996) yielding glucose and galactose. The glucose moiety is further metabolized via glycolysis, whereas the galactose moiety follows different pathways depending on the particular LAB. While some thermophilic strains of LAB (e.g., Lactobacillus bulgaricus and Streptococcus thermophilus) are known to release the galactose moiety of lactose into the medium, other LAB (e.g., Lactobacillus helveticus, Leuconostoc lactis and Streptococcus salivarius) metabolize this saccharide via the Leloir pathway (de Vos, 1996; Poolman, 1993; Vaughan et al., 2001). This pathway was one of the first central metabolic pathways to be discovered, by L. F. Leloir and coworkers in the early 1950s. It includes the key enzyme galactokinase (GalK), and hexose-1-P uridylyltransferase (GalT) plus UDP-glucose 4-epimerase (GalE), all of which are involved in the conversion of galactose to glucose-1P. The generated glucose-1P, after conversion to glucose-6P by phosphoglucomutase, enters the glycolytic pathway. Aldose-1-epimerase, a mutarotase (GalM), is an additional, more recently characterized enzyme required for rapid galactose metabolism (Bouffard et al., 1994; Mollet & Pilloud, 1991; Poolman et al., 1990). GalM catalyses the interconversion of the α- and β-anomers of galactose. This enzyme was found to be essential for efficient lactose utilization in E. coli since cleavage of this β-galactoside by β-galactosidase yields glucose and β-D-galactose, the latter being the sole substrate for GalK (Bouffard et al., 1994).

The existence of genes encoding components of the lactose permease-β-galactosidase system seems to be limited among the L. lactis strains as they have been identified only in the genomes of the dairy-derived strain IL1403 (Bolotin et al., 2001), non-dairy NCDO2054 (Vaughan et al., 1998) and KF147 isolated from mung bean sprouts (Siezen et al., 2010). Remarkably, in addition to galactose genes of the Leloir pathway cluster, these strains contain genes needed for lactose assimilation, such as lacZ (β-galactosidase) and lacA (thiogalactoside acetyltransferase), arranged in an identical layout. Directly upstream of the aforementioned genes required for lactose hydrolysis and subsequent galactose conversion, there is the gene encoding the LacS permease for sugar uptake.

Some details concerning the role of the lactose permease-β-galactosidase system in lactose utilization have been reported for the slow lactose fermenter - L. lactis NCDO2054 (Vaughan et al., 1998), and for the devoid of the lac-plasmid, essentially lactose-negative L. lactis IL1403 strain (starts to utilize lactose slowly after approximately 40 h of incubation) (Aleksandrzak-Piekarczyk et al., 2005). Since these strains possess the complete lactose permease-β-galactosidase system and an active Leloir pathway, it seems odd that they are barely capable of lactose metabolism. In the case of L. lactis NCDO2054, which can accumulate a high intracellular concentration of lactose-6-phosphate by using an efficient lac-PTS and possesses low-level P-β-galactosidase activity, it has been suggested that the slow fermentation of lactose may be due to this rate-limiting P-β-galactosidase activity and the inhibitory effect of the accumulated lactose-6-phosphate (Bissette & Anderson 1974; Crow & Thomas, 1984). However, other explanations of lactose fermentation problem can be envisaged: (i) lactose transport is inefficient due to low affinity of LacS for lactose or (ii) the strains lack a functional β-galactosidase. Indeed, the *lacS* gene of *L. lactis* IL1403 is almost identical to that of *L. lactis* NCDO2054, but also to *galP* of the lactose-negative *L. lactis* MG1363 strain (Grossiord et al., 2003). These permeases belong to the same subfamily (TC No. 2.A.2.2.3 according to the Transporter Classification Database: http://www.tcdb.org/; Saier, 2000), which includes transporters specific for galactose uptake, in contrast to LacS permeases of another subfamily (TC No. 2.A.2.2.1) with a proven high lactose-transport rate. The lack of LacS involvement in lactose transport is confirmed by the fact that disruption of *lacS* in *L. lactis* IL1403 had a minor effect on lactose assimilation (Aleksandrzak-Piekarczyk et al., 2005). Another indispensable factor in lactose assimilation, the β-galactosidase enzyme, is also encoded by the genomes of *L. lactis* IL1403 and NCDO2054 strains. In spite of the high similarity in the protein level of both enzymes, β-galactosidase of *L. lactis* NCDO2054, in contrast to the one of *L. lactis* IL1403 (Aleksandrzak-Piekarczyk et al., 2005), seems to be highly active and strongly regulated (Griffin et al., 1996). It has been suggested that the *lacZ* gene of *L. lactis* IL1403 may not be expressed or the encoded enzyme may be inactive since this strain does not exhibit β-galactosidase activity (Aleksandrzak-Piekarczyk et al., 2005). Furthermore, the *in trans* complementation of chromosomal *lacZ* by an active β-galactosidase in *L. lactis* IL1403 did not improve its ability for lactose assimilation, indicating that the lack of β-galactosidase activity is not the only obstacle in its ability to efficiently ferment lactose (unpublished personal observations).

Taken together, it seems that in *L. lactis* strains lactose permease-β-galactosidase systems play a minor role in lactose assimilation or function under certain environmental conditions. It appears that the major obstacle is the galactose-specific LacS permease, which shows only weak affinity for lactose and functions almost only in transport of galactose (Fig. 1). This thesis is confirmed by the study of Solem et al. (2008), in which an efficient lactose transporter (LacS; TC No. 2.A.2.2.1) and β-galactosidase (LacZ), encoded by the *lacSZ* operon, were introduced from lactose-positive *S. thermophilus* into the lactose-negative strain *L. lactis* MG1363, devoid of lactose permease-β-galactosidase system. As a result, fast-growing lactose-positive mutant strains were obtained. This shows that addition of the LacSZ system containing LacS with a proven high lactose-transport rate can strongly increase the lactose-transport capacity in *L. lactis*.

3. Metabolism of β–glucosides

In addition to dairy environment, plant surfaces and fermenting plant material are also important ecosystems occupied by *L. lactis*. With regard to fermentation, lactococcal strains usually occur there only at the beginning of this process, to be later replaced by microorganisms more resistant to low pH values (Kelly & Ward, 2002; Kelly et al., 1998). The majority of plant-associated strains belong to *L. lactis* subsp. *lactis*, whereas *L. lactis* subsp. *cremoris* is typical for dairy fermentations (Kelly & Ward, 2002; Kelly et al., 1998). In comparison to the dairy environment, fermenting plant material differs highly with respect to chemical composition, exhibiting, for instance, much lower protein concentration and wider availability of carbohydrates other than lactose. The ability of plant-associated *L. lactis* subsp. *lactis* strains to utilize such a large variety of plant carbohydrates is reflected in their genomes and sugar fermentation capabilities. Comparison between milk- and plant-associated lactococcal strains clearly shows that the latter possess a larger number of genes involved in transport and metabolism of carbohydrates, resulting in their increased sugar fermentation capabilities (Siezen et al., 2008).

Besides lactose, the PTS systems can also transport various other carbohydrates, including sugars widely distributed in plants, namely β-glucosides, like e.g. amygdalin, arbutin, cellobiose, esculin, gentobiose and salicin (Tobisch et al., 1997). Except for amygdalin, these sugars are composed of two molecules joined by the β-glucosidic bond, of which at least one is glucose. The best known example of this group is cellobiose, the structural unit of one of the most abundant renewable polymers on earth – cellulose, and also the main product in its enzymatic hydrolysis (Teeri, 1997). Unlike most of other β-glucosides (aryl-β-glucosides e.g., arbutin, amygdalin, esculin, and salicin), which are composed of a single glucose molecule and respective aglycone, cellobiose consists of two glucose molecules linked via a β(1-4) bond.

It is well known from sugar fermentation characteristics that *L. lactis* strains of different origin can utilize a variety of β-glucosides (e.g., Aleksandrzak-Piekarczyk et al., 2011; Bardowski et al., 1995; Fernández et al., 2011; Siezen et al., 2008). The metabolic potential for catabolism of these sugars can be chromosomally encoded by more than one genetic system, as was shown for *L. lactis* IL1403. Eight genes, which encode proteins homologous to EII proteins of β-glucoside-dependent PTS, involved in the uptake and phosphorylation of β-glucosides have been found throughout the *L. lactis* IL1403 chromosome (Bolotin et al., 2001). Three of them encode the three-domain EIIABC PTS components (PtbA, YedF and YleE), another three, EIIC permeases (CelB, PtcC and YidB), one an EIIA component (PtcA) and one an EIIB component (PtcB). CelB, PtcA, PtcB, PtcC and YidB are members of the Lac family (TC No. 4.A.3), which includes several lactose porters of Gram-positive bacteria as well as the *E. coli* and *Borrelia burgdorferi* N,N'-diacetylchitobiose (Chb) porters (according to http://www.tcdb.org/). The involvement of CelB and CelB/PtcC permeases in cellobiose transport has been experimentally confirmed in *L. lactis* IL1403 and MG1363, respectively (Aleksandrzak-Piekarczyk et al., 2011; Campelo et al., 2011). Although *L. lactis* IL1403 has such a large number of β-glucosides-specific PTS systems, CelB is the only permease operative in cellobiose uptake in this strain (Aleksandrzak-Piekarczyk et al., 2011) (Fig. 1), whereas in *L. lactis* MG1363 also another PTS permease, namely PtcC, seems to participate in the transport of this sugar, albeit to a much lesser extent than CelB (Campelo et al., 2011). It has been proposed that the observed low expression of the *ptcC* gene may be the result of repression by carbon catabolite control protein A (CcpA) as mutations in its binding site (catabolite responsive element - *cre*) in the *ptcC* promoter region led to high upregulation of this gene in strain NZ9000 compared to strain MG1363, even under repressive conditions (Linares et al., 2010).

On the other hand, the EIIAB components, namely PtcA and PtcB, seem to be more versatile, being involved in the metabolism of numerous sugars (arbutin, cellobiose, glucose, lactose, salicin) in *L. lactis* (Aleksandrzak-Piekarczyk et al., 2011; Castro et al., 2009; Pool et al., 2006). No other PTS systems dedicated to transport of other β-glucosides have yet been described in detail in any *L. lactis* strain. However, according to unpublished preliminary data, the PtbA protein appears to be involved in the transport of arbutin, esculin and salicin, but not cellobiose, in *L. lactis* IL1403 (unpublished personal observation) (Fig. 1). In this strain, inactivation of the *ptbA* gene led to serious defects in growth in medium supplemented with each of these sugars (unpublished).

After translocation by PTS through the bacterial membrane, the P-β-glucoside sugar is cleaved by P-β-glucosidase into glucose and glucose-6P or the respective aglycon (Tobisch et al., 1997). There are plenty of genes encoding P-β-glucosidases present in *L. lactis* chromosomes sequenced so far. Their large number is probably the result of adaptation of these bacteria to life on plants with abundant where β-glucosides. However, the data concerning their involvement in β-glucosides assimilation are rather scarce in scientific literature. It has only been demonstrated that a P-β-glucosidase, BglS, is responsible for hydrolysis of cellobiose, but not of salicin in *L. lactis* IL1403 (Aleksandrzak-Piekarczyk et al., 2005) (Fig. 1). On the other hand, no function has been attributed to another P-β-glucosidase encoded by the *bglA* gene, and forming one operon with *ptcC*. According to unpublished results, the disruption of *bglA* did not alter growth of the IL1403 mutant strain in medium supplemented with a wide array of sugars (unpublished personal analysis).

Expression of β-glucosides' catabolic genes can be controlled by various regulatory mechanisms. Among them, catabolite repression (Aleksandrzak-Piekarczyk et al., 2005, 2011; Zomer at al., 2007) and transcriptional antitermination through the BglR protein (Bardowski et al., 1994) were shown to be operational in *L. lactis*. The antitermination mechanism allows for expression of β-glucoside-specific genes in the absence of a metabolically preferred carbon source, such as glucose (Rutberg, 1997). It is believed that antiterminator proteins act by binding to a ribonucleic antiterminator (RAT) site at a specific mRNA secondary structure to prevent the formation of a hairpin terminator structure that would otherwise terminate transcription (Aymerich & Steinmetz, 1992; Rutberg, 1997). The binding of the antiterminator protein to the mRNA permits transcription through the sequestered terminator sequence into a β-glucoside-specific operon that is not normally transcribed. The function of BglR has been studied earlier in *L. lactis* IL1403, and it was shown to be involved in the activation of assimilation of β-glucosides such as arbutin, esculin and salicin, except for cellobiose (Bardowski et al., 1994; 1995) (Fig. 1). Inspection of the *L. lactis* IL1403 genome sequence downstream of *bglR* revealed the presence of two genes, *ptbA* and *bglH*, encoding proteins homologous to a putative three-domain EIIABC PTS component specific for the assimilation of β-glucosides, and P-β-glucosidase, respectively. Upstream of *bglR*, a putative *cre*-box (differing from the *cre* consensus by one nucleotide), a putative promoter sequence and a RAT sequence were identified. This RAT sequence has been reported previously (Bardowski et al., 1994, 1995) to be involved in the autoregulation of BglR. This sequence partially overlapped a putative *rho*-independent terminator, which comprised six nucleotides at the 3' end of the RAT. The *ptbA* gene is located 141 nt downstream of *bglR*. *In silico* sequence analysis revealed that the *ptbA* gene is also preceded by a DNA sequence highly similar to the RAT consensus sequence, suggesting that the regulation of *ptbA* expression may involve the BglR-mediated antitermination mechanism (unpublished personal analysis). Moreover, the short intergenic DNA region (47 nt) between *ptbA* and the next gene (*bglH*), plus the lack of an obvious hairpin structure or a promoter sequence strongly suggest that these two genes might be cotranscribed, and thus undergo common BglR-mediated regulation (unpublished) (Fig. 1).

4. Alternative lactose utilization system and its interconnection with cellobiose assimilation

The existence in several lactococcal strains devoid of *lac*-plasmids of cryptic lactose transport and catabolism systems has already been suggested in earlier studies (Anderson & McKay, 1977; Cords & McKay, 1974; de Vos & Simons, 1988; Simons et al., 1993). The presence in *L. lactis* of chromosomally-encoded lactose permease has been proposed since introduction of the *E. coli lacZ* gene into a lactose-deficient *L. lactis* strain restored its ability to utilize lactose (de Vos & Simons, 1988). Moreover, P-β-galactosidase activities have also been detected in strains cured of their lactose plasmids, suggesting the presence of chromosomally-encoded cryptic *lac*-PTS(s) (Anderson & McKay, 1977; Cords & McKay, 1974). However, it was suggested that these PTSs are not specific for lactose, but rather for the translocation of other sugars (e.g., β-glucosides), and lactose could be transported alternatively. This hypothesis was supported by observations suggesting that a putative P-β-glucosidase, involved in cellobiose hydrolysis, is probably also involved in lactose-6-P cleavage in *L. lactis* strain ATCC7962 (Simons et al., 1993). This seems reasonable, as according to http://www.tcdb.org/, PTS lactose transporters belong to the Lac family (TC No. 4.A.3) and porters of this family have broad substrate specificity. Besides lactose, they can also transport aromatic β-glucosides and cellobiose.

Until recently (Aleksandrzak et al., 2000; Aleksandrzak-Piekarczyk et al., 2005, 2011; Kowalczyk et al., 2008), little information on the organization in *L. lactis* strains of chromosomal alternative lactose utilization genes has been available. It was shown that in *lac*-plasmid-free, and thus lactose-negative *L. lactis* IL1403, the ability to assimilate lactose can be induced in two ways: (i) by the presence of cellobiose or (ii) by inactivation of CcpA (Aleksandrzak et al., 2000; Aleksandrzak-Piekarczyk et al., 2005). The CcpA protein is a member of the LacI-GalR family of bacterial repressors and exists only in Gram-positive bacteria (Weickert & Adhya, 1992). It exerts its regulatory role in carbon catabolite repression (CCR) by binding to DNA sites called *cres*, which occur in the vicinity of CcpA-regulated genes (Weickert & Chambliss, 1990). In *L. lactis* the known targets of CcpA are the *gal* operon for galactose utilization (Luesink et al., 1998), the *fru* operon for fructose utilization (Barrière et al., 2005), the *ptcABC* operon for cellobiose utilization (Zomer et al., 2007), and *cel-lac* genes for cellobiose and lactose utilization (Aleksandrzak-Piekarczyk et al., 2011). Thus, one could speculate that in *L. lactis* IL1403 cellobiose-inducible chromosomal alternative lactose utilization genes are under the negative control of CcpA, and, therefore, inactivation of the *ccpA* gene could result in their derepression and ability to assimilate lactose by the IL1403 *ccpA* mutant.

Further studies of Aleksandrzak-Piekarczyk et al. (2005, 2011) and Kowalczyk et al. (2008) provided details on interconnected metabolism of β-glucosides (cellobiose) and β-galactosides (lactose) and its variable regulation in *L. lactis* IL1403. Several genes have been implicated in coupled cellobiose and lactose assimilation in *L. lactis* IL1403, such as *bglS* and *celB*, *ptcA* and *ptcB*, encoding proteins homologous to P-β-glucosidase and EII components of cellobiose-specific PTS, respectively (Fig. 1). It has been shown that in *L. lactis* IL1403 the cellobiose-specific PTS system, comprising of *celB*, *ptcB* and *ptcA*, is also able to transport lactose because cellobiose-specific permease CelB has also an affinity for lactose, and, moreover, is the only permease involved in lactose uptake (Aleksandrzak-Piekarczyk et al., 2011). Furthermore, internalized lactose-P is hydrolyzed exclusively by BglS – an enzyme with dual P-β-glucosidase and P-β-galactosidase activity, and high affinity for cellobiose (Aleksandrzak-Piekarczyk et al., 2005) (Fig. 1). Thus, BglS activity generates glucose and galactose-P molecules. Glucose enters the Embden-Meyerhof-Parnas glycolytic pathway through phosphorylation by glucokinase, whereas galactose-P requires dephosphorylation performed by an unidentified phosphatase or phosphohexomutase, before entering the Leloir pathway (Neves et al., 2010) (Fig. 1). Moreover, this alternative lactose utilization system has been shown to be tightly controlled by CcpA-directed negative regulation (Fig. 1), since inactivation of the *ccpA* gene led to derepression of *bglS*, *celB*, *ptcA* and *ptcB* and *L. lactis* IL1403 *ccpA* mutant ability to assimilate lactose (Aleksandrzak-Piekarczyk et al., 2011). In addition to CcpA-mediated repression, the *celB* and *bglS* genes are specifically activated by cellobiose, as its presence leads to an increase in their transcription. This phenomenon has not been observed when other sugars, such as glucose, galactose or salicin, were used as carbon sources (Aleksandrzak-Piekarczyk et al., 2011). Preliminary results suggest that a hypothetical transcriptional regulator, namely YebF, could be engaged in this cellobiose-dependent activation of *celB* and *bglS* (Aleksandrzak-Piekarczyk et al., 2011; unpublished personal analysis) (Fig. 1). The YebF protein belongs to the RpiR family of phosphosugar binding proteins (Sorensen & Hove-Jensen, 1996), and, in addition to its sugar binding domain (SIS), it has a putative helix-turn-helix (HTH) DNA-binding domain. In addition to *yebF* mutant ferment lactose inability (Aleksandrzak-Piekarczyk et al., 2005), inactivation of the *yebF* gene in IL1403 resulted in inability to grow on cellobiose (unpublished personal analysis), suggesting the gene's requirement in both cellobiose and lactose assimilation. Further studies on this phenomenon in *L. lactis* are needed to address it in greater detail.

When cellobiose is available, it activates the cellobiose-specific PTS transport system, comprising CelB, PtcB and PtcA proteins, and *L. lactis* IL1403 is able to grow on cellobiose and lactose. This growth is supported by the activity of cellobiose-inducible BglS protein, which splits lactose-P into galactose-P and glucose. Then, after the dephosphorylation step, galactose is further metabolized through the Leloir pathway, while glucose enters glycolysis. Therefore, inactivation of the *ccpA* gene results in derepression of the cellobiose-specific PTS transport system and also of the *bglS* gene, which in turn enable the IL1403 strain to grow on lactose.

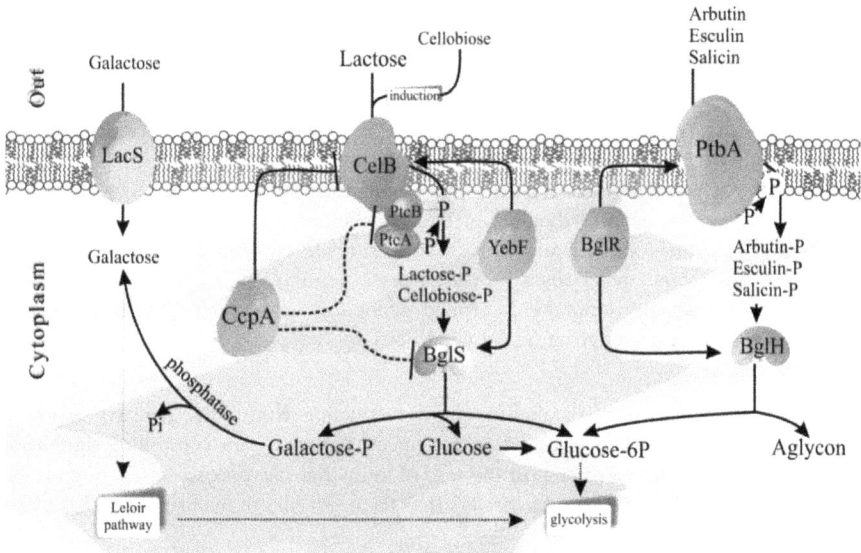

Figure 1. Schematic representation of the proposed mechanism of chromosomally-encoded lactose, cellobiose-inducible lactose and β-glucosides metabolism and of its regulation in *L. lactis* IL1403. In this model the key elements are the CelB, PtcB, PtcA, BglS and PtbA proteins. In the presence of glucose, IL1403 is unable to assimilate either lactose or β-glucosides. Under these conditions, these catabolic systems are either repressed by the CcpA protein and/or are not induced by the BglR activator.

Besides cellobiose, other β-glucosides like arbutin, esculin and salicin are transported by the PtbA-mediated PTS system. In the absence of any of these three sugars, *ptbA* expression is not induced by the inactive the phosphorylated BglR antiterminator protein. Once a β-glucoside is available, BglR becomes dephosphorylated and active, inducing the expression of the *ptbA* gene. The PtbA protein transports, with concomitant phosphorylation, arbutin, esculin and salicin, which are then probably hydrolyzed by BglH, a P-β-glucosidase, encoded by a gene located downstream of and in the same operon as the *ptbA* gene.

It is also proposed in this model that LacS is not engaged in lactose internalization and its function is limited to galactose transport.

5. Conclusions

Despite the fact that the metabolism of lactose and β-glucosides is very important for the biotechnological processes catalysed by *L. lactis*, thorough studies of the chromosomally encoded features enabling use of these carbon sources were so far rather scarce. The reason for this could be the fact that *L. lactis* demonstrates a very large and complex metabolic capability towards carbohydrates used as carbon and energy sources, and, moreover, that this genetic potential is tightly regulated by various environmental and intracellular factors. It seems that the main obstacle in studies on the complicated mechanisms involved in assimilation of β–glycoside sugars was the lack of complex data specifying the sequences of genes potentially involved in the metabolism of these sugars and its regulation. Indeed, recent access to the genomic sequences of some these bacteria greatly advanced the research on the metabolism of various β–glycosides. As expected, the results of sequencing of lactococcal genomes and genes annotations confirmed that there are numerous genes encoding potential β-glucosides-specific transport systems and β-glucosidases, sometimes with dual activities. And, to complicate the matter even further, the analysis of the list of genes annotated in *L. lactis* leads to over a hundred transcriptional regulators. A relatively large number of them may be related to carbon metabolism control. These regulators, together with signals modulating their activity, and the controlled genes form a regulatory network that is necessary for sensing the environmental conditions and adjusting the catabolic capacities of the cell.

Detailed knowledge of sugar metabolism and the regulators controlling gene expression in *Lactococcus lactis* may contribute to the improvement of mechanisms controlling significant cellular processes in these bacteria. In the case of industrial microorganisms, acting on the defined regulatory network may drastically affect the properties of the bacteria and have an impact on bioprocesses.

Lastly, is shown as an example that by the use of a simple microbiological screen, it is possible and worthwhile to modify the metabolic potential of lactococcal strains initially unable to assimilate lactose. By inactivation of the *ccpA* gene or induction of particular genes by supplementation of the medium with cellobiose and thus activation of YebF, it is possible to turn on an alternative lactose assimilation pathway in *L. lactis* IL1403. In contrast to plasmid-located *lac*-operons, the *cel-lac* system is within the chromosome, resulting in a stable, highly adapted strain, potentially valuable for the dairy industry.

Author details

Tamara Aleksandrzak-Piekarczyk
Institute of Biochemistry and Biochemistry, Polish Academy od Sciences, Warsaw, Poland

Acknowledgement

Some of the data presented were funded in part by the NCN grant UMO-2011/01/B/NZ2/05377.

6. References

Aleksandrzak, T., Kowalczyk, M., Kok, J. & Bardowski, J. (2000). Regulation of carbon catabolism in *Lactococcus lactis*. In: *Food biotechnology, Proceedings of an International Symposium organized by the Institute of Technical Biochemistry, Technical University of Lodz*.

Bielecki, S., Tramper, J., Polak, J. (eds), Vol. 17, pp. 61–66. Elsevier Science BV. ISBN 978-0-444-50519-4, Zakopane, Poland, May 1999

Aleksandrzak-Piekarczyk, T., Polak, J., Jezierska, B., Renault, P. & Bardowski, J. (2011). Genetic characterization of the CcpA-dependent, cellobiose-specific PTS system comprising CelB, PtcB and PtcA that transports lactose in *Lactococcus lactis* IL1403. *International Journal of Food Microbiology*, Vol. 145, No. 1, pp. 186-94, ISSN 1879-3460

Aleksandrzak-Piekarczyk, T., Kok, J., Renault, P. & Bardowski, J. (2005). Alternative lactose catabolic pathway in *Lactococcus lactis* IL1403. *Applied and Environmental Microbiology*, Vol. 71, No. 10, pp. 6060-9, ISSN 0099-2240

Anderson, DG. & McKay, LL. (1977). Plasmids, loss of lactose metabolism, and appearance of partial and full lactose-fermenting revertants in *Streptococcus cremoris* B1. *Journal of Bacteriology*, Vol. 129, No. 1, pp. 367-77, ISSN 0021-9193

Aymerich, S. & Steinmetz, M. (1992). Specificity determinants and structural features in the RNA target of the bacterial antiterminator proteins of the BglG/SacY family. *Proceedings of the National Academy of Sciences of the United States of America*, Vol. 89, No. 21, pp. 10410-4, ISSN 0027-8424

Bardowski, J., Ehrlich, SD. & Chopin, A. (1994). BglR protein, which belongs to the BglG family of transcriptional antiterminators, is involved in beta-glucoside utilization in *Lactococcus lactis*. *Journal of Bacteriology*, Vol. 176, No. 18, pp. 5681-5, ISSN 0021-9193

Bardowski, J., Ehrlich, SD. & Chopin, A. (1995). A protein, belonging to a family of RNA-binding transcriptional anti-terminators, controls beta-glucoside assimilation in *Lactococcus lactis*. *Developments in Biological Standardization*, Vol. 85, pp. 555-9, ISSN 0301-5149

Barrière, C., Veiga-da-Cunha, M., Pons, N., Guédon, E., van Hijum, SA., Kok, J., Kuipers, OP., Ehrlich, DS. & Renault, P. (2005). Fructose utilization in *Lactococcus lactis* as a model for low-GC gram-positive bacteria: its regulator, signal, and DNA-binding site. *Journal of Bacteriology*, Vol. 187, No. 11, pp. 3752-61, ISSN 0021-9193

Bauer, S., Tholen, A., Overmann, J. & Brune, A. (2000). Characterization of abundance and diversity of lactic acid bacteria in the hindgut of wood- and soil-feeding termites by molecular and culture-dependent techniques. *Archives of Microbiology*, Vol. 173, No. 2, pp. 126-37, ISSN 0302-8933

Bayjanov, JR., Wels, M., Starrenburg, M., van Hylckama Vlieg, JE., Siezen, RJ. & Molenaar, D. (2009). PanCGH: a genotype-calling algorithm for pangenome CGH data. *Bioinformatics (oxford, England)*, Vol. 25, No. 3, pp. 309-14, ISSN 1367-4811

Bissett, DL. & Anderson, RL. (1974). Lactose and D-galactose metabolism in group N streptococci: presence of enzymes for both the D-galactose 1-phosphate and D-tagatose 6-phosphate pathways. *Journal of Bacteriology*, Vol. 117, No. 1, pp. 318-20, ISSN 0021-9193

Bolotin, A., Wincker, P., Mauger, S., Jaillon, O., Malarme, K., Weissenbach, J., Ehrlich, SD. & Sorokin, A. (2001). The complete genome sequence of the lactic acid bacterium

Lactococcus lactis ssp. *lactis* IL1403. *Genome Research*, Vol. 11, No. 5, pp. 731-53, ISSN 1088-9051

Bouffard, GG., Rudd, KE. & Adhya, SL. (1994). Dependence of lactose metabolism upon mutarotase encoded in the *gal* operon in *Escherichia coli*. *Journal of Molecular Biology*, Vol. 244, No. 3, pp. 269-78, ISSN 0022-2836

Campelo, AB., Gaspar, P., Roces, C., Rodríguez, A., Kok, J., Kuipers, OP., Neves, AR. & Martínez, B. (2011). The Lcn972 bacteriocin-encoding plasmid pBL1 impairs cellobiose metabolism in *Lactococcus lactis*. *Applied and Environmental Microbiology*, Vol. 77, No. 21, pp. 7576-85, ISSN 1098-5336

Castro, R., Neves, AR., Fonseca, LL., Pool, WA., Kok, J., Kuipers, OP. & Santos, H. (2009). Characterization of the individual glucose uptake systems of *Lactococcus lactis*: mannose-PTS, cellobiose-PTS and the novel GlcU permease. *Molecular Microbiology*, Vol. 71, No. 3, pp. 795-806, ISSN 1365-2958

Chopin, A., Chopin, MC., Moillo-Batt, A. & Langella, P. (1984). Two plasmid-determined restriction and modification systems in *Streptococcus lactis*. *Plasmid*, Vol. 11, No. 3, pp. 260-3, ISSN 0147-619X

Cocaign-Bousquet, M., Even, S., Lindley, ND. & Loubière, P. (2002). Anaerobic sugar catabolism in *Lactococcus lactis*: genetic regulation and enzyme control over pathway flux. *Applied Microbiology and Biotechnology*, Vol. 60, No. 1-2, pp. 24-32, ISSN 0175-7598

Cords, BR. & McKay, LL. (1974). Characterization of lactose-fermenting revertants from lactose-negative *Streptococcus lactis* C2 mutants. *Journal of Bacteriology*, Vol. 119, No. 3, pp. 830-9, ISSN 0021-9193

Crow, VL. & Thomas, TD. (1984). Properties of a *Streptococcus lactis* strain that ferments lactose slowly. *Journal of Bacteriology*, Vol. 157, No. 1, pp. 28-34, ISSN 0021-9193

Daly, C. (1983). The use of mesophilic cultures in the dairy industry. *Antonie Van Leeuwenhoek*, Vol. 49, No. 3, pp. 297-312, ISSN 0003-6072

David, S., Stevens, H., van Riel, M., Simons, G. & de Vos, WM. (1992). *Leuconostoc lactis* beta-galactosidase is encoded by two overlapping genes. *Journal of Bacteriology*, Vol. 174, No. 13, pp. 4475-81, ISSN 0021-9193

Davidson, BE., Kordias, N., Dobos, M. & Hillier, AJ. (1996). Genomic organization of lactic acid bacteria. *Antonie Van Leeuwenhoek*, Vol. 70, No. 2-4, pp. 161-83, ISSN 0003-6072

De Vos, WM. & Simons, G. (1988). Molecular cloning of lactose genes in dairy lactic streptococci: the phospho-beta-galactosidase and beta-galactosidase genes and their expression products. *Biochimie*, Vol. 70, No. 4, pp. 461-73, ISSN 0300-9084

de Vos, WM. & Vaughan, EE. (1994). Genetics of lactose utilization in lactic acid bacteria. *FEMS Microbiology Reviews*, Vol. 15, No. 2-3, pp. 217-37, ISSN 0168-6445

de Vos, WM. (1996). Metabolic engineering of sugar catabolism in lactic acid bacteria. *Antonie Van Leeuwenhoek*, Vol. 70, No. 2-4, pp. 223-42, ISSN 0003-6072

Fernández, E., Alegría, A., Delgado, S., Martín, MC. & Mayo, B. (2011). Comparative phenotypic and molecular genetic profiling of wild *Lactococcus lactis* subsp. *lactis* strains of the *L. lactis* subsp. *lactis* and *L. lactis* subsp. *cremoris* genotypes, isolated from starter-free cheeses made of raw milk. *Applied and Environmental Microbiology*, Vol. 77, No. 15, pp. 5324-35, ISSN 1098-5336

Franz, CMAP., Cho, G-S., Holzapfel, WH. & Gálvez, A. (2010). Safety of Lactic Acid Bacteria, In: *Biotechnology of Lactic Acid Bacteria: Novel Applications*, Mozzi, F., Raya, RR. & Vignolo, GM., pp. 341-59, Wiley-Blackwell, ISBN 978-0-8138-1583-1, Oxford, UK

Gasson, MJ. (1983). Plasmid complements of *Streptococcus lactis* NCDO 712 and other lactic streptococci after protoplast-induced curing. *Journal of Bacteriology*, Vol. 154, No. 1, pp. 1-9, ISSN 0021-9193

Gasson, MJ. (1990). In vivo genetic systems in lactic acid bacteria. *FEMS Microbiology Reviews*, Vol. 7, No. 1-2, pp. 43-60, ISSN 0168-6445

Gilliland, SE. (1989). Acidophilus milk products: a review of potential benefits to consumers. *Journal of Dairy Science*, Vol. 72, No. 10, pp. 2483-94, ISSN 0022-0302

Godon, JJ., Delorme, C., Ehrlich, SD. & Renault, P. (1992). Divergence of Genomic Sequences between *Lactococcus lactis* subsp. *lactis* and *Lactococcus lactis* subsp. *cremoris*. *Applied and Environmental Microbiology*, Vol. 58, No. 12, pp. 4045-7, ISSN 0099-2240

Górecki, RK., Koryszewska-Bagińska, A., Gołębiewski, M., Żylińska, J., Grynberg, M. & Bardowski, JK. (2011). Adaptative potential of the *Lactococcus lactis* IL594 strain encoded in its 7 plasmids. *Plos One*, Vol. 6, No. 7, pp. e22238, ISSN 1932-6203

Griffin, HG., MacCormick, CA. & Gasson, MJ. (1996). Cloning, DNA sequence, and regulation of expression of a gene encoding beta-galactosidase from *Lactococcus lactis*. *DNA Sequence : the Journal of DNA Sequencing and Mapping*, Vol. 6, No. 6, pp. 337-46, ISSN 1042-5179

Grossiord, BP., Luesink, EJ., Vaughan, EE., Arnaud, A. & de Vos, WM. (2003). Characterization, expression, and mutation of the *Lactococcus lactis galPMKTE* genes, involved in galactose utilization via the Leloir pathway. *Journal of Bacteriology*, Vol. 185, No. 3, pp. 870-8, ISSN 0021-9193

Gunnewijk, MG. & Poolman, B. (2000a). HPr(His approximately P)-mediated phosphorylation differently affects counterflow and proton motive force-driven uptake via the lactose transport protein of *Streptococcus thermophilus*. *The Journal of Biological Chemistry*, Vol. 275, No. 44, pp. 34080-5, ISSN 0021-9258

Gunnewijk, MG. & Poolman, B. (2000b). Phosphorylation state of HPr determines the level of expression and the extent of phosphorylation of the lactose transport protein of *Streptococcus thermophilus*. *The Journal of Biological Chemistry*, Vol. 275, No. 44, pp. 34073-9, ISSN 0021-9258

Gunnewijk, MG., Postma, PW. & Poolman, B. (1999). Phosphorylation and functional properties of the IIA domain of the lactose transport protein of *Streptococcus thermophilus*. *Journal of Bacteriology*, Vol. 181, No. 2, pp. 632-41, ISSN 0021-9193

Itoi, S., Abe, T., Washio, S., Ikuno, E., Kanomata, Y. & Sugita, H. (2008). Isolation of halotolerant *Lactococcus lactis* subsp. *lactis* from intestinal tract of coastal fish. *International Journal of Food Microbiology*, Vol. 121, No. 1, pp. 116-21, ISSN 0168-1605

Itoi, S., Yuasa, K., Washio, S., Abe, T., Ikuno, E. & Sugita, H. (2009). Phenotypic variation in *Lactococcus lactis* subsp. *lactis* isolates derived from intestinal tracts of marine and freshwater fish. *Journal of Applied Microbiology*, Vol. 107, No. 3, pp. 867-74, ISSN 1365-2672

Kelly, W. & Ward, L. (2002). Genotypic vs. phenotypic biodiversity in *Lactococcus lactis*. *Microbiology (reading, England)*, Vol. 148, No. Pt 11, pp. 3332-3, ISSN 1350-0872

Kelly, WJ., Davey GP., & Ward, LJ. (1998). Characterization of lactococci isolated from minimally processed fresh fruit and vegetables. *International Journal of Food Microbiology*, Vol. 45, No. 2, pp. 85-92, ISSN 0168-1605

Kelly, WJ., Ward, LJ. & Leahy, SC. (2010). Chromosomal diversity in *Lactococcus lactis* and the origin of dairy starter cultures. *Genome Biology and Evolution*, Vol. 2, pp. 729-44, ISSN 1759-6653

Klijn, N., Weerkamp, AH. & de Vos, WM. (1995). Detection and characterization of lactose-utilizing Lactococcus spp. in natural ecosystems. *Applied and Environmental Microbiology*, Vol. 61, No. 2, pp. 788-92, ISSN 0099-2240

Kowalczyk, M., Cocaign-Bousquet, M., Loubiere, P. & Bardowski, J. (2008). Identification and functional characterisation of cellobiose and lactose transport systems in *Lactococcus lactis* IL1403. *Archives of Microbiology*, Vol. 189, No. 3, pp. 187-96, ISSN 0302-8933

Kuipers, OP., de Ruyter, PG., Kleerebezem, M. & de Vos WM. (1998). Quorum sensing-controlled gene expression in lactic acid bacteria. *Journal of Biotechnology*, Vol. 64, No. 1, pp. 15-21

Latorre-Guzman, BA., Kado, CI. & Kunkee, R. (1977). *Lactobacillus hordniae*, a new species from the leafhopper (*Hordnia circellata*). *International Journal of Systematic Bacteriology*, Vol. 27, No. 4, pp. 362–70, ISSN 1466-5026

Le Bourgeois, P., Lautier, M., van den Berghe, L., Gasson, MJ. & Ritzenthaler, P. (1995). Physical and genetic map of the *Lactococcus lactis* subsp. *cremoris* MG1363 chromosome: comparison with that of *Lactococcus lactis* subsp. *lactis* IL 1403 reveals a large genome inversion. *Journal of Bacteriology*, Vol. 177, No. 10, pp. 2840-50, ISSN 0021-9193

Linares, DM., Kok, J. & Poolman, B. (2010). Genome sequences of *Lactococcus lactis* MG1363 (revised) and NZ9000 and comparative physiological studies. *Journal of Bacteriology*, Vol. 192, No. 21, pp. 5806-12, ISSN 1098-5530

Lorca, G., Reddy, L., Nguyen, A., Sun, EI., Tseng, J., Yen, M-R. & Saier, MH. (2010). Lactic Acid Bacteria: Comparative Genomic Analyses of Transport Systems, In: *Biotechnology of Lactic Acid Bacteria: Novel Applications*, Mozzi, F., Raya, RR. & Vignolo, GM., pp:73-87. ISBN 978-0-8138-1583-1 Oxford, UK

Luesink, EJ., van Herpen, RE., Grossiord, BP., Kuipers, OP. & de Vos, WM. (1998). Transcriptional activation of the glycolytic las operon and catabolite repression of the gal operon in *Lactococcus lactis* are mediated by the catabolite control protein CcpA. *Molecular Microbiology*, Vol. 30, No. 4, pp. 789-98, ISSN 0950-382X

McClelland, M., Sanderson, KE., Spieth, J., Clifton, SW., Latreille, P., Courtney, L., Porwollik, S., Ali, J., Dante, M., Du, F., Hou, S., Layman, D., Leonard, S., Nguyen, C., Scott, K., Holmes, A., Grewal, N., Mulvaney, E., Ryan, E., Sun, H., Florea, L., Miller, W.,

Stoneking, T., Nhan, M., Waterston, R. & Wilson, RK. (2001). Complete genome sequence of *Salmonella enterica* serovar Typhimurium LT2. *Nature*, Vol. 413, No. 6858, pp. 852-6, ISSN 0028-0836

McKay, LL. (1983). Functional properties of plasmids in lactic streptococci. *Antonie Van Leeuwenhoek*, Vol. 49, No. 3, pp. 259-74, ISSN 0003-6072

Mollet, B. & Pilloud, N. (1991). Galactose utilization in *Lactobacillus helveticus*: isolation and characterization of the galactokinase (galK) and galactose-1-phosphate uridyl transferase (galT) genes. *Journal of Bacteriology*, Vol. 173, No. 14, pp. 4464-73, ISSN 0021-9193

Neves, AR., Pool, WA., Kok, J., Kuipers, OP. & Santos, H. (2005). Overview on sugar metabolism and its control in *Lactococcus lactis* - the input from in vivo NMR. *FEMS Microbiology Reviews*, Vol. 29, No. 3, pp. 531-54, ISSN 0168-6445

Neves, AR., Pool, WA., Solopova, A., Kok, J., Santos, H. & Kuipers, OP. (2010). Towards enhanced galactose utilization by *Lactococcus lactis*. *Applied and Environmental Microbiology*, Vol. 76, No. 21, pp. 7048-60, ISSN 1098-5336

Nomura, M., Kimoto, H., Someya, Y. & Suzuki, I. (1999). Novel characteristic for distinguishing *Lactococcus lactis* subsp. *lactis* from subsp. *cremoris*. *International Journal of Systematic Bacteriology*, Vol. 49 Pt 1, pp. 163-6, ISSN 0020-7713

Nomura, M., Kobayashi, M. & Okamoto, T. (2002). Rapid PCR-based method which can determine both phenotype and genotype of *Lactococcus lactis* subspecies. *Applied and Environmental Microbiology*, Vol. 68, No. 5, pp. 2209-13, ISSN 0099-2240

Orla-Jensen, S. (1919). The lactic acid bacteria, pp. 1-196. Host & Son, Copenhagen, Denmark

Pérez, T., Balcázar, JL., Peix, A., Valverde, A., Velázquez, E., de Blas, I. & Ruiz-Zarzuela, I. (2011). *Lactococcus lactis* subsp. *tructae* subsp. nov. isolated from the intestinal mucus of brown trout (*Salmo trutta*) and rainbow trout (*Oncorhynchus mykiss*). *International Journal of Systematic and Evolutionary Microbiology*, Vol. 61, No. Pt 8, pp. 1894-8, ISSN 1466-5034

Pool, WA., Neves, AR., Kok, J., Santos, H. & Kuipers, OP. (2006). Natural sweetening of food products by engineering *Lactococcus lactis* for glucose production. *Metabolic Engineering*, Vol. 8, No. 5, pp. 456-64, ISSN 1096-7176

Poolman, B. (1993). Energy transduction in lactic acid bacteria. *FEMS Microbiology Reviews*, Vol. 12, No. 1-3, pp. 125-47, ISSN 0168-6445

Poolman, B., Royer, TJ., Mainzer, SE. & Schmidt, BF. (1990). Carbohydrate utilization in *Streptococcus thermophilus*: characterization of the genes for aldose 1-epimerase (mutarotase) and UDPglucose 4-epimerase. *Journal of Bacteriology*, Vol. 172, No. 7, pp. 4037-47, ISSN 0021-9193

Rademaker, JL., Herbet, H., Starrenburg, MJ., Naser, SM., Gevers, D., Kelly, WJ., Hugenholtz, J., Swings, J. & van Hylckama Vlieg, JE. (2007). Diversity analysis of dairy and nondairy *Lactococcus lactis* isolates, using a novel multilocus sequence analysis scheme and (GTG)5-PCR fingerprinting. *Applied and Environmental Microbiology*, Vol. 73, No. 22, pp. 7128-37, ISSN 0099-2240

Rutberg, B. (1997). Antitermination of transcription of catabolic operons. *Molecular Microbiology*, Vol. 23, No. 3, pp. 413-21, ISSN 0950-382X

Saier, MH. (2000). Families of transmembrane sugar transport proteins. *Molecular Microbiology*, Vol. 35, No. 4, pp. 699-710, ISSN 0950-382X

Salama, M., Sandine, W. & Giovannoni, S. (1991). Development and application of oligonucleotide probes for identification of *Lactococcus lactis* subsp. *cremoris*. *Applied and Environmental Microbiology*, Vol. 57, No. 5, pp. 1313-8, ISSN 0099-2240

Sandine, WE. (1988). New nomenclature of the non-rod-shaped lactic acid bacteria. *Biochimie*, Vol. 70, No. 4, pp. 519-21, ISSN 0300-9084

Schleifer, KH., Kraus, J., Dvorak, C., Kilpper-Bälz, R., Collins MD. & Fischer, W. (1985). Transfer of *Streptococcus lactis* and related streptococci to the genus *Lactococcus* gen. nov. *Systematic and Applied Microbiology*, Vol. 6, No. 2, pp. 183-95, ISSN 07232020

Schultz, JE. & Breznak, JA. (1978). Heterotrophic bacteria present in hindguts of wood-eating termites [*Reticulitermes flavipes* (Kollar)]. *Applied and Environmental Microbiology*, Vol. 35, No. 5, pp. 930-6, ISSN 0099-2240

Siezen, RJ., Bayjanov, J., Renckens, B., Wels, M., van Hijum, SA., Molenaar, D. & van Hylckama Vlieg, JE. (2010). Complete genome sequence of *Lactococcus lactis* subsp. *lactis* KF147, a plant-associated lactic acid bacterium. *Journal of Bacteriology*, Vol. 192, No. 10, pp. 2649-50, ISSN 1098-5530

Siezen, RJ., Starrenburg, MJ., Boekhorst, J., Renckens, B., Molenaar, D. & van Hylckama Vlieg, JE. (2008). Genome-scale genotype-phenotype matching of two *Lactococcus lactis* isolates from plants identifies mechanisms of adaptation to the plant niche. *Applied and Environmental Microbiology*, Vol. 74, No. 2, pp. 424-36, ISSN 1098-5336

Simons, G., Nijhuis, M. & de Vos, WM. (1993). Integration and gene replacement in the *Lactococcus lactis* lac operon: induction of a cryptic phospho-beta-glucosidase in LacG-deficient strains. *Journal of Bacteriology*, Vol. 175, No. 16, pp. 5168-75, ISSN 0021-9193

Smit, G., Smit, BA. & Engels, WJ. (2005). Flavour formation by lactic acid bacteria and biochemical flavour profiling of cheese products. *FEMS Microbiology Reviews*, Vol. 29, No. 3, pp. 591-610, ISSN 0168-6445

Solem, C., Koebmann, B. & Jensen, PR. (2008). The extent of co-metabolism of glucose and galactose by *Lactococcus lactis* changes with the expression of the *lacSZ* operon from *Streptococcus thermophilus*. *Biotechnology and Applied Biochemistry*, Vol. 50, No. Pt 1, pp. 35-40, ISSN 1470-8744

Sørensen, KI. & Hove-Jensen, B. (1996). Ribose catabolism of *Escherichia coli*: characterization of the *rpiB* gene encoding ribose phosphate isomerase B and of the *rpiR* gene, which is involved in regulation of *rpiB* expression. *Journal of Bacteriology*, Vol. 178, No. 4, pp. 1003-11, ISSN 0021-9193

Stiles, ME. & Holzapfel, WH. (1997). Lactic acid bacteria of foods and their current taxonomy. *International Journal of Food Microbiology*, Vol. 36, No. 1, pp. 1-29, ISSN 0168-1605

Tailliez, P., Tremblay, J., Ehrlich, SD. & Chopin, A. (1998). Molecular diversity and relationship within *Lactococcus lactis*, as revealed by randomly amplified polymorphic

DNA (RAPD). *Systematic and Applied Microbiology*, Vol. 21, No. 4, pp. 530-8, ISSN 0723-2020

Tanigawa, K., Kawabata, H. & Watanabe, K. (2010). Identification and typing of *Lactococcus lactis* by matrix-assisted laser desorption ionization-time of flight mass spectrometry. *Applied and Environmental Microbiology*, Vol. 76, No. 12, pp. 4055-62, ISSN 1098-5336

Teeri, TT. (1997) Crystalline cellulose degradation: new insight into the function of cellobiohydrolases. *Trends in Biotechnology*, Vol. 15, No. 5, pp. 160–67, ISSN 0167-7799

Tobisch, S., Glaser, P., Krüger, S. & Hecker, M. (1997). Identification and characterization of a new beta-glucoside utilization system in *Bacillus subtilis*. *Journal of Bacteriology*, Vol. 179, No. 2, pp. 496-506, ISSN 0021-9193

van Hylckama Vlieg, JE., Rademaker, JL., Bachmann, H., Molenaar, D., Kelly, WJ. & Siezen, RJ. (2006). Natural diversity and adaptive responses of *Lactococcus lactis*. *Current Opinion in Biotechnology*, Vol. 17, No. 2, pp. 183–190, ISSN 0958-1669

van Rooijen, RJ., van Schalkwijk, S. & de Vos, WM. (1991). Molecular cloning, characterization, and nucleotide sequence of the tagatose 6-phosphate pathway gene cluster of the lactose operon of *Lactococcus lactis*. *The Journal of Biological Chemistry*, Vol. 266, No. 11, pp. 7176-81, ISSN 0021-9258

Vaughan, EE., David, S. & de Vos, WM. (1996). The lactose transporter in *Leuconostoc lactis* is a new member of the LacS subfamily of galactoside-pentose-hexuronide translocators. *Applied and Environmental Microbiology*, Vol. 62, No. 5, pp. 1574-82, ISSN 0099-2240

Vaughan, EE., Pridmore, RD. & Mollet, B. (1998). Transcriptional regulation and evolution of lactose genes in the galactose-lactose operon of *Lactococcus lactis* NCDO2054. *Journal of Bacteriology*, Vol. 180, No. 18, pp. 4893-902, ISSN 0021-9193

Vaughan, EE., van den Bogaard, PT., Catzeddu, P., Kuipers, OP. & de Vos, WM. (2001). Activation of silent *gal* genes in the *lac-gal* regulon of *Streptococcus thermophilus*. *Journal of Bacteriology*, Vol. 183, No. 4, pp. 1184-94, ISSN 0021-9193

Wegmann, U., O'Connell-Motherway, M., Zomer, A., Buist, G., Shearman, C., Canchaya, C., Ventura, M., Goesmann, A., Gasson, MJ., Kuipers, OP., van Sinderen, D. & Kok, J. (2007). Complete genome sequence of the prototype lactic acid bacterium *Lactococcus lactis* subsp. *cremoris* MG1363. *Journal of Bacteriology*, Vol. 189, No. 8, pp. 3256-70, ISSN 0021-9193

Weickert, MJ. & Adhya, S. (1992). A family of bacterial regulators homologous to Gal and Lac repressors. *The Journal of Biological Chemistry*, Vol. 267, No. 22, pp. 15869-74, ISSN 0021-9258

Weickert, MJ. & Chambliss, GH. (1990). Site-directed mutagenesis of a catabolite repression operator sequence in *Bacillus subtilis*. *Proceedings of the National Academy of Sciences of the United States of America*, Vol. 87, No. 16, pp. 6238-42, ISSN 0027-8424

Williams, SG., Greenwood, JA. & Jones, CW. (1992). Molecular analysis of the *lac* operon encoding the binding-protein-dependent lactose transport system and beta-galactosidase in *Agrobacterium radiobacter*. *Molecular Microbiology*, Vol. 6, No. 13, pp. 1755-68, ISSN 0950-382X

Zomer, AL., Buist, G., Larsen, R., Kok, J. & Kuipers, OP. (2007). Time-resolved determination of the CcpA regulon of *Lactococcus lactis* subsp. *cremoris* MG1363. *Journal of Bacteriology*, Vol. 189, No. 4, pp. 1366-81, ISSN 0021-9193

Exopolysaccharides of Lactic Acid Bacteria for Food and Colon Health Applications

Tsuda Harutoshi

Additional information is available at the end of the chapter

1. Introduction

Lactic acid bacteria (**LAB**) are used in many fermented foods, particularly fermented dairy products such as cheese, buttermilk, and fermented milks. LAB produce lactic acid, carbon dioxide, and diacetyl/acetoin that contribute to the flavor, texture, and shelf life of fermented foods. Some LAB produce exopolysaccharide (EPS), and generally, EPS play a major role as natural texturizer in the industrial production of yoghurt, cheese, and milk-based desserts. Recently, EPS produced by LAB have received increasing attention, mainly because of their health benefits. In particular, immune stimulation, antimutagenicity, and the antitumor activity of fermented dairy products prepared with EPS-producing LAB or EPS themselves have been investigated [1-4].

EPS are polysaccharides secreted from the cell, or produced on the outer cell by extracellular enzymes. EPS from LAB are divided into two classes, homo- and hetero-EPS. Homo-EPS are composed of one type of monosaccharide, whereas hetero-EPS consist of regular repeating units of 3-8 different carbohydrate moieties synthesized from intracellular sugar nucleotide precursors [5]. The biosynthesis of homo-EPS and hetero-EPS are different. Homo-EPS are made from sucrose using glucansucrase or levansucrase [6-7], and the synthesis of hetero-EPS involves four major steps, sugar transportation, sugar nucleotide synthesis, repeating unit synthesis, and polymerization of the repeating units [8]. The major physiological function of EPS is believed to be biological defenses against various stresses such as phage attack, toxic metal ions, and desiccation [9], and it is very unlikely that bacteria use EPS as an energy source. However, some potentially probiotic LAB strains have been reported to degrade EPS produced by the other LAB strains [10-11].

The term "probiotic" was first proposed by Fuller [12], and its definition was further refined to "Live microorganisms which when consumed in adequate amounts as part of food confer a health benefit on the host" [13]. Probiotic LAB thus represent a class of live food ingredients that exert a beneficial effect on the health of the host. Beneficial microorganisms in the intestine are enhanced by "prebiotics," which are defined as "nondigestible food ingredients that beneficially affect the host by selectively stimulating the growth and activity of one or a limited number of bacterial species already resident in the colon, and thus improving host health" [14].

Most of the current prebiotics are low molecular weight except for inulin. As long carbohydrate chains are metabolized more slowly than the short ones, and polysaccharides thus exert prebiotic effects in more distal colonic regions compared to oligosaccharides, which are more rapidly digested in the proximal colon [15]. Therefore, EPS produced by LAB can be used as prebiotics. This chapter reviews the physicochemical properties, genetics, and bioactivities of the EPS produced by LAB.

2. Chemical composition of EPS

2.1. Homo-EPS

Some LAB can produce EPS that are either secreted to the environment or attached to the cell surface forming capsules. EPS are classified into two groups: homo-EPS, consisting of a single type of monosaccharide (α-D-glucans, β-D-glucans, fructans, and others represented by polygalactan) and hetero-EPS, composed of different types of monosaccharides, mainly D-glucose, D-galactose, L-rhamnose, and their derivatives [16].

The differences arise between the homopolysaccharides mainly because of the features of their primary structure such as the pattern of main chain bonds, molecular weight, and branch structure. Two important groups of homo-EPS are produced by LAB; (i) α-glucans, mainly composed of α-1,6- and α-1,3-linked glucose residues, namely dextrans, produced by *Leuconostoc mesenteroides* subsp. *mesenteroides* and *Leuconostoc mesenteroides* subsp. *dextranicum* and mutans produced by *Streptococcus mutans* and *Streptococcus sobrinus*; and (ii) fructans, mainly composed of β-2,6-linked fructose molecules, such as levan produced by *Streptococcus salivarius* [17].

The formation of dextran from sucrose has been recorded for *Leuc. mesenteroides* subsp. *mesenteroides*. However, the ability to form dextran is often lost when serial transfers are made in media with increasing salt concentrations. Nevertheless, non-dextran-producing strains of *Leuconostoc* sp. can revert to dextran production when they are inoculated into medium containing tomato or orange juice [18]. In the 1950s, the use of a cell-free enzyme solution permitted dextran synthesis under controlled conditions yielding a polymer of greater purity. A common feature of all dextrans is the preponderance of α-1,6-linkages with branch points at positions 2, 3, or 4 [17]. Some strains of *Leuconostoc amelibiosum* [19] and *Lactobacillus curvatus* [20] are reported to be dextran-producing strains.

Mutan is the glucan synthesized by various serotypes of *Str. mutans*, and differs from dextran in that it contains a high percentage of α-1,3 linkages. Differences in solubility result

Homo-EPS	Main linkage (branching linkage)	Organism
Glucans		
Dextran	α-1,6 (α-1,3)	*Leuc. mesenteroides* subsp. *mesenteroides, Leuc. mesenteroides* subsp. *dextranicum, Leuc. amelibiosum, Lb. curvatus*
Mutan	α-1,3 (α-1,6)	*Str. mutans*
Alternan	α-1,3 and α-1,6	*Leuc. mesenteroides*
Fructans		
Levan	β-2,6 (β-2,1)	*Leuc. mesenteroides, Lb. reuteri*
Inulin	β-2,1 (β-2,6)	*Str. mutans*

Table 1. Homo EPS produced by LAB

from the proportions of different types of linkages; water-soluble glucans are rich in α-1,6 linkages, while water-insoluble glucans are rich in α-1,3 linkages [17]. Ingestion of mutan has been linked with dental caries, as insoluble mutans can adhere to teeth, thus helping microorganisms adhere to the surface of teeth.

Alternan has alternate α-1,6 and α-1,3 linkages, and this structure is thought to be responsible for its distinctive physical properties including high solubility and low viscosity. These characteristics provide this glucan with a potential commercial application as a low viscosity texturizer in foods. *Leuc. mesenteroides* NRRL B-1355 was first reported to be an alternan-producing strain [21]

Levan is an EPS produced from sucrose. It is fructan composed of β-2,6-linked fructose molecules with some β-2,1-linked branches. Incidentally, inulin is a fructan composed of β-2,1-linked fructose molecules with some β-2,6-linked branches. *Str. salivarius, Leuc. mesenteroides,* and *Lactobacillus reuteri* are known to be levan-producing LAB [22-23]. In addition, the EPS produced by *Lactobacillus sanfranciscensis* TMW 1.392 has been reported to be fructan [11].

2.2. Hetero-EPS

The chemical composition of hetero-EPS shows wide variablity. Hetero-EPS are polymerized repeating units mainly composed of D-glucose, D-galactose, and L-rhamnose. The composition of the monosaccharide subunits and the structure of the repeating units are considered not to be species-specific, except in case of *Lactobacillus kefiranofaciens* subsp. *kefiranofaciens*. This species, isolated from kefir grain, a fermented dairy food from the North Caucasus region, produces large amounts of polysaccharides [24]. Hetero-EPS-producing strains of *Streptococcus thermophilus, Lactococcus lactis, Lactobacillus delbrueckii,* and *Lactobacillus helveticus,* among others have been identified (Table 2) [25-49]. Heterofermentative LAB such as *Leuc. dextranicum* are well known homo-EPS producers, while homofermentative LAB are well-studied hetero-EPS producers. Heterofermentative in addition to homofermentative LAB can produce EPS. *Lactobacillus fermentum* is an EPS-producing heterofermentative LAB for which the EPS structure has been determined [50]. Figueroa et al. reported that *Lactobacillus brevis* and *Lactobacillus buchneri* showed ropiness on glucose- or sucrose-containing media, although they did not investigate whether such ropiness derived from hetero-EPS or from other slimy substances [51].

The quantities of hetero-EPS produced by LAB vary greatly. EPS production is 50-350 mg/l for *Str. thermophilus,* 80-600 mg/l for *Lc. lactis* subsp. *cremoris,* 60-150 mg/l for *Lb. delbrueckii* subsp. *bulgaricus,* 50-60 mg/l for *Lactobacillus casei* [52], and approximately 140 mg/l for *Lactobacillus plantarum* [45, 53]. The highest recorded yields of hetero-EPS are 2775 mg/l for *Lactobacillus rhamnosus* RW-9595M [54] and 2500 mg/l for *Lb. kefiranofaciens* WT-2B [55]. However, the quantities of EPS produced by LAB are much lower than the yields from other industrially important microorganisms such as *Xanthomonas campestris,* which produces 30-50 g/l xanthan gum [56]. Even so, amounts of EPS produced by LAB are sufficient to exploit for in situ applications. LAB are 'generally recognized as safe' (GRAS) microorganisms, and

LAB strain culture would be a useful method to produce EPS for food applications if the LAB could be grown in edible and safe culture media such as whey, and if fermentation conditions were optimized to obtain a high yield.

Fermentation conditions using undefined media have been improved to maximize yields. However, a chemically defined medium containing a carbohydrate source, mineral salts, amino acids, vitamins, and nucleic acid bases is more suitable for investigating the influence of different nutrients on LAB growth and EPS biosynthesis. The total yield of EPS produced by LAB depends on the composition of the medium (carbon and nitrogen sources) and the growth conditions, i.e., temperature, pH, and incubation time.

Under conditions of higher temperatures and slower growth, the production of the polymer per cell in *Lb. delbrueckii* subsp. *bulgaricus* NCFB 2772 was greater in milk [57]. Another study investigated the optimum culture conditions for EPS production by *Lb. delbrueckii* subsp. *bulgaricus* RR in semidefined medium [58], and determined the optimum temperature and pH conditions for EPS production to be 36°C - 39°C and pH 4.5 - 5.5. The optimal temperature for EPS production was approximately 40°C for thermophilic LAB strains, and around 25°C for mesophilic LAB. Gamar et al. [59] reported increased slime production at lower incubation temperatures, and an increase in the final EPS concentration in *Lb. rhamnosus* following incubation at 25°C instead of 30°C. The effects of temperature on EPS production in whey were investigated in *Lb. plantarum* [53], and the yield was found to be higher at 25°C than at either 30°C or 37°C. Moreover, an inverse relationship was observed between EPS production per cell and the growth temperature for *Lactobacillus sake* [49], i.e. the lower the temperature, the higher the EPS production per cell. However, the growth rate in the exponential phase decreased at low temperatures. Therefore, the temperature for the maximal production of EPS is based on a balance of cell density and EPS production per cell. Maximal EPS production by *Lb. sake* was obtained under anaerobic conditions at 20°C, although EPS production per cell was higher at 10°C. Therefore, it is possible that severe environmental conditions trigger EPS production as a protective mechanism.

The effects of alterations to the nitrogen and carbon sources used in EPS production have also been investigated. According to early reports, neither LAB growth nor EPS production was specifically linked to the presence of casein or whey proteins in the growth medium. Garcia et al. [57] reported that EPS production by *Lb. delbrueckii* subsp. *bulgaricus* NCFB 2772 increased during the early growth pase in the presence of hydrolyzed casein in milk, while the addition of hydrolyzed casein to MRS medium did not increase EPS production. This strain produced 25 mg/l EPS when grown on fructose in a defined medium, and 80 mg/l EPS when grown on glucose [60]. The optimum Bacto-casitone concentration for EPS production by *Lb. delbrueckii* subsp. *bulgaricus* RR was investigated in semidefined medium [58]. In this study, there was a significant relationship between the Bacto-casitone concentration and EPS production; the higher the casitone concentration, the higher the EPS yield that was obtained. For *Lb. plantarum* grown in whey, yeast extract was a more effective nitrogen source for EPS production than soy peptide, tryptone, peptone, and Lab-Lemco powder, and glucose was a more effective carbon source than galactose, sucrose, maltose, fructose, and raffinose [53]. EPS production by *Lb. casei* CG11 was investigated in basal minimum medium containing galactose, glucose, lactose, sucrose, maltose, and melibiose; glucose was the most efficient carbon source, and lactose and galactose were the least efficient ones [61]. EPS production by *Lb. rhamnosus* C83 was investigated in a chemically defined medium containing different carbon sources (glucose, fructose, mannose, and maltose) at different concentrations. Mannose at 40 g/l was by far the most efficient carbon source. Furthermore, increased Mg, Mn and Fe concentrations stimulated EPS production in synthetic media [59]. In addition, Macedo et al. [54] reported about the importance of salts in culture media and the strong positive effect of salts and amino acids on *Lb. rhamnosus* RW-9595M growth and EPS production. The addition of salts and amino acids largely increased EPS production (to 2775 mg/l) in whey permeate supplemented with yeast extract, although the addition of amino acids alone had no effect on EPS production.

It has been shown that an optimal ratio between the carbon and nitrogen is absolutely necessary to achieve high EPS yields [62]. The production of EPS by *Str. thermophilus* LY03 is modulated by both the absolute quantities and the ratio of carbon to nitrogen (C/N ratio). The carbon source is converted into lactic acid to produce energy as well as to synthesize the cell wall and EPS, and nitrogen is necessary for the synthesis of essential cell components. Therefore, a higher C/N ratio and sufficient quantities of both carbon and nitrogen increase EPS production.

3. EPS biosynthesis by LAB

3.1. Homo EPS biosynthesis

Homo EPS are synthesized outside the cell by specific glycosyltransferase (GTF) or fructosyltransferase (FTF) enzymes (commonly named glucansucrases or fructan-sucrases). Homo-EPS producing LAB also use extracellular GTF enzymes to synthesize high-molecular mass α-glucans from sucrose. This process uses sucrose as a specific substrate, and the energy required for the process comes from sucrose hydrolysis. There is no energy requirement for EPS-production other than for enzyme biosynthesis because EPS synthesis by GTF or FTF does not involve active transport processes or the use of activated carbohydrate precursors. Therefore, large amounts of sucrose can easily be converted to EPS. *Lb. sanfranciscensis* produces up to 40 g/l levan and 25 g/l 1-kestose during growth in the presence of 160 g/l sucrose [63].

Glucan synthesis reactions catalysed by GTF can be written as follows (Fig. 1):
sucrose + H_2O → glucose + fructose
sucrose + acceptor carbohydrate → oligosaccharide + fructose
sucrose + glucan (n) → glucan (n+1) + fructose

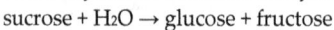

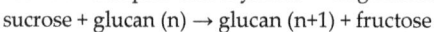

Although GTF enzymes have a high degrees of similarity, lactobacilli produce a broad spectrum of glucans, including polymers with α-1,6 linkages (dextran), α-1,3 linkages (mutan), and both α-1,6 and α-1,4 linkages (alternan). The relative molecular weight of glucans from lactobacilli range from 1×10^6 Da to 5×10^7 Da [6]. In addition, GTF enzymes are not saturated by their substrate, and transfer reactions exceed the sucrose hydrolysis under sucrose concentrations above 100 mM [64].

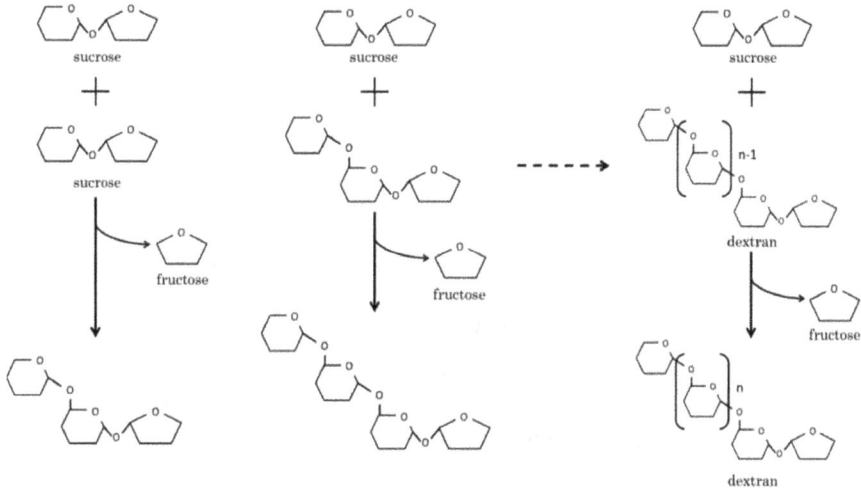

Figure 1. The dextran synthesis by GTF (dextran sucrase).

The GTF enzymes of streptococci are generally produced constitutively. In contrast, the GTF enzymes of *Leuconostoc* species are specifically induced by sucrose. For example, GTF expression in *Leuc. mesenteroides* is low in the presence of carbon sources other than sucrose and is increased by the addition of sucrose [5]. GTF expression during sucrose fermentation is 10-15-fold higher than that measured during glucose fermentation in *Leuc. mesenteroides* Lcc4. In fed-batch fermentation with both glucose and sucrose, GTF activity was similar to that obtained with sucrose alone. These results show that GTF expression is low in the presence of glucose alone, and that GTF activity is significantly induced by sucrose. A sucrose concentration of 20 g/l is sufficient to ensure the induction of enzyme synthesis, and higher concentrations (up to 60 g/l) do not lead to a further increase in enzyme synthesis [65].

The fructan synthesis reaction catalyzed by FTF can be written as follows:
sucrose + H_2O → fructose + glucose
sucrose + acceptor carbohydrate → oligosaccharide + glucose
sucrose + fructan (n) → fructan (n+1) + glucose

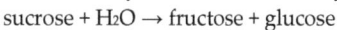

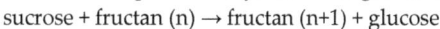

Fructans generally have a relative molecular weight exceeding 5×10^6 Da. Similar to GTFs, FTFs are not saturated by their substrate, namely, sucrose, and transfer reactions exceed the rate of sucrose hydrolysis for sucrose concentrations above 200 mM [5]. FTFs such as Lev, Inu, and LevS from lactobacilli exhibit pH optima of between 5.0 and 5.5. The optimum temperature for enzymes from the thermophilic *Lb. reuteri* is higher (50°C) than that of the *Lb. sanfranciscensis* enzyme (35°C – 40°C) [5].

3.2. Hetero EPS biosynthesis

Hetero EPS are not synthesized by extracellular enzymes, but are instead synthesized by a complex sequence of interactions involving intracellular enzymes. EPS are made by polymerization of repeating units, and these repeating units are built by a series of addition of sugar nucleotides at the cytoplasmic membrane. Sugars are the starting materials for the synthesize sequence. LAB strains can utilize various monosaccharides and disaccharides as energy sources, via some well-studied sugar uptake systems include primary transport systems, direct coupling of sugar translocation to ATP hydrolysis via a transport-specific ATPase; secondary sugar transport systems, coupling of sugar transport to the transport of ions or other solutes, both as symport and antiport transport systems; and group translocation systems, coupling of sugar transport to phosphorylation via the phosphoenolpyruvate (PEP)-dependent phosphotransferase system (PTS; Fig. 2) [8]. Polysaccharides must be hydrolyzed before uptake. For example, starch is hydrolyzed by α-amylase, and the raction products are subsequently hydrolyzed by the enzymes described above.

Lc. lactis strains possess a lactose-specific PEP-PTS sugar transport system that imports extracellular lactose, resulting in increased intracellular lactose-6-phosphate. Lactose-6-phosphate is then hydrolyzed, and the galactose-6-phosphate moiety is metabolized by the tagatose-6P pathway (Fig. 2).

Lb. delbrueckii subsp. *bulgaricus* and *Str. thermophilus* are generally galactose-negative and take up lactose via a lactose/galactose antiport transport system. The glucose moiety of imported lactose is fermented by these strains, while the galactose moiety is excreted via the lactose/galactose antiport system.

After the addition of a hetero-EPS repeating unit, the unit is exported through the cell membrane and becomes polymerized into the final hetero-EPS. Hence, several enzymes and proteins are involved in the biosynthesis and secretion of heterotype EPS, and the enzymes and proteins involved in these processes may not be unique to hetero-EPS anabolism.

Sugars taken into the cell are converted into sugar nucleotides. Iintracellular monosaccharides are converted to sugar nucleotide substrates for polymerization reactions, including UDP (uridine diphosphate), dNTP (thymidine diphosphate), and GDP (guanosine diphosphate). Such polymerization reactions are catalyzed by glycosyl pyrophosphorylases.

Glu-1P (Gal-1P) + UTP → UDP-Glu (UDP-Gal) + pyrophosphate
UDP-glucose is then converted to UDP-galactose by epimerases such as UDP-glucose-4-epimerase. This reaction is reversible.
UDP-glucose ↔ UDP-galactose

Glycosidic linkages are formed on membranes in the cytoplasm. A sugar moiety is transferred to C55-polyprenyl phosphate, a carrier lipid and component of the membrane, by priming glycosyl transferases. This transfer triggers the addition of a repeating unit to the hetero-EPS molecule. Disruption of the priming glycosyl transferase gene generates non-EPS-producing mutants [66]. Thus, priming glycosyl transferases are thought to be crucial

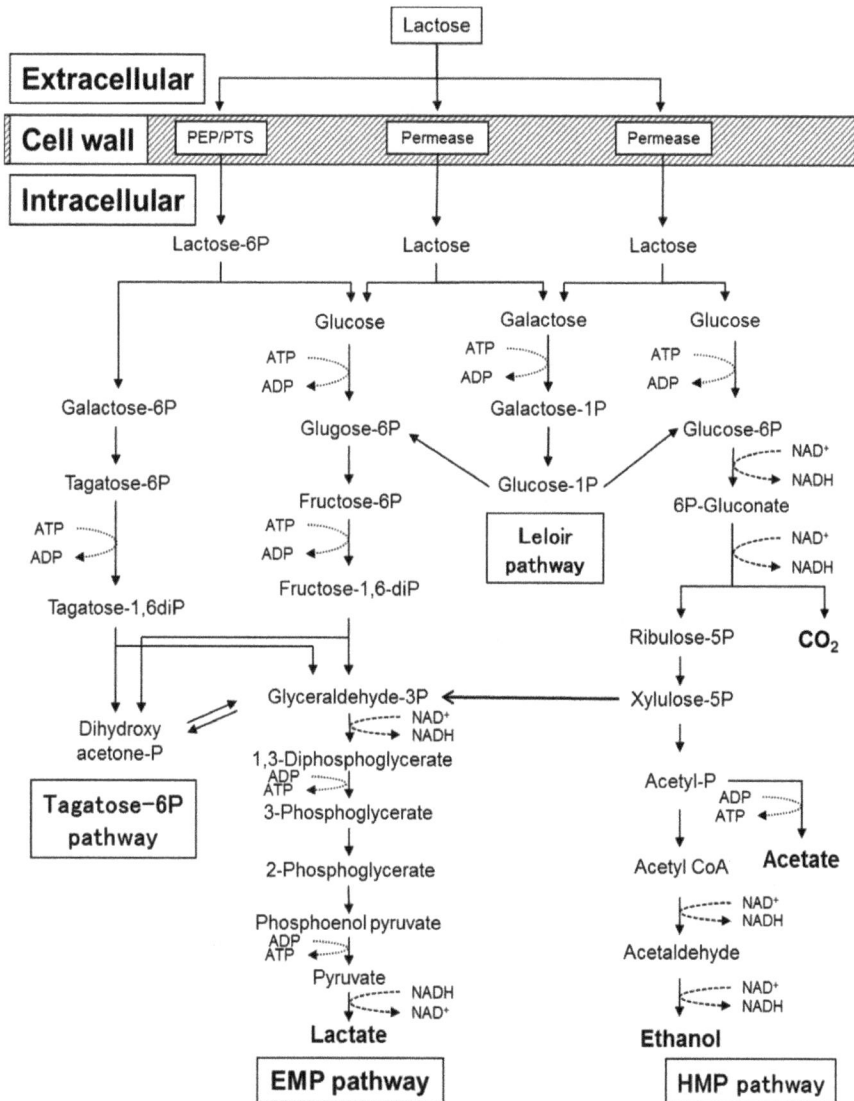

Figure 2. Pathway of lactose fermentation in lactic acid bacteria.

for EPS biosynthesis. The addition of the repeating unit is completed by the action of glycosyl transferase on the sugar residue attached to C55-polyprenyl phosphate. Therefore, the type and number of glycosyl transferases available determine the range of repeating units in hetero-EPS. C55-polyprenyl phosphate is also involved in bacterial cell wall biosynthesis, and therefore, cell wall biosynthesis and EPS synthesis compete for this substrate. The repeating unit is exported through the bacterial membrane, and is polymerized to become a hetero-EPS (Fig. 3).

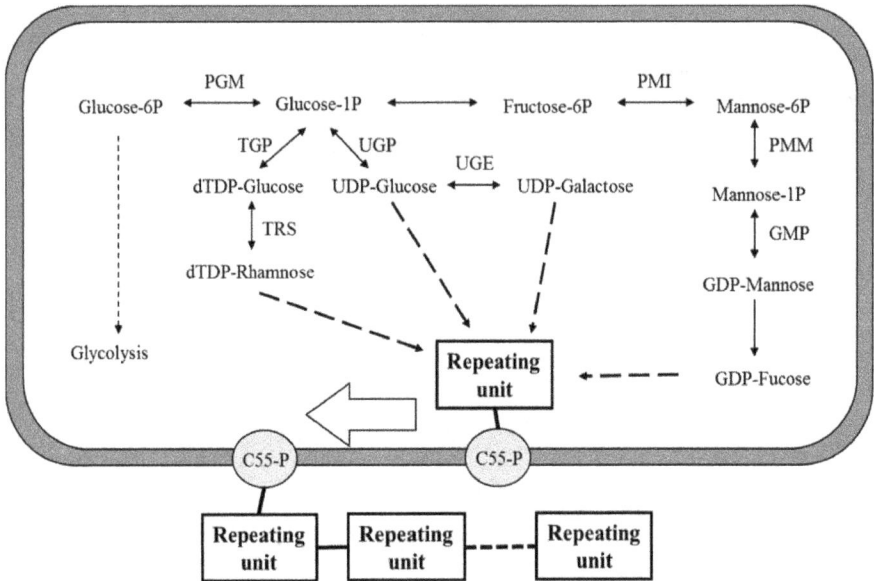

Figure 3. Outline of biosynthesis of hetero EPS.
PGM: α-phosphoglucomutase, UGP: UDP-glucose pyrophospholyrase
UGE: UDP-galactose 4-epimerase, TGP: dTDP-glucose pyrophospholyrase
TRS: dTDP-rhamnose synthetic enzyme system, PMI: phosphomannoisomerase
PMM: phosphomannomutase, GMP: GDP-mannose pyrophospholyrase

3.3. Instability of EPS production

The instability of hetero-EPS production has been reviewed by de Vuyst et al. [8]. Briefly, a loss in the ability to produce slime may be caused by repeated subculture of bacterial strains or incubation at high temperatures. The loss of plasmids from ropy mesophilic LAB strains is generally the reason for loss of slime production. On the other hand, thermophilic LAB, namely, *Lb. delbrueckii* subsp. *bulgaricus* and *Str. thermophilus*, have been shown to lack plasmids encoding components required for slime production. These species can usually recover the ability to produce slime following loss due to culture conditions. Thus, genetic instability could be a consequence of the actions of mobile genetic elements such as insertion sequences. Recently, the EPS gene cluster in *Lb. fermentum* TDS030603 was reported to be located in chromosomal DNA [67].

Priming glycosyl transferases are thought to be crucial for EPS biosynthesis and disruption of the priming glycosyl transferase gene generates non-EPS-producing mutants. Tsuda et al. generated the EPS-producing mutant strain 301102S from the non-EPS-producing *Lb. plantarum* 301102 following exposure to the mutagens acridine orange and novobiocin [4]. The activities of α-phosphoglucomutase (PGM), UDP-glucose pyrophosphorylase (UGP), and UDP-galactose 4-epimerase (UGE) were measured in parental and mutant strains by using the method of Mozzi [68], and were found to be almost the same for both [Tsuda & Miyamoto, unpublished data]. Next, priming glycosyl transferase genes in parental and mutant strains were amplified with the thermal cycler. Primers were designed to amplify a priming glycosyl transferase gene referring to complete *Lb. plantarum* WCFS1 genome sequenced [69]. PCR products were subjected to restriction digestion, which allowed identification of putative priming glycosyl transferase gene. PCR products were also applied to single strand conformation polymorphism (SSCP) analysis for detecting point mutations. However, both parental and mutant strains had the same priming glycosyl transferase gene sequence, and similar levels of activities of the PGM, UGP, and UGE enzymes. Thus, although priming glycosyl transferases are essential, other factors may also be necessary for EPS production, and a mutation affecting EPS production may occure in another gene. Morona et al. reported that an autophosphorylating protein-tyrosine kinase is essential for encapsulation in *Streptococcus pneumoniae* [70]. A point mutation in the gene encoding the autophosphorylating protein-tyrosine kinase affecting the ATP-binding domain resulted in loss of EPS production.

4. Polysaccharides and oligosaccharides for colon health

EPS produced by LAB have various functional roles in human or animal health including immunomodulatory properties, antiviral activity, antioxidant activity, and antihypertensive activity [1, 55, 71, 72], and have also been used as food additives for texture improvement. These properties have been extensively reviewed [8, 9, 56, 73, 74]. Besides these properties, prebiotics based on LAB and oligosaccharides have other health benefits. Prebiotics are usually non-digestible oligosaccharides that selectively stimulate the growth and activity of a limited number of bacterial species in the colon, such as bifidobacteria and lactobacilli, and therefore, improve host health. Detrimental bacteria may form substances such as ammonia, hydrogen sulfide, indles, and amines that are noxious to the host. However, beneficial bacteria such as bifidobacteria and lactobacilli inhibit the proliferation of detrimental bacteria, and their cell components stimulate the host immune system [75]. Gastrointestinal microflora consist of approximately 10^{14} colony forming units (cfu)/g of various types of both detrimental and beneficial bacteria, and the numbers and composition vary greatly along the gastrointestinal tract. The balance of the gastrointestinal micro flora influences different aspects of host health such as bowel movement, tympanites flatulence, and the absorption of nutrients. Many factors may upset this balance, including stress, consumption of antibiotics, infection, food poisoning, and the natural ageing process. To redress this balance, the growth and activities of beneficial bacteria may be enhanced by specific ingredients in foods.

Speceis	Strain	Glc	Gal	Rha	Fuc	NAc Gal	GlcA	Gly	Reference
Streprococcus thermophilus	CNCMI 733	1	2			1			[25]
	SFi39	1	1						[26]
	SFi12	1	3	2					[26]
	LY03	1	4						[27]
	OR901		5	2					[28]
	MR-1C		5	2	1				[29]
Lactococcus lactis subsp. *cremoris*	NIZO B891	3	2						[30]
	Ropy352	2	3						[31]
	NIZO B39	2	3	2					[32]
	SBT 0495	2	2	1					[33]
Lactobacillus delbrueckii subsp. *bulgaricus*	OLL 1073R-1	1	1.6						[34]
	NCFB 2772	1	2.4						[35]
	Lb18	1	1						[36]
	EU23	1		1					[37]
	rr	1	5	1					[38]
	NCFB 2772	1	7	0.8					[35]
Lb. helveticus	TN-4	1	1						[39]
	766	2	1						[40]
	2091	1	2						[41]
	Lb161	5	2						[42]
Lb. rhamnosus	RW-9595M	2	1	4					[43]
	GG	1	4	1					[44]
Lb. plantarum	EP56	3	1			1			[45]
	EP56	3	1	1					[45]
Lb. pentosus	LPS26	1		2			2		[46]
Lb. paracasei	34-1		3			1		1	[47]
Lb. kefiranofaciens	K₁	1	1						[48]
Lb. sake	0-1	3		2					[49]

Table 2. Monosaccharide ratio in hetero EPS
Glc: glucose, Gal: galactose, Rha: rhamnose, Fuc: fucose, NAc Glu: N-acetyl glucosamine, NAc Gal: N-acetyl galactosamine, GlcA: glucuronic acid.

Various oligosaccharides have been identified as prebiotics, that can increase the number of *Bifidobacterium* in the host colon. Galacto-oligosaccharides (GOS) and fructo-oligosaccharides (FOS) are considered important prebiotics. Other carbohydrates including gluco-oligosaccharides, isomalto-oligosaccharides, lactulose, mannan-oligosaccharides, and nigero-oligosaccharides are also considered prebiotics. Increased numbers of bifidobacteria and/or lactobacilli in the colon have been shown to have beneficial effects, although the specific mixtures of populations of these genera necessary to provide health-promoting effects has not yet been determined. This is because the beneficial effects are likely to be due to improvement in the balance of coloni micro flora. However, difference do exist in the micro flora among individuals. To function most effectively, prebiotics must be resistant to digestive processes in the stomach and small bowel, so that they can come into contact with the bacteria growing in the large intestine.

The food for specified health use (FOSHU) system was introduced in Japan in 1991. FOSHU refers to foods containing ingredients that provide health benefits and have officially approved physiological effects on the human body. FOSHU is intended to be consumed for the maintenance or promotion of health or for special health uses, for example, to control conditions such as blood pressure or blood cholesterol. To be defined as FOSHU, it is important to assess the safety of the food as well as the effectiveness of health promotion, and this assessment must be approved by the Ministry of Health, Labour and Welfare in Japan. At present (2012), 990 foods are recognized as FOSHU, and of these, 86 provide gastrointestinal health benefit. Foods for balancing gastrointestinal micro flora contain galactosylsucrose, soy oligosaccharides, lactulose, GOS, FOS, isomalto-oligosaccharides, raffinose, xylo-oligosaccharides, mannobiose, and brewer's yeast cell wall as functional ingredients.

4.1. GOS

GOS are well-known type of prebiotic oligosaccharides found in human milk. The concentration of oligosaccharides is 100 times higher in human breast milk than in bovine milk [76]. Many studies have shown that breast-fed infants have intestinal microflora dominated by bifidobacteria. The reason for this phenomenon is thought to be that the oligosaccharides in breast milk, including GOS, can reach the upper gut without being digested where the bifidobacteria can utilize them. At present, GOS is produced by the enzymatic treatment of lactose by β-galactosidase. GOS produced in this manner usually have degrees of polymerization (DP) between 2 and 10. Furthermore, the type of glycosidic linkage is determined by the reaction conditions: final products usually possess β-1,2, β-1,3, or β-1,4 linkages. GOS is given a caloric value of 2 kcal/g in Japan and Europe for food-labelling purpose.

The effect of GOS on defecation has been studied in healthy volunteers. Defecation frequency was significantly increased, and faeces became significantly softer after the subjects drank a beverage containing 5.0 g of GOS, on a daily basis. Therefore, consumption of a beverage containing 5.0 g of GOS can improve defecation in individuals with a tendency for constipation [77]. Ishikawa et al. reported that the number of faecal bifidobacteria increased significantly after subjects consumed 2.5 g of GOS/day for 3 weeks [78]. GOS utilization by enterobacteria was further investigated in vitro. The trisaccharide forms of GOS were utilized by *Bifidobacterium, Lactobacillus acidophilus, Lb. reuteri, Bacteroides, Clostridium perfringens, Klebsiella pnumoniae, Enterococcus faecium,* and the tetra-saccharide forms were utilized by *Bifidobacterium adolescentis, Bifidobacterium breve, Bifidobacterium infantis,* and *Ent. faecium.* These results suggest that a higher DP of GOS enhanced selectivity, and that the tetrasaccharide forms of GOS are specifically utilized by bifidobacteria. Similarly, *Bifidobacterium lactis* DR10 utilizes trisaccharide and tetra-saccharide forms of GOS, whereas *Lb. rhamnosus* DR20 prefers disaccharides and monosaccharides [79]. Barboza reported that *Bif. breve* and *Bif. longum* subsp. *infantis* can consume GOS with a DP ranging from 3 to 8 [80]. Furthermore, *Bif. longum* subsp. *infantis* preferentially consume GOS with a DP of 4, and *Bif. adolescentis* utilizes GOS with DP of 3. In addition, the structure of GOS influences its utilization by lactobacilli and bifidobacteria [81]. Trisaccharides of 4'-GOS (β-1,4 linkage) and 6'-GOS (β-1,6 linkage) can be used as the sole carbon source. Almost all lactobacilli and bifidobacteria tested preferred to utilize 4'-GOS, while *Lb. acidophilus, Lb. reuteri,* and *Lb. casei* could utilize both 4'- and 6'-GOS. GOS are used to stimulate beneficial bacteria, but can also be utilized by bacteroides and clostridia [82]. GOS selectivity may be enhanced by altering the structure and increasing the DP.

The use of beneficial bacteria or their enzymes in the synthesis of prebiotics may be a good way to produce prebiotics with high specificity. Rabiu reported that five different GOS were produced using β-galactosidase extracted from five different *Bifidobacterium* species, and that each GOS showed an increased growth rate in producer strains, except for *Bif. adolescentis* [83]. The utilization of these GOS by faecal bacteria was investigated using commercial GOS as control. The number of *Bacteroides* was decreased with GOS from bifidobacteria, whereas both GOS extracts and commercial GOS increased the number of bifidobacteria, lactobacilli, and clostridia.

4.2. FOS

FOS is used as a generic term for all β-2,1 linear fructans with a variable DP. Inulin and oligofructose are common forms of FOS that are widely found in nature. Chicory inulin has a DP of 2-60, and the product of its partial enzymatic hydrolysis is oligofructose or FOS with a DP of 2-10.

The effect of FOS intake on intestinal microflora was studied in humans. The number of bifidobacteria in faeces was significantly increased during the FOS intake (1 g/d) period, and a significant increase in stool frequency and a softening effect on stool were observed [84]. FOS increased the level of bifidobacteria in faeces, whereas that of bacteroides, clostridia, and fusobacteria decreased in subjects that were fed FOS (15 g/d) for 15 days [85]. Another study measured the increase in number of *Bifidobacterium* species in faeces by using real-time PCR [86]. The composition of bifidobacteria in the gut microflora was studied by clone library analysis in ten volunteers. All ten volunteers carried *Bif. longum*, and nine of these also carried *Bif. adolescentis*. The consumption of inulin (10 g/d) increased the number of bifidobacteria in faeces with *Bif. adolescentis* showing the highest increase response among *Bifidobacterium* species. Rossi et al. reported that only 8 of 55 *Bifidobacterium* strains fermented inulin in pure cultures, although inulin increased the number of bifidobacteria in faecal culture [87]. They, therefore, suggested that most bifidobacteria were not able to utilize long fructans in the absence of other intestinal bacteria that can hydrolyze fructans, and that fermentation of oligosaccharides in the colon is the result of a complex metabolic sequence carried out by numerous species.

4.3. Selection of high-efficiency prebiotics

It is not clear which oligosaccharides are the most suitable substrates for the selective growth of specific beneficial species or strains. Several research group have suggested useful methods to investigate the potential prebiotic activity of oligosaccharides [88-92]. Potential prebiotic activities were determined on the basis of the changes in the growth of beneficial and undesirable bacteria, such as bifidobacteria, lactobacilli, clostridia, and bacteroides. Such methods can evaluate the ability of specific strains to utilize a particular prebiotic, and a comparison of the prebiotic activities of oligosaccharides by using these methods could help in the choice of prebiotics for improving the gastrointestinal microflora on an individual basis. However, it is important to understand that only a limited group of bacteria can be chosen from the gastrointestinal microflora by using these methods, and that polysaccharides and oligo-saccharides are fermented by numerous species in the gastrointestinal tract.

Oligosaccharides produced by beneficial bacteria or their enzymes may enhance the growth of beneficial bacteria. A novel GOS mixture produced using *Bif. longum* NCIBM 41171 galactosidases increased the proportion of bifidobacteria in faeces relative to commercial GOS [93]. In the above-described study, oligosaccharides synthesized by the enzymes from *Bifidobacterium* strains were favored by the producer strains [83]. These studies suggest that the oligosaccharides produced by beneficial bacteria are selectively utilized by the producer strain, because the enzymes required for their degradation are already available. In addition, glycosyltransferases may possess both hydrolytic and transglycosylation activities [94], and glycosidases and glycosyltransferases may coexist in the same strains. Schwab et al. reported the production of novel oligosaccharides [95]. Hetero-oligosaccharides were produced from lactose, mannose, fucose, and N-acetylglucosamine by using crude cell extracts and whole cells of LAB and bifidobacteria. These hetero-oligosaccharides contained mannose, fucose, and N-acetylglucosamine, and could be digested by LAB strains. The prebiotic activities of these oligosaccharides were not investigated; however, a similar approach using probiotic and intestinal beneficial bacteria may lead to the production of highly selective prebiotics.

The dietary fiber, arabinoxylan is the predominant hemicellulose from cereals and exhibits prebiotic activity [96]. The addition of water-unextractable arabinoxylans increased the population of bifidobacteria and bacteroides in a medium inoculated with faecal slurry. Polysaccharides are not usually utilized by microorganisms. Remarkably, however, *Bifidobacterium bifidum* DSM20456 can utilize the EPS produced by *Pediococcus pentosaceus, Lb. plantarum, Weissella cibaria,* and *Weissella confusa,* and some growth is observed in cas of *Bif. longum, Bif. adolescentis,* and *Lb. acidophilus* [97]. For EPS production by LAB, reduced yields were frequently observed after the maximal level had been reached, which might be caused by the enzymes produced by the bacteria [98]. Tsuda and Miyamoto investigated the prebiotic activity of EPS produced by *Lb. plantarum* 301102S [52], a mutant strain derived from *Lb. plantarum* 301102. Oral administration of the parental strain 301102 showed the survivability and proliferation in porcine gastrointestinal tract [99]. The potential prebiotic activities of EPS, GOS, and inulin were measured in 37 LAB strains, and the activity scores of EPS in the strains 301102 and 301102S were highest. This suggests that the EPS produced by the mutant strain is utilized by the same strain 301102S and the parental strain, and that the parental strain has enzymes that can degrade the EPS.

5. Conclusion

Poly- and hetero-oligosaccharides produced by LAB may be potential prebiotics. Studies on the production of polysaccharides and oligosaccharides by enzymes in beneficial microorganisms may lead to the production of highly selective prebiotics, although in vitro evaluation may be difficult because of degradation and utilization of polysaccharides by various microorganisms in the gastrointestinal tract. Administration of synbiotic food containing a combination of a probiotic bacterial strain and the prebiotic sugar produced by that strain could be effective in improving human health.

Author details

Tsuda Harutoshi
National Institute of Health and Nutrition, Tokyo, Japan

6. References

[1]   Chabot S, Yu H.L, de Leseleuc L, Cloutier D, Van Calsteren M.R, Roy D, Lacroix M, Oth D (2001) Exopolysaccharides from *Lactobacillus rhamnosus* RW-9595M stimulate TNF, IL-6 and IL-12 in human and mouse cultured immunocompetent cells, and IFN-γ in mouse splenocytes. Lait 81: 683-698.

[2]   Kitazawa H, Harata T, Uemura J, Saito T, Kaneko T, Itoh T (1998) Phosphate group requirement for mitogenic activation of lymphocytes by an extracellular phosphopolysaccharide from *Lactobacillus delbrueckii* ssp. *bulgaricus.* Int. J. Food Microbiol. 40: 169-175.

[3] Sreekumar O, Hosono A (1998) The antimutagenic properties of a polysaccharide produced by *Bifidobacterium longum* and its cultured milk against some heterocyclic amines. Can. J. Microbiol. 44: 1029-1036.

[4] Tsuda H, Hara K, Miyamoto T (2008) Binding of mutagens to exopolysaccharide produced by *Lactobacillus plantarum* mutant strain 301102S. J. Dairy Sci. 91: 2960-2966.

[5] Ganzle M. Michael G., Schwab C (2005) Exopolysaccharide production by intestinal lactobacilli. In: Tannock G. W, editors. Probiotics & Prebiotics: Scientific Aspects. Norfolk: Caister Academic Press. pp. 83-96.

[6] Kralj S., van Geel-Schutten G. H., Dondorff M. M. G., Kirsanovs S, van der Maarel M. J. E. C., Dijkhuizen L (2004) Glucan synthesis in the genus *Lactobacillus*: isolation and characterization of glucansucrase genes, enzymes and glucan products from six different strains. Microbiology 150: 3681-3690.

[7] van Hijum S. A. F. T., Szalowska E., van der Maarel M. J. E. C., Dijkhuizen L (2004) Biochemical and molecular characterization of a levansucrase from *Lactobacillus reuteri*. Microbiology 150: 621-630.

[8] de Vuyst L., de Vin F., Vaningelgem F., Degeest B (2001) Recent developments in the biosynthesis and applications of heteropolysaccharides from lactic acid bacteria. Int. Dairy J. 11: 687-707.

[9] Ruas-Madiedo P., Hugenholtz J., Zoon P (2002) An overview of the functionality of exopolysaccharides produced by lactic acid bacteria. Int. Dairy J. 12:163-171.

[10] Ruijssenaars H, Stingele F, Hartmans S (2000) Biodegradability of food-associated extracellular polysaccharides. Curr. Microbiol. 40: 194-199.

[11] Korakli M, Ganzle M.G, Vogel R.F (2002) Metabolism by bifidobacteria and lactic acid bacteria of polysaccharides from wheat and rye, and exopolysaccharides produced by *Lactobacillus sanfranciscensis*. J. Appl. Microbiol. 92: 958-965.

[12] Fuller R (1989) Probiotics in man and animals. J. Appl. Bacteriol. 66: 365-378.

[13] FAO/WHO (2002) Guidelines for the evaluation of probiotics in food. Report of a joint FAO/WHO working group on drafting guidelines for the evaluation of probiotics in food. London, Ontario, Canada.

[14] Gibson G.R, Roberfroid M.B (1995) Dietary modulation of the human colonic microbiota: Introducing the concept of prebiotics. J. Nutr. 125: 1401-1412.

[15] Rastall R (2003) Enhancing the functionality of prebiotics and probiotics. In: Mattila-Sandholm T, Saarela M, editors. Functional Dairy Products. Florida: CRC Press LLC. pp. 301-315.

[16] Mayo B, Aleksandrzak-Piekarczyk T, Fernandez M, Kowalczyk M, Alvarez-Martin P, Bardowski J (2010) Updates in the metabolism of lactic acid bacteria. In: Mozzi F, Raya R. R, Vignolo G. M, editors. Biotechnology of Lactic Acid Bacteria. Iowa: Blackwell Publishing. pp. 3-33.

[17] Cerning J (1990) Exocellular polysaccharides produced by lactic acid bacteria. FEMS Microbiol. Lett. 87: 113-130.

[18] Pederson C. S, Albury M (1955) Variation among the heterofermentative lactic acid bacteria. J. Bacteriol. 70: 702-708

[19] Dellaglio F., Dicks L. M. T., Torriani S (1995) The genus *Leuconostoc*. In: Wood B. J. B., Holzapfel W. H, editors. The genera of lactic acid bacteria. Glasgow: Blackie Academic and Professional. pp. 235-278.

[20] Minervini F, Angelis M. D, Surico R. F, Cagno R. D, Ganzle M, Gobbetti M (2010) Highly efficient synthesis of exopolysaccharides by *Lactobacillus curvatus* DPPMA10 during growth in hydrolyzed wheat flour agar. Int. J. Food Microbiol. 141: 130-135.

[21] Cote G. L, Robyt F. J (1982) Isolation and partial characterization of an extracellular glucansucrase from *Leuconostoc mesenteroides* NRRl B-1355 that synthesizes an alternating (1→6), (1→3)-α-D-glucan. Carbohydr. Res. 101:57-74.

[22] Uchida K (1996) Nyuusannkinn no kouzou to kinntaiseibunn. In: Nyuusannkinn kennkyuu syuudannkai, editors. Nyuusannkinn no kagaku to gijutsu. Tokyo: Gakkai Syuppann Center. pp. 59-88. (In Japanese)

[23] Van Geel-Schutten G. H, Faber E. J, Smit E, Bonting K, Smith M. R, Ten Brink B, Kamerling J. P, Vliegenthart J. F. G, Dijkhuizen L (1999) Biochemical and structural characterization of the glucan and fructan exopolysaccharides synthesized by the *Lactobacillus reuteri* wild-type strain and by mutant strains. Appl. Environ. Microbiol. 65: 3008-3014.

[24] De Vos P, Garrity G. M, Jones D, Krieg N. R, Ludwig W, Rainey F. A, Schleifer K. H, Whitman W. B editors (2009) Bergey's Manual of Systematic Bacteriology Second Edition Volume Three. New York: Springer.

[25] Doco T, Wieruszeski J. M, Fournet B (1990) Structure of an exocellular polysaccharide produced by *Streptococcus thermophilus*. Carbohydr. Res. 198: 313-321.

[26] Lemoine J, Chirat F, Wieruszeski J. M, Strecker G, Favre N, Neeser J. R (1997) Structural characterization of the exocellular polysaccharides produced by *Streptococcus thermophilus* SFi39 and SFi12. Appl. Environ. Microbiol. 63: 3512-

[27] Degeest B, de Vuyst L (2000) Correlation of activities of the enzymes a-phosphoglucomutase, UDP-galactose 4-epimerase, and UDP-glucose pyrophosphorylase with exopolysaccharide biosynthesis by *Streptococcus thermophilus* LY03. Appl. Environ. Microbiol. 66: 3519-3527.

[28] Bubb W. A, Urashima T, Fujiwara R, Shinnai T, Ariga H (1997) Structural characterization of the exocellular polysaccharide produced by *Streptococcus thermophilus* OR901. Carbohydr. Res. 301: 41-50.

[29] Low D, Ahlgren J. A, Horne D, McMahon D. J, Oberg C. J, Broadbent J. R (1998) Role of *Streptococcus thermophilus* MR-1C capsular exopolysaccharide in cheese moisture retention. Appl. Environ. Microbiol. 64: 2147-2151.

[30] Van Casteren W. H. M, de Waaed P, Dijkema C, Schols H. A, Voragen A. G. J (2000) Structural characterization and enzymic modification of the exopolysaccharide produced by *Lactococcus lactis* subsp. *cremoris* B891. Carbohydr. Res. 327: 411-422.

[31] Knoshaug E. P, Ahlgren J. A, Trempy J. E (2007) Exopolysaccharide expression in *Lactococcus lactis* subsp. *cremoris* Ropy352: Evidence for novel gene organization. Appl. Environ. Microbiol. 73: 897-905.

[32] van Casteren W. H. M, Dijkema C, Schols H. A, Beldman G, Voragen A. G. J (2000) Structural characterization and enzymic modification of the exopolysaccharide produced by *Lactococcus lactis* subsp. *cremoris* B39. Carbohydr. Res. 324: 170-181.

[33] Nakajima H, Hirota T, Toba T, Itoh T, Adachi S (1992) Structure of the extracellular polysaccharide from slime-forming *Lactococcus lactis* subsp. *cremoris* SBT 0495. Carbohydr. Res. 224: 245-253.

[34] Kitazawa H, Ishii Y, Uemura J, Kawai Y, Saito T, Kaneko T, Noda K, Itoh T (2000) Augmentation of macrophage functions by an extracellular phosphopolysaccharide from *Lactobacillus delbrueckii* ssp. *bulgaricus*. Food Microbiol. 17:109-118.

[35] Grobben G. J, Smith M. R, Sikkema J, de Bont J. A. M (1996) Influence of fructose and glucose on the production of exopolysaccharides and the activities of enzymes involved in the sugar metabolism and the synthesis of sugar nucleotides in *Lactobacillus delbrueckii* subsp. *bulgaricus* NCFB 2772. Appl. Microbiol. Biotechnol. 46:279-284.

[36] Petry S, Furlan S, Waghorne E, Saulnier L, Cerning J, maguin E (2003) Comparison of the thickening properties of four *Lactobacillus delbrueckii* subsp. *bulgaricus* strains and physicochemical characterization of their exopolysaccharides. FEMS Microbiol. Lett. 221: 285-291.

[37] Harding L. P, Marshall V. M, Elvin M, Gu Y, Laws A. P (2003) Structural characterization of a perdeuteriomethylated exopolysaccharide by NMR spectroscopy: characterization of the novel exopolysaccharide produced by *Lactobacillus delbrueckii* subsp. *bulgaricus* EU23. Carbohydr. Res. 338:61-67.

[38] Gruter M, Leeflang B. R, Kuiper J, Kamerling J. P, Vliegenthart F. G (1993) Structural characterization of the exopolysaccharide produced by *Lactobacillus delbrueckii* subspecies *bulgaricus* rr grown in skimmed milk. Carbohydr. Res. 239:209-226.

[39] Yamamoto T, Nunome T, Yamauchi R, Kato K, Sone Y (1995) Structure of an exocellular polysaccharide of *Lactobacillus helveticus* TN-4, a spontaneous mutant strain of *Lactobacillus helveticus* TY1-2. Carbohydr. Res. 275: 319-332.

[40] Robijn G. W, Thomas J. R, Haas H, van den Berg D. J. C, Kamerling J. P, Vliengenthart J. F. G (1995) The structure of the exopolysaccharide produced by *Lactobacillus helveticus* 766. Carbohydr. Res. 276: 137-154.

[41] Staaf M, Widmalm G, Yang Z, Huttunen E (1996) Structural elucidation of an extracellular polysaccharide produced by *Lactobacillus helveticus*. Carbohydr. Res. 291: 155-164.

[42] Staaf M, Yang Z, Huttunen E, Widmalm G (2000) Structural elucidation of the viscous exopolysaccharide produced by *Lactobacillus helveticus* Lb161. Carbohydr. Res. 326: 113-119.

[43] Van Calsteren M. R, Pau-Roblot C, Begin A, Roy D (2002) Structure determination of the exopolysaccharide produced by *Lactobacillus rhamnosus* strains RW-9595M and R. Biochem. J. 363: 7-17.

[44] Landersjo C, Yang Z, Huttunen E, Widmalm G (2002) Structural studies of the exopolysaccharide produced by *Lactobacillus rhamnosus* strain GG (ATCC 53103). Biomacromolecules 3: 880-884.

[45] Tallon R, Bressollier P, Urdaci M. C (2003) Isolation and characterization of two exopolysaccharides produced by *Lactobaciilus plantarum* EP56. Res. Microbiol. 154: 705-712.

[46] Rodriguez-Carvajal M. A, Sanchez J. I, Campelo A. B, Martinez B, Rodriguez A, Gil-Serrano A. M (2006) Structure of the high-molecular weight exopolysaccharide isolated from *Lactobacillus pentosus* LPS26. Carbohydr. Res. 343: 3066-3070.

[47] Robijn G. W, Wienk H. L. J, van den Berg D. J. C, Haas H, Kamerling J. P, Vliengenthart J. F. G (1996) Structural studies of the exopolysaccharide produced by *Lactobacillus paracasei* 34-1. Carbohydr. Res. 285: 129-139.

[48] Mukai T, Toba T, Itoh T, Adachi S (1990) Structural investigation of the capsular polysaccharide from *Lactobacillus kefiranofaciens* K1. Carbohydr. Res. 204: 227-232.

[49] van den Berg D. J. C, Robijn G. W, Janssen A. C, Giuseppin M. L. F, Vreeker R, Kamerling J. P, Vliegenthart J. F. G, Ledeboer A. M, Verrips C. T (1995) Production of a novel extracellular polysaccharide by *Lactobacillus sake* 0-1 and characterization of the polysaccharide. Appl. Environ. Microbiol. 61: 2840-2844.

[50] Leo F, Hashida S, Kumagai D, Uchida K, Motoshima H, Arai I, Asakuma S, Fukuda K, Urashima T (2007) Studies on a neutral exopolysaccharide of *Lactobacillus fermentum* TDS030603. J. Appl. Glycosci. 54: 223-229.

[51] Figueroa C, Davila A. M, Pourquie J (1995) Lactic acid bacteria of the sour cassava starch fermentation. Lett. Appl. Microbiol. 21: 126-130.

[52] Cerning J (1995) Production of exopolysaccharides by lactic acid bacteria and dairy propionibacteria. Lait, 75: 463-472.

[53] Tsuda H, Miyamoto T (2010) Production of exopolysaccharide by *Lactobacillus plantarum* and the prebiotic activity of the exopolysaccharide. Food Sci. Technol. Res. 16: 87-92.

[54] Macedo M.G, Lacroix C, Gardner N.J, Champagne C.P (2002). Effect of medium supplementation on exopolysaccharide production by *Lacctobacillus rhamnosus* RW-9595M in whey permeate. Int. Dairy J. 12: 419-426.

[55] Maeda H, Zhu X, Suzuki S, Suzuki K, Kitamura S (2004) Structural characterization and biological activities of an exopolysaccharide kefiran produced by *Lactobacillus kefiranofaciens* WT-2B[T]. J. Agric. Food Chem. 52: 5533-5538.

[56] de Vuyst L, Degeest B (1999) Heteropolysaccharides from lactic acid bacteria. FEMS Microbiol. Rev. 23: 153-177.

[57] Garcia-Garibay M, Marshall V. M. E (1991) Polymer production by *Lactobacillus delbrueckii* ssp. *bulgaricus*. J. Appl. Bacteriol. 70: 325-328.

[58] Kimmel S. A, Roberts R. F, Ziegler G. R (1998) Optimization of exopolysaccharide production by *Lactobacillus delbrueckii* subsp. *bulgaricus* RR grown in a semidefined medium. Appl. Environ. Microbiol. 64: 659-664.

[59] Gamar L, Blondeau K, Simonet J. M (1997) Physiological approach to extracellular polysaccharide production by *Lactobacillus rhamnosus* strain C83. J. Appl. Microbiol. 83: 281-287.

[60] Grobben G. J, Chin-Joe I, Kitzen V. A, Boels I. C, Boer F, Sikkema J, Smith M. R, de Bont J. A. M (1998) Enhancement of exopolysaccharide production by *Lactobacillus delbrueckii* subsp. *bulgaricus* NCFB 2772 with a simplified defined medium. Appl. Environ. Microbiol. 64: 1333-1337.

[61] Cerning J, Renard C. M. G. C, Thibault J. F, Bouillanne C, Landon M, Desmazeaud M, Topisirovic L (1994) Carbon source requirements for exopolysaccharide production by *Lactobacillus casei* CG11 and partial structure analysis of the polymer. Appl. Environ. Microbiol. 60: 3914-3919.

[62] de Vuyst L, Vanderveken F, Van de Ven S, Degeest B (1998) Production by and isolation of exopolysaccharides from *Streptococcus thermophilus* grown in a milk medium and evidence for their growth-associated biosynthesis. J. Appl. Microbiol. 84: 1059-1068.

[63] Korakli M, Pavlovic M, Ganzle M. G, Vogel R. F (2003) Exopolysaccharide and kestose production by *Lactobacillus sanfranciscensis* LTH2590. Appl. Environ. Microbiol. 69: 2073-2079.

[64] Kralj S, van Geel-Schutten G. H, van der Maarel M. H. E. C, Dijkhuizen L (2004) Biochemical and molecular characterization of *Lactobacillus reuteri* 121 reuteransucrase. Microbiology 150: 2099-2112.

[65] Neubauer H, Bauche A, Mollet B (2003) Molecular characterization and expression analysis of the dextransucrase DsrD of *Leuconostoc mesenteroides* Lcc4 in homologous and heterologous *Lactococcus lactis* cultures. Microbiology 149: 973-982.

[66] Dabour N, LaPointe G (2005) Identification and molecular characterization of the chromosomal exopolysaccharide biosynthesis gene cluster from *Lactococcus lactis* subsp. *cremoris* SMQ-461. Appl. Environ. Microbiol. 71: 7414-7425.

[67] Dan T, Fukuda K, Sugai-Bannai M, Takakuwa N, Motoshima H, Urashima T (2009) Characterization and expression analysis of the exopolysaccharide gene cluster in *Lactobacillus fermentum* TDS030603. Biosci. Biotechnol. Biochem. 73: 2656-2664.

[68] Mozzi F, Savoy de Giori G, Font de Valdez G (2003) UDP-galactose 4-epimerase: a key enzyme in exopolysaccharide formation by *Lactobacillus casei* CRL 87 in controlled pH batch cultures. J. Appl. Microbiol. 94: 175-183.

[69] Kleerebezem M. et al. (2003) Complete genome sequence of *Lactobacillus plantarum* WCFS1. Proc. Natl. Acad. Sci. USA. 100: 1990-1995.

[70] Morona J. K, Paton J. C, Miller D. C, Morona R (2000) Tyrosine phosphorylation of CpsD negatively regulates capsular polysaccharide biosynthesis in *Streptococcus pneumonia*. Mol. Microbiol. 35: 1431-1442.

[71] Nagai T, Makino S, Ikegami S, Itoh H, Yamada H (2012) Effects of oral administration of yogurt fermented with *Lactobacillus delbrueckii* ssp. *bulgaricus* OLL1073R-1 and its exopolysaccharides against influenza virus infection in mice. Int. Immunopharmacolgy 11: 2246-2250.

[72] Kodali V. P, Sen R (2008) Antioxidant and free radical scavenging activities of an exopolysaccharide from a probiotic bacterium. Biotechnol. J. 3: 245-251.

[73] Kleerebezem M, van Kranenburg R, Tuinier R, Boels I. C, Zoon P, Looijesteijn E, Hugenholtz J, de Vos W. M (1999) Exopolysaccharides produced by *Lactococcus lactis*: from genetic engineering to improved rheological properties? Ant. van Leeuwenhoek 76: 357-365.

[74] O'Connor E. B, Barrett E, Fitzgerald G, Hill C, Stanton C, Ross R. P (2005) Production of vitamins, exopolysaccharides and bacteriocins by probiotic bacteria. In: Tamime A. Y, editors. Probiotic Dairy Products. Oxford: Blackewell Publishing Ltd. pp. 167-194.

[75] Mitsuoka T (1992) Intestinal flora and aging. Nutr. Rev. 50: 438-446.

[76] Kunz C, Rudloff S, Baier W, Kein N, Strobel S (2000) Oligosaccharides in human milk: structure, functional, and metabolic aspects. Ann. Rev. Nutr. 20: 699-722.

[77] Deguchi Y, Matsumoto K, Ito A, Watanuki M (1997) Effects of β 1-4 galactooligosaccharides administration on defecation of healthy volunteers with constipation tendency. Eiyougakuzasshi 55: 13-22. (in Japanese)

[78] Ishikawa F, Takayama H, Matsumoto K, Ito M, Chonan O, Deguchi Y, Kikuchi-Hayakawa H, Watanuki M (1995) Effects of β 1-4 linked galactooligosaccharides on human fecal microflora. Bifizusu 9: 5-18. (in Japanese)

[79] Gopal P. K, Sullivan P. A, Smart J. B (2001) Utilisation of galacto-oligosaccharides as selective substrates for growth by lactic acid bacteria including *Bifidobacterium lactis* DR10 and *Lactobacillus rhamnosus* DR20. Int. Dairy J. 11: 19-25.

[80] Barboza M, Sela D. A, Pirim C, LoCascio R. G, Freeman S. L, German J. B, Mills D. A, Lebrilla C. B (2009) Glycoprofiling bifidobacterial consumption of galacto-oligosaccharides by mass spectrometry reveals strain-specific, preferential consumption of glycans. Appl. Environ. Microbiol. 75: 7319-7325.

[81] Cardelle-Cobas A, Corzo N, Olano A, Pelaez C, Requena T, Avila M (2011) Galactooligosaccharides derived from lactose and lactulose: Influence of structure on *Lactobacillus, Streptococcus* and *Bifidobacterium* growth. Int. J. Food Microbiol. 149: 81-87.

[82] Macfarlane G.T, Steed H, Macfarlane S (2008). Bacterial metabolism and health-related effects of galacto-oligosaccharides and other prebiotics. J. Appl. Microbiol. 104: 305-344.

[83] Rabiu B. A, Jay A. J, Gibson G. R, Rastall R. A (2001) Synthesis and fermentation properties of novel galacto-oligosaccharides by β-galactosidases from *Bifidobacterium* species. Appl. Environ. Microbiol. 67: 2526-2530.

[84] Tokunaga T, Nakada Y, Tashiro Y, Hirayama M, Hidaka H (1993) Effects of fructooligosaccharides intake on the intestinal microflora and defecation in healthy volunteers. Bifizusu 6: 143-150. (in Japanese)

[85] Gibson G. R, Beatty E. R, Wang X, Cummings J. H (1995) Selective stimulation of bifidobacteria in the human colon by oligofructose and inulin. Gastroenterology 108: 975-982.

[86] Ramirez-Farias C, Slezak K, Fuller Z, Duncan A, Holtrop G, Louis P (2009) Effect of inulin on the human gut microbiota: stimulation of *Bifidobacterium adolescentis* and *Faecalibacterium prausnitzii*. Br. J. Nutr. 101: 541-550.

[87] Rossi M, Corradini C, Amaretti A, Nicolini M, Pompei A, Zanoni S, Matteuzzi D (2005) Fermentation of fructooligosaccharides and inulin by bifidobacteria: a comparative study of pure and fecal cultures. Appl. Environ. Microbiol. 71: 6150-6158.

[88] Olano-Martin E, Gibson G. R, Rastall R. A (2002) Comparison of the in vitro bifidogenic properties of pectins and pectic-oligosaccharides. J. Appl. Microbiol. 93: 505-511.

[89] Palframan R, Gibson G. R, Rastall R. A (2003) Development of a quantitative tool for the comparison of the prebiotic effect of dietary oligosaccharides. Lett. Appl. Microbiol. 37: 281-284.

[90] Vulevic J, Rastall R. A, Gibson G. R (2004) Developing a quantitative approach for determining the in vitro prebiotic potential of dietary oligosaccharides. FEMS Microbiol. Lett. 236: 153-159.

[91] Sanz M. L, Gibson G. R, Rastall R. A (2005) Influence of disaccharide structure on prebiotic selectivity in vitro. J. Agric. Food Chem. 53: 5192-5199.

[92] Huebner J, Wehling R. L, Hutkins R. W (2007) Functional activity of commercial prebiotics. Int. Dairy J. 17: 770-775.

[93] Depeint F, Tzortzis G, Vulevic J, I'Anson K, Gibson G. R (2008) Prebiotic evaluation of a novel galactooligosaccharide mixture produced by the enzymatic activity of *Bifidobacterium bifidum* NCIMB 41171, in healthy humans: a randomized, double-blind, crossover, placebo-controlled intervention study. Am. J. Clin. Nutr. 87: 785-791.

[94] van den Broek L. A. M, Hinz S. W. A, Beldman G, Vincken J. P, Voragen A. G. J (2008) *Bifidobacterium* carbohydrases – their role in breakdown and synthesis of (potential) prebiotics. Mol. Nutr. Food Res. 52: 146-163.

[95] Schwab C, Lee V, Sorensen K. I, Ganzle M. G (2011) Production of galactooligosaccharides and heterooligosaccharides with disrupted cell extracts and whole cells of lactic acid bacteria and bifidobacteria. Int. Dairy J. 21: 748-754.

[96] Vardakou M, Palop C. M, Christakopoulos P, Faulds C. B, Gasson M. A, Narbad A (2008) Evaluation of the prebiotic properties of wheat arabinoxylan fractions and induction of hydrolase activity in gut microflora. Int. J. Food Microbiol. 123: 166-170.

[97] Hongpattarakere T, Cherntong N, Wichienchot S, Kolida S, Rastall R. A (2012) In vitro prebiotic evaluation of exopolysaccharides produced by marine isolated lactic acid bacteria. Carbohydr. Polym. 87: 846-852.

[98] Pham P.L, Dupont I, Roy D, Lapointe G, Cerning J (2000) Production of exopolysaccharide by *Lactobacillus rhamnosus* R and analysis of its enzymatic

degradation during prolonged fermentation. Appl. Environ. Microbiol. 66: 2302-2310.

[99] Tsuda H, Hara K, Miyamoto T (2008) Survival and colonization of orally administered *Lactobacillus plantarum* 301102 in porcine gastrointestinal tract. Anim. Sci. J. 7: 274-278.

Chapter 8

Lactic Acid Bacteria in Philippine Traditional Fermented Foods

Charina Gracia B. Banaay, Marilen P. Balolong and Francisco B. Elegado

Additional information is available at the end of the chapter

1. Introduction

The Philippine archipelago is home to a diverse array of ecosystems, organisms, peoples, and cultures. Filipino cuisine is no exception as distinct regional flavors stem from the unique food preparation techniques and culinary traditions of each region. Although Philippine indigenous foods are reminiscent of various foreign influences, local processes are adapted to indigenous ingredients and in accordance with local tastes. Pervasive throughout the numerous islands of the Philippines is the use of fermentation to enhance the organoleptic qualities as well as extend the shelf-life of food.

Traditional or indigenous fermented foods are part and parcel of Filipino culture since these are intimately entwined with the life of local people. The three main island-groups of the Philippines, namely – Luzon, Visayas, and Mindanao, each have their own fermented food products that cater to the local palate. Fermentation processes employed in the production of these indigenous fermented foods often rely entirely on natural microflora of the raw material and the surrounding environment; and procedures are handed down from one generation to the next as a village-art process. Because traditional food fermentation industries are commonly home-based and highly reliant on indigenous materials without the benefit of using commercial starter cultures, microbial assemblages are unique and highly variable per product and per region. Hence the possibility of discovering novel organisms, products, and interactions are likely.

Various microorganisms are involved in common food fermentation processes. In particular, lactic acid bacteria (LAB) in food is a type of biopreservation system. They not only contribute to the flavor of the food but LAB are also able to control pathogenic and spoilage microorganisms through various ways that include, but are not limited to, production of peroxidases, organic acids, and bacteriocins. Traditionally, identification of LAB in foods is largely dependent on culture-based methods; and properties of each isolate are evaluated under controlled conditions. However, with the advent of molecular techniques, the enumeration of microorganisms missed by culture-dependent methods is now possible. Also, as more LAB metabolites, such as bacteriocins, are being reported, a wider database for identification and comparison with potential novel products are now available.

As the production and consumption of traditional fermented food products become increasingly relevant in the face of rapidly increasing population and food insecurity, more research and development to ensure the safety and nutritional quality of these fermented products is warranted. For a more extensive discussion of the principles and technology of Philippine fermented foods, the readers are directed to Sanchez (2008). This book is a detailed reference based on decades of research. Some data from the book will be presented again here in addition to other data from more recent studies. It is not the intention of this present paper to repeat what has been presented in the book, especially regarding fermentation processes, but only to present, as complete as possible, the data that are available regarding LAB present in indigenous/traditional fermented foods.

This paper aims to briefly review the various lactic acid-fermented indigenous fermented specialties in the different regions of the Philippines. Majority of the discussion will focus on recent data gathered from bacteriocin research and metagenomics studies of Philippine fermented specialties. Lastly, the health applications of the different fermented food products and their development as functional foods will be evaluated.

2. Regional fermented specialties in the Philippines

There are various lactic acid-fermented indigenous food products in the Philippines. Table 1 gives a summary of these different fermented specialties found in the different regions. Although a particular product type can be seen throughout the whole country, the texture, taste, and appearance would vary depending on the local taste, materials used, and process employed. For example, bagoong is a common fermented fish paste found all over the Philippines but the characteristic of the product found in Luzon is different from that found in the Visayas and Mindanao regions. Bagoong also takes on different names; there is bagoong na isda, bagoong alamang, bagoong na sisi, and guinamos (Sanchez, 2008). A product that is processed in a similar manner is dayok; it is made of brined fish entrails. Research indicates that this is also a lactic acid-fermented food but the LAB involved have not been identified yet (Besas and Dizon, 2012). Longanisa is sausage made of beef, pork, or chicken. It also takes on many forms depending on where it is made. The more famous ones are Vigan Longanisa in Northern Luzon, Pampanga Longanisa in Central Luzon, Lucban Longanisa in Southern Luzon, and Cebu Longanisa in the Visayas. The tastes vary from spicy, garlicky, sour, to sweet.

In lactic acid-fermented foods, LAB are important in preventing the growth of spoilage organisms, and altering flavor, aroma, and texture of the product. Although LAB are initially present in low numbers in the raw materials used, they soon proliferate as other organisms are inhibited by the initial addition of salt and as the continuous growth of LAB decreases the pH of the food making it less conducive for growth of other organisms. Recent studies, however, have shown that there are a lot more benefits that can be derived from LAB in traditional fermented foods.

CATEGORY	PRODUCT NAME	REGION	MAJOR INGREDIENTS	LACTIC ACID BACTERIA INVOLVED (as determined from culture-based methods)	APPEARANCE AND/OR USAGE
Fermented vegetables, fruits	Burong mustasa	Luzon	Mustard leaves, cooked rice and/or rice washings	*Leuconostoc mesenteroides, Enterococcus faecalis, Lactobacillus plantarum*	Side dish
	Burong pipino	Whole Phil	cucumber	*Leu. mesenteroides, L. brevis, Pediococcus cerevisiae, L. plantarum*	Side dish
	Burong mangga	Whole Phil	Immature mango	*Leu. mesenteroides, L. brevis, P. cerevisiae, L. plantarum*	Side dish
	Atchara	Whole Phil	Immature papaya or chayote, or turnip (singkamas)	Unknown	Side dish
Cheese	Kesong puti	Luzon, Visayas	Cow or carabao milk	*Lactococcus lactis*	White soft cheese
Fermented fish and fishery products	Balao-balao	Luzon	Cooked rice, shrimp, salt	*Leu. mesenteroides, P. cerevisiae, L. plantarum*	Side dish, condiment
	Burong-isda	Luzon	Freshwater fish, rice, salt	*Leu. mesenteroides, E. faecalis, P. cerevisiae, L. plantarum, P. acidilactici, Leu. paramesenteroides*	Side dish, condiment
	Tinabal	Visayas	Parrot fish (for tinabal molmol) and frigate fish (for tinabal mangko), salt	*P. pentosaceus, S. equinus, Leuconostoc* sp., *Lactobacillus* sp.	Side dish, viand
	Burong talangka	Luzon	Small shore crabs (*Varuna litterata*)	*Leu. mesenteroides, E. faecalis, P. cerevisiae, L. plantarum*	Side dish, viand
	Patis	Whole Phil	Small fish, salt	*P. halophilus* (in mixed fermentation)	Fish sauce (patis), fish paste (bagoong), used as condiment, sauce, flavoring agent, viand
	Bagoong isda	Whole Phil	Small fish, salt		
	Bagoong alamang	Whole Phil	Small shrimps, salt		
	Bagoong na sisi	Visayas	Shell fish, salt		
	Guinamos	Bagoong isda in Visayas, Mindanao	Salt water small fish (dilis/belabid – *Stolephorus* sp.), salt		Condiment, viand, side dish
	Dayok	Visayas, Mindanao	Fish entrails, salt	Unidentified LAB	Condiment, viand, side dish
Fermented meat, sausages	Longanisa	Whole Phil	Ground pork, beef, or chicken meat, spices and preservatives	*P. acidilactici, Lactococcus lactis* (together with *Micrococcus aurantiacus*)	Viand
	Agos-os	Visayas	Sweet potato and ground pig's head	*E. faecalis*	Viand
	Burong kalabi	Luzon	Cooked rice, ground carabao meat	*L. plantarum*	Side dish, viand
	Burong babi	Luzon	Cooked rice, ground pork	*L. plantarum*	Side dish, viand

	Puto	Whole Phil	Rice, sugar	*L. mesenteroides, E. faecalis, P. cerevisiae* (in mixed	Steamed rice cake
	Bibingka	Whole Phil	Rice, sugar	fermentation with *Saccharomyces cerevisiae)*	Baked rice cake
	Tapuy	Luzon	Rice, glutinous rice	*Leuconostoc, L. plantarum* (in mixed fermentation with molds and yeasts)	Wine; beer
	Pangasi	Mindanao	Rice	Unknown	Wine
	Landang	Visayas, Mindanao	Cassava, or buli palm flour	Unknown	Dried jelly pellets pellets, rice substitute
	Puto balanghoy	Mindanao	Cassava	Unknown	Steamed cake
Fermented rice, cassava, sugar cane, coconut, soya	Basi	Luzon	Sugar cane	Unknown	Wine
	Suka	Whole Phil	Sugar cane juice (for sukang Iloco), palm inflorescence sap (for sukang tuba)	*Leuconostoc, Lactobacillus, Streptococcus* in the initial fermentation phase only	Vinegar, condiment, flavoring
	Sinamak	Luzon	Sugar cane juice, spices (chilies, onions, garlic)	Unknown	Spiced vinegar, condiment, flavoring
	Pinakurat	Visayas, Mindanao	Coconut sap, chilies, salt, various spices	Unknown	Spiced vinegar, condiment, flavoring
	Tuba	Whole Phil	Coconut sap	Unknown	Wine
	Lambanog	Whole Phil	Coconut sap	Unknown	Wine
	Toyo	Whole Phil	Soybeans	*P. halophilus, E. faecalis, L. delbrueckii* (in mixed fermentation with *Aspergillus sojae* and *Saccharomyces rouxii)*	Condiment, flavoring agent, seasoning

(Sources: Banaay et al., 2004; Besas and Dizon, 2012; Lee, 1999; Olympia et al., 1995; Sanchez, 2008; Tan et al., 2007)

Table 1. Regional Lactic Acid-Fermented Specialties in the Philippines

3. Research initiatives on LAB from Philippine fermented foods

3.1. Bacteriocin research

Bacteriocins are antimicrobial proteins or peptides produced by certain bacterial strains. Unlike the peptide antibiotics they usually have a narrow spectrum of antimicrobial activity, usually inhibiting growth of closely related bacterial species or strains and lacking lethality to the producer strain (Riley and Wertz, 2002).

The bacteriocins of LAB are small, cationic, hydrophobic, or amphiphilic peptides or small proteins, composed of 20 to 60 amino acid residues (Chen & Hoover, 2003). The bactericidal mode of action and biochemical properties depend on the protein moiety that could be specific to a particular LAB strain, *i.e.* the N-terminal amino acids as determinant of receptors in the cell wall of the susceptible strains/species and C-terminal amino acids for the biochemical properties. LAB bacteriocin must have the following desirable properties: "(1) not active and nontoxic to eukaryotic cells, (2) become inactivated by digestive proteases, having little influence on the gut microbiota, (3) low pH and heat-tolerant, (4) have a relatively broad antimicrobial spectrum, against many food-borne pathogenic and spoilage bacteria, (5) show a bactericidal mode of action, usually acting on the bacterial cytoplasmic membrane: no cross resistance with antibiotics, and (6) have genetic determinants that are usually plasmid-encoded, facilitating genetic manipulation" (Apaga, 2012 as cited from Abriouel et al., 2007).

LAB bacteriocins have attracted attention in recent years because of their generally regarded as safe (GRAS) status and good value as natural biopreservatives which can find applications in the food and cosmetic industries (Cleveland et al., 2001; Daeschel, 1993; Riley and Wertz, 2002). Nisin, produced by strains of *Lactococcus lactis*, has been used in over 50 countries as anti-listerial and anti-clostridium substance. LAB bacteriocins with selective inhibition on food pathogens such as *Listeria monocytogenes*, but no inhibition on important lactic acid bacterial inocula such as the noted probiotic *Lactobacillus paracasei* or *Lactobacillus rhamnosus*; and yogurt-producing *Lactobacillus delbrueckii* subsp. *bulgaricus* and *Lactococcus thermophilus*, may provide advantage over those that have a wider spectrum of antimicrobial activity and would kill these beneficial organisms, including nisin (De Vos, 1993; Jack and Ray, 1995; Nielsen et al., 1990). Hence, efforts on the search for LAB bacteriocins and elucidation of their properties are actively being pursued by several research laboratories. The future holds a wide array of LAB bacteriocins available for various specific applications.

3.2. Isolation and identification of bacteriocin-producing LAB

Some efforts on the isolation of bacteriocin-producing LAB had been started for more than a decade now in two major research institutions in the country namely: University of the Philippines Los Banos (specifically, the National Institutes of Molecular Biology and Biotechnology or BIOTECH-UPLB and the Institute of Biological Sciences or IBS-UPLB) and the Philippine Root Crop Research and Training Center, Visayas State University (VSU). These two institutions branched out knowledge on bacteriocin research through affiliate tutorship, as thesis advisers and as trainors to students and staff from a few other academic institutions which also did bacteriocin researches like University of Santo Tomas (UST), University of the Philippines Manila (UPM), De La Salle University (DLSU) and Ateneo de Manila University (ADMU). BIOTECH-UPLB and IBS-UPLB jointly worked on bacteriocins of *Lactobacillus plantarum* or plantaricins and those of *Pediococcus acidilactici* or pediocins. On the other hand, VSU devoted some efforts on the enterocins of *Enterococcus* spp. (Tan et al., 2001). DLSU also tried isolation of bacteriocin-producing LAB for food applications. UST was able to isolate bacteriocin-like inhibitory substances against medically important pathogens like *K. pneumoniae* (Dedeles et al., 2011). UPM and ADMU worked on human and animal health applications of bacteriocins.

Various fermented food products with proteinaceous components were the major sources of isolated LAB for bacteriocin screening. Such fermented food products are home-grown or produced by small enterprises and are still commercially available from public markets in Luzon, Philippines and some parts of the Visayas like Leyte island. Examples of Philippine indigenous fermented foods that were good sources of bacteriocin-producing LAB are fermented rice and shrimp (*balao-balao*), fermented rice and fish mixture (*burong kanin at isda*), fermented pork (*burong babi*) in Central Luzon (Elegado et al., 2003; Gervasio and Lim, 2007) and fermented pork and sweet potato (*agos-os*) in Eastern Visayan region (Samar and Leyte). On the other hand, pickled vegetables like mustard leaf (*burong mustasa*) and green papaya (*achara*), fermenting fruits like pickled green mango, *bignay* or mango wine (Samnang 2010), fermented salted fish (*bagoong*), spicy sausages (*longganisa*) may contain some LAB but often times they are not bacteriocinogenic (Gervasio and Lim, 2007). The obvious reasons are the presence of inhibitory substances like salt, spices, alcohol or acid and of course the dearth of proteinaceous materials in the food material.

In one of the first isolation studies for bacteriocinogenic LAB, various proteinaceous fermented foods native to Central and Southern, Philippines were screened for bacteriocin-producing bacterial isolates. Seventy one out of several hundreds of colony-forming unit isolated by agar plate streaking were found antagonistic to the indicator microorganism, *Lactobacillus plantarum* ATCC 14917, through direct assay. By "spot-on-lawn" assay by pH-neutralized culture supernatant, nine (9) isolates were confirmed to be bacteriocin producers (Elegado et al., 2003). Banaay et al. in 2004 also reported on the isolation of 1,100 putative LAB from indigenous fermented foods in Luzon, Philippines. A strain of *Lactobacillus plantarum* was selected as the best bacteriocin producer. In another study, out of the 160 putative LAB obtained from 19 fermented food products from public markets in Central Luzon, 32 LAB isolates were found to be bacteriocinogenic (Gervasio and Lim, 2007). Santiago et al. (2008) were also able to find two LAB isolates, *Lactobacillus fermentum* LBA-19 and *Lactobacillus casei* LTI-21, screened from among several LAB isolates from various fermented food products from different regions in the Philippines.

Being pleomorphic, identification of LAB is quite challenging. A combination of various microbiological and molecular biology tools would help in finding the real identity. Banaay et al. (2004) did a thorough identification of the bacteriocinogenic LAB isolate using conventional morphological, biochemical and physiological methods, chemotaxonomic methods, as well as molecular methods. This is especially relevant to the identification of *Lactobacillus plantarum* which is a known pleomorphic bacteria. Most other Philippine LAB researchers often times directly apply 16S rRNA gene sequencing and homology search for LAB purified through repeated agar streaking and putatively identified as LAB just after determining its acid–forming, Gram positive and catalase negative properties. (Elegado et al., 2003; Gervasio and Lim, 2007; Santiago et al., 2008). Aside from 16S rRNA genes, other conserved genes were used for identification such as phenylalanyl-tRNA synthase (*pheS*) gene (Dedeles et al., 2011). Detection of bacteriocin genes through PCR may also be helpful in confirming the identity of the bacteriocinogenic LAB as well as the probability of producing the bacteriocin (Table 2).

ISOLATE/ STRAIN No.	IDENTIFICATION	(primer) HOMOLOGY to *P. acidilactici* type strain	REFERENCE	Bacteriocin gene by PCR; fingerprinting; HOMOLOGY
AA-5a	partial 16S rRNA gene ID: *P. acidilactici*	(1492R)98% *P. acidilactici* UL5; 99% *P. acidilactici* DSM20284 (27F) 99% *P. acidilactici* LAB 001; 99% *P. acidilactici* DSM20284	Elegado et al. 2003	ped [+]; REP and RAPD
4E2	partial 16S rRNA gene ID: *P. acidilactici*	(1492R) 98% *P. acidilactici* UL5; 99% *P. acidilactici* DSM20284 (27F) 99% *P. acidilactici* LAB 001; 99% *P. acidilactici* DSM20284	Apaga (2012)	ped[+]
4E4	partial 16S rRNA gene ID: *P. acidilactici*	(1492R) 97% *P. acidilactici* UL5; 99% *P. acidilactici* DSM20284 (27F) 98% *P. acidilactici* 8D2CCH01MX; 99% *P. acidilactici* DSM20284	Apaga (2012)	ped[+]
4E5	partial 16S rRNA gene ID: *P. acidilactici*	(1492R) 99% *P. acidilactici* DSM20284 (27F) 99% *P. acidilactici* DSM20284	Laxamana et al. (2011)	ped[+]; REP
4E6	partial 16S rRNA gene ID: *P. acidilactici*	(1492R) 98% *P. acidilactici* UL5; 99% *P. acidilactici* DSM20284; (27F) 99% *P. acidilactici* 8D2CCH01MX ; 99% *P. acidilactici* DSM20284	Apaga (2012)	ped[+];[99% *P. acidilactici* bacteriocin genes ; pSMB74]
4E10	partial 16S rRNA gene ID: *P. acidilactici*	(1492R) 96% *P. acidilactici* UL5; 99% *P. lolii* to NGRI0510Q (27F) 99% *P. acidilactici* LAB 001 ; 99% *P. lolii* NGRI 0510Q	Apaga (2012)	ped[-]
4BL7	partial 16S rRNA gene ID: *P. acidilactici*	(1492R) 98% *P. acidilactici* UL5; 99% *P. acidilactici* DSM20284 (27F) 99% *P. acidilactici* 8D2CCH01MX; 99% *P. acidilactici* DSM20284	Apaga (2012)	ped[+]
3G3	API CHL50 ID: *Lactobacillus pentosus*(doubtful) partial 16S rRNA gene ID: *P. acidilactici*	(1492R) 99% *P. acidilactici* IMAU20090 (27F) 98% *P. acidilactici* DSM20284	Elegado and Perez (2012)	ped[+] ; REP; ped[+]
3G8	partial 16S rRNA gene ID: *P. acidilactici*	(1492R) 99% *P. acidilactici* UL5 (27F) 98% *P. acidilactici* DSM20284	Elegado and Perez (2012)	ped[+]
3F3	partial 16S rRNA gene ID: *P. acidilactici*	(1492R) 95% *P. acidilactici* UL5 (27F) 98% *P. acidilactici* UL5; 99% *P. acidilactici* DSM20284	Apaga (2012)	ped[+]
3F8	partial 16S rRNA gene ID: *P.acidilactici*	(1492R) 98% *P. acidilactici* UL5; 99% *P. acidilactici* DSM20284 (27F) 99% *P. acidilactici* LAB 001; 99% *P. acidilactici* DSM20284	Apaga (2012)	ped[+]
3F10	partial 16S rRNA gene ID: *P. acidilactici*	(1492R) 97% *P. acidilactici* UL5; 99% *P. acidilactici* DSM20284 (27F) 97% *P. acidilactici* LAB 001; 99% *P. acidilactici* DSM20284	Apaga (2012)	ped[+] [99% *P. acidilactici* genomic scaffold];
IG7	partial 16S rRNA gene ID: *P. acidilactici*	(1492R) 97% *P.acidilactici* UL5 99% *P. acidilactici* 8D2CCH01MX; (27F) 98% *P. acidilactici* DSM20284	Apaga (2012)	ped[+] [100% pediocin operon;PSMB74];

K₂A₂-3	API: *Pediococcus pentosaceus* (good) partial 16S rRNA gene ID: *P. acidilactici*	(1492R) 97% *P. acidilactici* UL5; 99% *P. acidilactici* DSM20284 (27F) 99% *P. acidilactici* LAB 001; 99% *P. acidilactici* DSM20284	Villarante (2011); Elegado and Perez (2012)	ped⁺ ; plan⁺ ped⁺ ; REP
K₂A₂-1	API: *P. acidilactici* (doubtful)	-	Abuel (2007)	ped⁺ ; plan⁺ped⁺
K₂A₂-5	API: *P. acidilactici* (doubtful); partial 16S rRNA gene ID: *P. acidilactici*	(1492R) 97% *P. acidilactici* UL5; 99% *P. acidilactici* DSM20284 (27F) 99% *P. acidilactici* LAB 001; 99% *P. acidilactici* DSM20284	Apaga (2012)	ped⁺ [99% *P. acidilactici* genomic scaffold]; plan⁺
K₂A₁-1	partial 16S rRNA gene ID: *P. acidilactici*	(1492R) 99% *P. acidilactici* L94; 99% *P. acidilactici* DSM20284 (27F) 98% *P.acidilactici* JS-9-4; 99% *P. acidilactici* DSM20284	Apaga (2012)	ped⁺
K₃A₂-2	API: *Lactococcus lactis* (good) partial 16S rRNA gene ID: *P. pentosaceus*	-		ped⁺ ; plan⁺; ped⁺
K₃A₂-3	partial 16S rRNA gene ID: *P. acidilactici*	100% *P. acidilactici* UL5	Elegado and Perez (2012)	ped⁺
S3	partial 16S rRNA gene ID: *P. acidilactici*	(1492R) 98% *P. acidilactici* UL5; 99% *P. acidilactici* DSM20284 (27F) 97% *P. acidilactici* LAB 001; 99% *P. acidilactici* DSM20284	Apaga (2012)	ped⁺ [99% pediocin operon; pSMB72];

Table 2. Identification and bacteriocin gene determination of putative *Pediococcus acidilactici* through 16S rRNA and pediocin gene PCR amplification and sequencing.

3.3. Purification and characterization of bacteriocins

Purification of bacteriocin peptides or small proteins into homogeneity is necessary in order to fully characterize them, particularly the determination of molecular mass, the primary structure or amino acid sequence and secondary structure. For pediocin, it was found that a simple and rapid method is effective for its purification. This method involves adsorption of pediocin onto the cell wall of the producer cell at pH 6 and 0.05 M NaCl and then subsequent desorption at pH 2.0 and 1 M NaCl (Elegado et al., 1997; Yang et al., 1992). This method seemed more applicable to pediocin but not with the lactococcin, nisin or plantaricin. The reason is not clear but it could be related to variation in cell wall properties. The pH-adsorption/desorption method was able to provide materials for pH and temperature tolerance assays, estimation of molecular mass through SDS-PAGE, residual activity determination after protease, amylase and other enzyme actions (Laxamana et al., 2011). Enough amount of semi-purified bacteriocin from pediococci using this method was obtained for further purification through preparative reverse phase HPLC for various characterization studies, including the determination of secondary structures by circular dichroism and confirmation of double bonds through trypsin digestion and electrospray mass spectrometry (Elegado and Kwon, 1998). Other preparative purification methods prior to reverse phase HPLC and spectrometry included ion exchange chromatography and gel filtration chromatography (Elegado et al., 2003), and hydrophobic interaction chromatography (Villarante et al., 2011). This method could also be applied with bacteriocins of pediococci and lactobacilli. The properties obtained from well characterized bacteriocinogenic LAB are shown in Table 3.

Isolate	Identity	Bacteriocin	Purification mode	Properties
AA5a	*Pediococcus acidilactici*	pediocin	pH adsorption/desorption Reversed-phase HPLC	Tolerant to pH 2-9 and 121 °C
BS25	*Lactobacillus plantarum*	plantaricin	Gel filtration chromatography Reversed-phase HPLC	MW = 3,830 Da
K2a2-3	*Pediococcus acidilactici*	pediocin	Hydrophobic interaction and ion-exchange chromatographies Reversed-phase HPLC	MW = 4,626 Da
K2a2-1	*Pediococcus acidilactici*	pediocin	pH adsorption/desorption	Optimum pH = 5-7 Resistant to boiling but not to autoclaving
4E5	*Pediococcus acidilactici*	pediocin	pH adsorption/desorption	Tolerant to pH 2-9; slight loss of activity at 100 °C; loss of activity at 121 °C; tolerates high salt; est. MW = 6,500 Da by SDS-PAGE

Table 3. List of purified and characterized bacteriocins from LAB isolated from Philippine indigenous fermented foods.

3.4. Optimization of bacteriocin production through fermentation kinetics

Bacteriocin production is largely dependent on the nutrients and nitrogen content of the fermentation medium. For instance, increased yeast extract concentration and polypeptone amount increases bacteriocin production. Molasses, raw sugar and sago hydrolyzates of amylase digestion were found to be good carbon sources. Other possible substrate base and supplements are cheese whey, coconut water and rice bran extract. Initial sugar concentration of usually 2 to 3% and inoculation rate of 3% by volume of at least 10^8 cells/mL provides good bacteriocin production (Elegado et al., 2001).

Bacteriocin production is highly dependent on cell or biomass growth. LAB are microaerophilic and most are either mesophilic or slightly thermophilic. The following conditions are applicable to their production: pH= 5.5 to 6.0; temperature = 35 – 40 °C; agitation = 50 rpm; without aeration. Usually, bacteriocin is optimally produced or secreted in the culture broth during the early stationary phase of growth. For *Pediococcus acidilactici*, culturing at 40 °C promotes earlier optimum bacteriocin production of around 10-12 hours. At 37 °C, bacteriocin production is from 14-16 hours (Sagpao et al., 2007).

3.5. Applications

Pediocins and plantaricins are the commonly found bacteriocins in Philippine fermented foods so far studied. Their antimicrobial properties have been investigated in several studies (Banaay et al., 2004; Elegado et al., 2003, 2004, 2007; Marilao et al., 2007). Although pediocins and plantaricins show promise, their applications are limited at present because it is a well-known fact that other bacteriocins aside from nisin are not yet approved for food use. For pediocins and plantaricins, the most practical use for now would be dermatological and animal health care use. But since the bacteriocin-producing LAB are of GRAS status, those with probiotic properties such as tolerance to acidic pH (2.0 -3.0) and bile (0.3%) and adhesion properties to intestinal mucosa would be an advantage when used as adjunct inocula in fermented food products (Gervasio and Lim, 2007).

Perhaps another importance of bacteriocin-producing LAB is their effectiveness in biomedical applications. In one study, for example, partially-purified pediocin K2a2-3, through pH-mediated bacteriocin extraction method, was found cytotoxic against human colon adenocarcinoma (HT29) and human cervical carcinoma (HeLa) cells *in vitro* as determined by MTT [3-(4,5-dimethylthiazol-2-yl)-2,5-diphenyltetrazolium bromide] assay (Villarante et al., 2011). Other potential biomedical applications will be discussed in the succeeding section.

4. Probiotics and functional foods

An offshoot of the initial research on bacteriocins of LAB isolated from indigenous fermented foods is the emergence of probiotic research towards developing functional foods for biomedical applications. Probiotics refer to microorganisms that, when administered in adequate amounts, confers health benefits to the host. Although there are many microorganisms that can be considered as probiotics, LAB are the most common types because they produce antimicrobial compounds that inhibit other harmful microorganisms, they are able to tolerate acids and bile present in the digestive system, and they are able to adhere and establish themselves in the gut surfaces.

Many benefits have been ascribed to probiotics. For example, *Lactobacillus casei* (Shirota strain in Yakult®) have been shown effective in preventing diarrhea due to enterotoxigenic *Escherichia coli* (ETEC) and choleragenic vibrios (*V. cholerae* biotype E1 Tor and classical *V. cholerae*) using rats (Jacalne et al., 1990). This may be accounted for by its ability to kill the pathogens and inhibit further growth (Consignado et al., 1994). Because the probiotic used in the two studies mentioned is a commercial strain, current research on probiotics progressed to the search for indigenous LAB for use in the development of locally-produced functional food and investigation of their utility for biomedical applications. Metagenomic approaches to investigating LAB present in fermented foods have shown the diversity of potentially beneficial species present other than those that are readily detected by conventional culture-based methods. The development of functional food products shows potential in disease management. Research using metagenomic analysis in searching for microbial markers for use in functional foods to address certain lifestyle diseases as well as malnutrition is on the way.

4.1. Metagenomic and diversity studies

Traditional culture-based methods have been used for isolating LAB from fermented foods. These studies form the basis for the starter cultures used in food fermentation technologies employed for commercial production. Sanchez (2008) gives detailed information on the different technologies and cultures used for the production of some traditional as well as developed technologies that have arisen from the culture-based studies conducted in earlier years.

In recent years culture-based approaches in LAB isolation have become more targeted for detection of bacteriocin-producers and those that have potential as probiotics. In one initiative, LAB isolates from fermented foods were screened for bacteriocin production and a PCR-based assay was used to detect specific bacteriocin-encoding genes. Acid and bile tolerance were also determined. Among all the isolates tested, *Lactobacillus fermentum* 4B1 and *Lactobacillus pentosus* 3G3 (later identified as *Pediococcus acidilactici*) have been identified as most promising for the development of new probiotic food products, hence they were chosen for subsequent biomedical application assays (Lim and Gervacio, 2007). In another study, LAB from traditionally fermented wine and vinegar from Visayas and Mindanao were isolated, identified, and tested for inhibitory activity against *Enterococcus faecium*, *Listeria innocua*, and *Staphylococcus aureus*. Five *Lactobacillus paracasei* and one *Lactobacillus brevis* showed antimicrobial properties against the tester strains (Licaros and Bautista, 2009).

With the advent of molecular techniques, the existence of non-culturable microorganisms has been acknowledged especially since the occurrence of culture-bias is already well-accepted. Culture-independent approaches, therefore, have been gaining popularity in microbial diversity studies and this includes researches on microorganisms found in fermented foods. The microbial populations in selected Philippine fermented foods were assessed through Polymerase Chain Reaction followed by Denaturing Gradient Gel Electrophoresis (PCR-DGGE) in two recent studies (Dalmacio et al., 2011; Larcia, 2010). Food samples tested include *burong mustasa* (fermented mustard), *alamang* (fermented shrimp paste), *burong isda* (fermented rice-fish mixture), *balao-balao/burong hipon* (fermented rice-shrimp mixture), *tuba* (sugar cane wine), and *sinamak* (spiced vinegar). Analysis of the 16S rRNA gene sequences revealed the presence of several LAB that have not been reported in these food products before. *Weissella cibaria*, *Lactobacillus plantarum*, *Lactobacillus pontis*, *Lactobacillus panis*, and *Lactobacilus fermentum* were detected in *burong mustasa* (Larcia, 2010). *L. panis* and *L. fermentum* were present in *alamang*; *L. pontis* and *L. plantarum* in *burong isda*; *L. panis*, *L. pontis*, and *L. fermentum* in burong hipon; and *W. cibaria*, *L. pontis*, *L. panis*, *L. fermentum* and *L. plantarum* in *burong mustasa* (Dalmacio et al., 2011).

The results of the two studies using molecular approaches in defining diversity of LAB in Philippine fermented foods show that culture-independent approaches are efficient tools for the analysis of microbial populations in fermented foods. Majority of the identified bacteria (LAB and other bacterial groups) have not been reported in culture-dependent studies. As such, the isolated bacterial 16S rRNA genes were cloned to have an initial partial 16S rRNA gene library for Philippine fermented foods (Dalmacio et al., 2011).

4.2. Biomedical applications

1.   Anti-Obesity

Obesity is defined as an abnormal or excessive fat accumulation that presents risks to health. Probiotics can help in fighting obesity by reducing lipid absorption through its action on bile acid metabolism, and by assimilation of cholesterol thus eliminating it from the host's system. Several studies were conducted to examine anti-obesity properties of different probiotic strains.

In one study, oral administration of *Lactobacillus paracasei* K3-4C, isolated from a locally fermented food had significant effect on lowering blood glucose levels (by 46%) and body weight (by 13%) in female BALB/c mice induced to be diabetic and obese through a 28-day high-fat diet (Parungao et al., 2006). In another study, orally administered *L. fermentum* 4B1 reduced adipose cell size, and decreased adipose tissue weight and overall body weight of mice fed with a high-fat diet for 49 days (Bautista et al., 2008). Likewise, oral administration of *P. acidilactici* 3G3 reduced body weight in diet-induced obese female Swiss mice (Parungao et al., 2009). In the last two studies described, the effects of the probiotics were determined to be comparable with the effects of the commercial anti-obesity drug Orlistat based on the parameters measured.

Recently, it has been postulated that the development of obesity may be caused by a shift in the composition of the gut microbiota towards the Firmicutes population (Ley et al., 2005). Firmicutes characterize obese versus lean/non-obese individuals together with a drop or no change in Bacteroidetes (Delzenne and Cani, 2010). Interestingly, Ley et al. (2006) found that a low fat diet had an effect to reverse the shift of Firmicutes/Bacteroidetes proportion. Because of this, dietary manipulation has been seen as a potential means of changing bacterial populations in the colonic microbiota and perhaps treating or at least preventing diseases like obesity. Although the root cause of obesity is excessive caloric intake coupled with a sedentary lifestyle (Blaut and Bischoff, 2010), Ley et al. (2005) proposed in their findings that alteration in the populations of mice gut microflora may have caused or may have been an effect of obesity. Because of this, current researches aim in using probiotics in the treatment of diseases such as obesity.

In two related studies (Arroyo and Fabiculana, 2011; Parungao et al., 2012), the effect of a functional food containing *P. acidilactici* 3G3 on microbial community changes in the gut of obese and non-obese mice was determined through PCR-DGGE. Results of these two preliminary studies showed that obese and non-obese mice had different baseline colonic microbiota. There were also indications that treatment with probiotics shifts the microbiota of obese mice towards the normal non-obese type. As these are preliminary studies, more research is warranted to elucidate the nature of the changes in gut microbiota and how it is related to obesity and the anti-obesity effects of probiotics.

2.   Immuno-enhancement

A preliminary *in vitro* study to examine the immune-enhancing properties of viable and heat-killed preparations of two LAB previously isolated from traditional fermented foods (*L. fermentum* 4B1 and *P. acidilactici* 3G3) on murine peritoneal macrophage cells and spleenic T-cells showed that isolate 4B1 was able to induce NO production in murine macrophages but, like 3G3, was unable to stimulate murine T-cell proliferation (Tan et al., 2008). Furthermore, this study showed that preparations of *L. fermentum* 4B1 have the ability to induce NO production in murine macrophage cells and its effects were more potent when it was alive.

The study also showed that isolate 4B1 exhibited better immune-enhancing effect than the probiotic species found in a commercial probiotic drink. T-cell proliferation, however, was not observed in any of the treatments in this study and was attributed to the delayed stimulation in cells responding to a first-time exposure to the different probiotic strain preparations used.

3.   Reduction of blood glucose levels

A study by Ngo et al. (2008) showed that oral administration of kefir, a common fermented food consumed by the elderly, significantly decreased blood glucose levels and body weight of diabetic obese male Sprague Dawley rats. The results of the study showed lower blood glucose levels (from 198.5 to 105.6 mg/dL) and clinically lower body weights (from 342.9 to 311.5 g) of the treated diabetic-obese rats than the untreated diabetic-obese control group.

4.   Prevention of hypercholesterolemia

The effect of *P. acidilactici* 3G3 administration on hypercholesterolemic Swiss Albino mice was determined (Parungao et al., 2009). This strain was able to assimilate cholesterol in the *in vitro* plate assay and decrease HDL, LDL, and total cholesterol in the *in vivo* assay using mice. Strain 3G3 was also shown to adhere well to the duodenum and middle colon. Results suggest the potential of *P. acidilactici* 3G3 in preventing hypercholesterolemia.

4.3. Development of functional foods

The development of functional foods containing known probiotic strains stems from earlier researches on bacteriocins and isolation of potential probiotics from traditional fermented foods. The beneficial effects of probiotic-supplemented chocolate bars (Arroyo et al., 2010; Arroyo and Fabiculana, 2011), fermented mustard leaves (Calapardo et al., 2006), and coffee wine (Parungao, 2007) have been investigated. Initial studies on mango-milk and carrot juice drinks supplemented with probiotic strains have also been conducted (Bugarin et al., 2010; Elegado et al., 2005). These potential functional foods contain probiotic strains, previously isolated from traditional fermented foods such as *P. acidilactici* AA5a (Elegado et al., 2003), *L. plantarum* BS25 (Banaay et al., 2004), and *P. acidilactici* 3G3 (Lim and Gervacio, 2007). Research on functional foods is still in its infancy but this food category shows promise in disease management as well as in contributing to food security in the country. Commercial interest in probiotic food products is increasing due to the growing understanding of its health benefits. This growing industry can derive benefits from the researches conducted on this emerging food category.

5. Future perspectives

Aside from the research works presented earlier in this paper as well as on-going follow-up studies related to them, future goals may include research on a variety of other possible biomedical applications of LAB with potential probiotic properties. The effect of probiotics on *Helicobacter pylori* infections (that may cause peptic ulcers) may be determined. Their ability to modulate inflammatory and hypersensitivity responses as well as their effect on irritable bowel syndrome and colitis may be investigated. Further research on possible anti-cancer properties of probiotics is warranted as follow-up studies on the work done by Villarante et al. (2011). These studies are very important as these have the potential to address some of the more serious health concerns of our society.

Much is still to be learned about the existing probiotic strains. The molecular biology and genomics of these isolates may be pursued in order to further elucidate their properties and mechanisms of action.

Determination of factors affecting probiotic viability in foods is also important as these will determine if their survival in the food, and therefore their delivery into the host, is maintained. This will constitute a quality control for functional foods.

The potential physiological effects of multiple prebiotic strains, as opposed to a single strain, are also interesting areas of research. The delivery of multiple probiotic strains may help ensure its effectiveness in an environment that contains high diversity of resident microflora. The potential benefits of synbiotics, (combination of probiotic and prebiotic) which have synergistic interaction, may also be investigated. A good combination will greatly enhance the health benefits to humans.

Author details

Charina Gracia B. Banaay
Institute of Biological Sciences, College of Arts and Sciences,
University of the Philippines Los Baños, Laguna, Philippines

Marilen P. Balolong
Department of Biology, University of the Philippines Manila,
Padre Faura Street, Ermita, Manila, Philippines
Institute of Molecular Biology and Biotechnology, National Institutes of Health,
University of the Philippines Manila, Pedro Gil Street, Ermita, Manila, Philippines

Francisco B. Elegado
National Institute of Molecular Biology and Biotechnology,
University of the Philippines Los Baños, Laguna, Philippines

6. References

Abriouel H, Galvez A, Lopez RL, Omar NB (2007) Bacteriocin-based strategies for food biopreservation. *International Journal of Food Microbiology* 120: 51–70.

Apaga DLT (2012) Detection of bacteriocin structural genes and bacteriocinogenic activity of several pediococcus isolates. Undergraduate thesis. University of the Philippines Manila. 56 p.

Arroyo KZO, Garcia JAR, Elegado FB, Calapardo M, Parungao MM (2010) Survival of *Lactobacillus pentosus* 3G3 in home-made chocolate bars. 39th Annual Convention and Scientific Meeting,, PSM, Inc. April 29-30, 2010, City of Naga, Camarines Sur.

Arroyo MJJ, Fabiculana PRS (2011) Gut microflora assessment of obese and non-obese mice (*Mus musculus* L.) orally-fed with lp-3g3 chocolate. Department of Biology, University of the Philippines Manila, Manila, Philippines. Unpublished Undergraduate Thesis.

Banaay CGB, Elegado FB, Dalmacio IF (2004) Identification and characterization of bacteriocinogenic *Lactobacillus plantarum* BS25 isolated from balao-balao, a locally fermented rice-shrimp mixture from the Philippines. *The Philippine Agricultural Scientist* 87(4): 427-438.

Bautista RLS, Ecarma NCA, Balolong ECJr, Hallare AV, Parungao MM (2008) The Effects of Orally-Administered *Lactobacillus* sp. 4B1 on the Adipose Tissues of Diet-Induced Obese Mice (*Mus musculus* L.). International Symposium on Probiotic from Asian Traditional Fermented Foods for Healthy Gut Function. August 19-20, 2008. Sari-Pan Pacific Hotel, Jakarta, Indonesia.

Besas JR, Dizon EI (2012) Influence of salt concentration on histamine formation in fermented tuna viscera (Dayok). *Food and Nutrition Sciences* 3:201-206.

Blaut M and Bischoff SC (2010) Probiotics and Obesity. *Ann Nutr Metab* 2010; 57(Suppl.1):20-23.

Bugarin MA, Sison AAD, Elegado FB, Calapardo M, Parungao MM (2010) Survival of *Lactobacillus plantarum* BS25 on Varying Ratio of Mango-Milk Substrates and Storage Temperature. 39th Annual Convention and Scientific Meeting, PSM, Inc. April 29-30, 2010, City of Naga, Camarines Sur.

Calapardo MR, Bueno MOV, Guillermo MKB, Saguibo JD, Parungao MM, Elegado FB (2006) Bile and Acid Tolerance of Bacteriocinogenic *Lactobacillus plantarum* and its Use as Adjunct Inoculum in Pickled Mustard Leaves. Proceedings of the 5th Asia-Pacific Biotechnology Congress and 35th Annual Convention of the PhilippineSociety for Microbiology Inc. (PSM). "Microbiology and Biotechnology: Roadmaps and Milestones for Enhancing Sustainable Productivity in the Asia-Pacific Region." Bohol Tropics Resort. Tagbilaran City, Bohol, Philippines.

Chen H, Hoover DG (2003) Bacteriocins and their food applications. http://www.ift.org/publications/crfsfs (accessed 22 July 2006)

Cleveland J, Chikindas ML, Montville TJ, Nes IF (2001) Bacteriocins: safe, natural antimicrobials for food preservation. *International Journal of Food Microbiology* 71: 1–2 0.

Consignado GO, Pena AC, Jacalne AV (1994) *In vitro* study on the bacterial activity of *Lactobacillus casei* (commercial Yakult drink) against four diarrhea-causing organisms: enterotoxigenic *E. coli*, *Salmonella enteritidis*, *Shigella dysenteriae*, *Vibrio cholerae*. Philipp J Microbiol Infect Dis . 23(2): 50-55.

Daeschel MA (1993) Applications and interactions of bacteriocins of lactic acid bacteria in food and beverages. In: Hoover, D.G., L.R. Steenson, eds. Bacteriocins of Lactic Acid Bacteria. New York: Academic Press Inc. P. 63-70.

Dalmacio LMM, Angeles AKJ, Larcia LLH, Parungao-Balolong MM, Estacio RC (2011) Assessment of microbial diversity in Philippine fermented food products through polymerase chain reaction- denaturing gradient gel electrophoresis (PCR-DGGE). Beneficial Microbes, 2(4): 273-281.

De Vos WM (1993) Future prospects for research and applications of nisin and other bacteriocins. In: Hoover, D.G., L.R. Steenson, eds. Bacteriocins of Lactic Acid Bacteria. New York: Academic Press Inc. P. 249-258.

Dedeles GR, Caranza MAE, Elegado FB (2011) Bacteriocin-like inhibitory substances (BLIS) from lactic acid bacteria and their potential as biopreservatives in foods. Proccedings of the 6th Asian Conference on Lactic Acid Bacteria and XIII International Congress of Bacteriology and Applied Microbiology. September 6-10, 2011. Sapporo, Japan. 120 p.

Delzenne NM and Cani PD (2010) Nutritional modulation of gut microbiota in the context of obesity and insulin resistance: potential interest of prebiotics. *Int. Dairy J.* 20: 277-280.

Elegado FB, Kim WJ, Kwon DY (1997) Rapid purification, partial characterization and antimicrobial spectrum of the bacteriocin, Pediocin ACM from *Pediococcus acidilactici* M. *International Journal of Food Microbiology* 37(1): 1-11.

Elegado FB, Kwon DY (1998) Primary structure and conformational studies of Pediocin AcM, a bacteriocin from *Pediococcus acidilactici* M. *The Philippine Journal of Biotechnology* 9(1): 19-26.

Elegado FB, Sonomoto K, Ishizaki A (2001) Molecular characterization of a bacteriocin of *Pediococcus acidilactici* and its production in sago and sugarcane-based substrate. *Biotechnology for Sustainable Utilization of Biological Resources in the Tropics.* Murooka, Y., T. Yoshida, T. Seki, P. matangkasombut, T.M. Espino, U. Soetisna and M.I.A. Karim. eds. JSPS-NRCT/DOST/LIPI/VCC Joint Seminar. Nov. 7-9, 2001. Bangkok, Thailand. 15:173-180.

Elegado FB, Opina ACL, Banaay CGB, Dalmacio IF (2003) Purification and characterization of novel bacteriocins from lactic acid bacteria isolated from Philippine fermented rice-shrimp or rice-fish mixtures. *The Philippine Agricultural Scientist* 86(1): 65-74.

Elegado FB, Guerra MARV, Macayan RA, Estolas MT, Lirazan MB (2004) Antimicrobial activity and DNA fingerprinting of bacteriocinogenic *Pediococcus acidilactici* through RAPD-PCR. *The Philippine Agricultural Scientist* 87(2):229-237.

Elegado FB, Calapardo MR, Fabregas JF, Ona SEN, Parungao MM (2005) Antilisterial Efficacy of Bacteriocinogenic *Pediococcus acidilactici* AA5a in Carrot Juice Fermentation. Proceedings of the 3rd Asian Conference on Lactic Acid bacteria. August 25-27, 2005. Sanur Paradise Hotel, Bali, Indonesia.

Elegado FB, Abuel BJA, Te JT Jr.,Calapardo MR, Parungao MM (2007) Antagonism against *Listeria* spp. and *Staphylococuus aureus* by bacteriocin-producing lactic acid bacteria screened from the intestine of Philippine carabao using polymerase chain reaction. *The Philippine Agricultural Scientist* 90(4): 305-314.

Elegado FB and Perez MTM (2011) Genetic identification of lactic acid bacteria and bacteriocin structural gene elucidation. National Academy of Science and Technology (NAST) Scientific Meeting. July 11-12, 2012. Manila Hotel, Manila, Philippines.

Gervasio ATR, Lim VMT (2007) Probiotic characterization of bacteriocinogenic lactic acid bacteria isolated from fermented foods of selected areas. Undergraduate thesis. University of the Philippines Manila. 105 p.

Jack RJT, Ray B (1995) Bacteriocin of gram positive bacteria. *Microbiological Reviews.* 59(2): 171–200.

Jacalne AV, Jacalne RR, Hirano H, Suetomi T, Villahermosa CG, Castaneda I (1990) *In-vivo* studies on the use of *Lactobacillus casei* (Yakult Strain) as biological agent for the prevention and control of diarrhea. Acta Med Philipp. 26(2): 116-122

Larcia, LLH (2010) A study of the bacterial profile of Philippine fermented mustard (*burong mustasa*) through culture-independent methods / Levi Letlet Larcia, II. Thesis (MS Biochemistry)--University of the Philippines Manila. 158pp. LG995 2010 B3 L37

Laxamana FLM, Carillo MCO, Elegado FB (2011) Characterization of a bacteriocin of lactic acid bacteria isolated from fermented rice-fish mixture. Proceedings of the 40th Annual Convention and Scientific Meeting of the Philippine Society for Microbiology, Inc. May 10-14, 2011. Manila, Philippines.

Lee C (1999) Cereal fermentations in countries of the Asia-Pacific Region. In: FAO, Fermented Cereals: A Global Perspective, FAO Agricultural Services Bulletin Issue 138. Food and Agriculture Organization 144 p.

Ley RE, Bäckhed F, Turnbaugh P, Lozupone C A, Knight RD, Gordon JI (2005) Obesity alters gut microbial ecology. Proc Natl Acad Sci USA, 102(31): 11070-11075.

Ley RE, Turnbaugh PJ, Klein S, Gordon JI (2006) *"Human gut microbes associated with obesity."* Nature 444:1022-1023.

Licaros A, Bautista A (2009) Culture-Dependent and Culture-Independent Analysis of Vinegars from Visayas and Mindanao. Department of Biology, University of the Philippines Manila, Manila, Philippines. Unpublished Undergraduate Thesis.

Lim V, Gervacio TA (2007) Bacteriocin-producing LAB from fermented foods of Central Luzon. Department of Biology, University of the Philippines Manila, Manila, Philippines. Unpublished Undergraduate Thesis.

Marilao CG, Calapardo MR, Ciron CE, Elegado FB (2007) Antilisterial action of *Pediococcus acidilactici* AA5a in pork sausage fermentation. *The Philippine Agricultural Scientist* 90(1): 40-45.

Ngo RE, Estrada MZ, Balolong ECJr., Parungao MM (2008) Orally-administered kefir on lowering blood glucose levels and body weight in diet-induced diabetic obese rats. Project Terminal Report. National Institutes of Health University of the Philippines Manila.

Nielsen JW, Dickson JS, Crouse JD (1990) Use of a bacteriocin produced by *Pediococcus acidilactici* to inhibit *Listeria monocytogenes* associated with fresh meat. *Applied and Environmental Microbiology* 54: 2349-2353.

Olympia M, Fukuda H, Ono H, Kaneko Y, Takano M (1995) Characterization of starch-hydrolyzing lactic acid bacteria isolated from a fermented fish and rice food, "Burong Isda", and its amylolytic enzyme. *Journal of Fermentation and Bioengineering* 80(2):124-130.

Parungao MM (2007) Probiotic Properties of Philippine Coffee Wine. The 4th Asian Conference on Lactic Acid Bacteria & the 3rd International Symposium on Lactic Acid Bacteria and Health. October 17-19, 2007. Shanghai, China.

Parungao MM, Castillo RL, Mortel MRA (2006) Effects of Oral Administration of *Lactobacillus paracasei* K3-4C on Blood Glucose Levels and Body Weight in Diabetic

Obese Mice (*Mus musculus* L.). December 15-17, 2006. Proceedings of the 11th Biological Sciences Graduate Congress, Chulalongkorn University, Bangkok, Thailand.

Parungao MM, Ong C, Laluces N, De Torres K, Usisa J, Calapardo M, Trinidad L, Elegado FB (2009) The Cholesterol-Reducing Ability and Adhesion Properties of *Lactobacillus pentosus* 3G3. Proceedings of the 5th Conference of Federation for Societies of Lactic Acid Bacteria. July 1-4, 2009. National University of Singapore, Singapore.

Parungao-Balolong M, Libed AAO, Loma K, Villena JPDS, Villafuerte AR, Balolong EC Jr., Dalmacio LMM (2012) Influence of probiotic drinks on the distal gut bacterial flora of mice (*Mus musculus* L.) fed with standard diet or high-fat diet. Terminal report, National Institutes of Health University of the Philippines Manila.

Riley MA, Wertz JE (2002) Bacteriocins: evolution, ecology and applications. *Annual Review of Microbiology* 56:117-137.

Sagpao SMN, Elegado FB, Zamora A (2007) Optimization of production and partial purification of pediocin from *Pediococcus acidilactici* PNCM 10289. Poster paper presented at the 29th National Academy of science and Technology (NAST) Philippines Annual Scientific Meeting, Manila Hotel, July 11-12, 2007.

Samnang (2010) Isolation and identification of lactic acid bacteria from fermenting bignay (*Antidesma bunius* (L.) Spreng) and mango (*Mangifera indica* L.) wines. M.S. thesis. University of the Philippines Los Banos. 242 p.

Sanchez PC (2008) Philippine fermented foods: principles and technology. The University of the Philippines Press. Diliman, Quezon City, Philippines. 516 p.

Santiago MR, Lopez CM, Tenorio EL, de Guzman EE (2007) Detection and partial charcaterization of two bacteriocins produced by *Lactobacillus casei* and *Lactobacillus fermentum* isolated from fermented rice-fish mixtures indigenous in the Philippines. Proceedings of the 37th Annual Convention and Scientific Meeting of the Philippine Society for Microbiology, Inc. May 7-9, 2008. Boracay Philippines. P. 32.

Tan JD, Galvez FCF, Asano K, Tomita F (2001) Isolation and partial purification of bacteriocin produced by microorganisms from *agos-os*. Biotechnology for Sustainable Utilization of Biological Resources in the Tropics. Murooka, Y., T. Yoshida, T. Seki, P. matangkasombut, T.M. Espino, U. Soetisna and M.I.A. Karim. eds. JSPS-NRCT/DOST/LIPI/VCC Joint Seminar. Nov. 7-9, 2001. Bangkok, Thailand. 15: 104-110.

Tan JAS, Yalung PM, Evangelista KV, Parungao MM (2008) *In vitro* Study on the Effect of *Lactobacillus* sp. 4B1 and *Lactobacillus* sp. 3G3 on Murine Macrophage Nitric Oxide Production and Splenic T-cell Proliferation. Proceedings of the International Symposium on Probiotic from Asian Traditional Fermented Foods for Healthy Gut Function.August 19-20, 2008. Sari-Pan Pacific Hotel, Jakarta, Indonesia.

Villarante KI, Elegado FB, Iwatani S, Zendo T, Sonomoto K, de Guzman EE (2010) Purification, characterization and *in vitro* cytotoxicity of the bacteriocin from *Pediococcus acidilactici* K2a2-3 against human colon adenocarcinoma (HT29) and human cervical carcinoma (HeLa) cells. *World Journal of Microbiology and Biotechnology* 27: 975-980.

Yang R, Johnson MC, Ray B (1992) Novel method to extract large amounts of bacteriocins from lactic acid bacteria. *Appl. Environ. Microbiol.* 58: 3355-3359.

Chapter 9

The Current Status and Future Expectations in Industrial Production of Lactic Acid by Lactic Acid Bacteria

Sanna Taskila and Heikki Ojamo

Additional information is available at the end of the chapter

1. Introduction

Conversion of carbohydrates to lactic acid is one of the most employed fermentation processes in food industry. Applications of lactic acid fermentation are found in dairy industry, production of wine and cider, production of fermented vegetable products and meat industry.

The main markets for lactic acid have been in food, pharmaceutical and cosmetics industry, but presently the main growing application of lactic acid is in the production of biodegradable and renewable raw material based poly lactic acid (PLA) polymers. Production of lactate esters (*e.g.* butyl lactate) is another growing application as environmentally friendly solvents [1]. Lactic acid has two optical isomers, L-(+)-lactic acid and D-(−)-lactic acid. Lactic acid is classified as GRAS (generally recognized as safe) for use as a food additive, although D(-)-lactic acid can be harmful to human metabolism and result in *e.g.* acidosis [2]. The optical purity of lactic acid is required for the production of PLA. The properties of PLA may however be adjusted by the ratio of the L- and D-PLA in a copolymer D-form increasing the melting point of the copolymer [3]. Optically pure L- or D-lactic acid can be obtained by microbial fermentation and presently more than 95 % of industrial production of lactic acid is based on fermentation.

Production figure of 260,000 t as 100 % lactic acid for conventional (excluding PLA) markets in 2008 and forecast over 1 million ton annual production of lactic acid for conventional markets and PLA by 2020 has been presented in 2010 [4]. DuPont patented PLA already in 1954 but it took almost 50 years before first large-scale production was started. The US-based NatureWorks is the largest producer of PLA having lactic acid production capacity of 180,000 t/a. The sustainability of the PLA product Ingeo® from NatureWorks has been evaluated [5]. Greenhouse gas emissions and nonrenewable energy consumption for Ingeo from cradle to factory-gate are 1.3 kg CO_2 eq./kg polymer and 42 MJ/kg polymer. These compare favorably with *e.g.* fossil-based PET (polyethylene terephtalate) with 3.2 kg CO_2 eq./kg polymer and 80 MJ/kg polymer, respectively. There is a huge potential for biodegradable and renewable raw materials based polymers if and when the economics for these become competitive. It is estimated that altogether 140 million tons of petroleum-based synthetic polymers are produced annually [6]. It should be emphasized that also many petroleum-based synthetic polymers (*e.g.* polyesters) are biodegradable. However at the moment there are only three commercial synthetic polymers replacing petroleum-based ones and produced on renewable raw materials: PLA, PTT (polytetramethylene terephtalate which is partly renewable) and PHA (polyhydroxyalkanoates). Natural polymers such as starches and celluloses are biodegradable and based on renewable raw materials, but their applications are limited by their properties. Reliance Life Sciences is producing copolymers of PLA and glycolic acid mainly for high-value medical applications. Lactic acid in this case is produced by bacterial fermentation.

The price of PLA is ca. 2.2 $/kg, the target being half of that [7]. This means that the price of lactic acid in captive use should be less than 0.8 $/kg. A major cost factor is the raw material used in fermentation medium. This is especially the case with fastidious lactic acid bacteria. Processes based on cheap polymeric waste and side stream materials are indeed widely studied. So far research on alternative fermentation modes and reactor systems has been mainly academic. PLA production requires both optically and chemically pure lactic acid. Optical purity can be guaranteed with several microbial strains under optimized fermentation conditions. Chemical purity is mainly dependent on the constituents in the fermentation medium especially when cheap materials are being used. Contrary to many other fermentation products lactic acid yield on monosaccharides is usually very high (> 90 %) the main impurity being the cell mass itself, which is easily separated from the product. The key economic drivers in the fermentative production of lactic acid are optimization of the production medium, high product yields, productivity, and the concentration of products formed, which influences the down-stream processing costs [8].

Lactic acid bacteria (LAB) are a group of Gram-positive bacteria belonging to genera Aerococcus, Alloiococcus, Atopobium, Bifidobacterium, Carnobacterium, Enterococcus, Lactobacillus (Lb.), Lactococcus (L.), Leuconostoc (Leuc.), Oenococcus, Pediococcus, Streptococcus (S.), Tetragenococcus, Vagococcus and Weissella (W.). LAB are non-sporulating rods or cocci which produce lactic acid as the main fermentation product under suitable substrates. LAB are oxidase and benzidine negative, lack cytochromes, and do not reduce nitrates to nitrite [9]. Most of the LAB are anaerobic, but some of them can shift to oxygen-dependent metabolism in aerobic conditions [10-11]. Lactic acid bacteria have complex nutrient requirements, including specific minerals, B vitamins, several amino acids, and purine and pyrimidine bases.

LAB ferment sugars via homo-, hetero-, or mixed acid fermentation. Homofermentative LAB produce lactic acid as main product from sugars, while hetero- or mixed acid fermentations produce also ethanol and/or acetic acid, formic acid and carbon dioxide.

Although it is a common practice to divide LAB into homo- and heterofermentative strains, the division is not that straightforward as the actual metabolism is dependent on both the nature of the C/energy substrate (*e.g.* hexose vs. pentose sugars) and fermentation conditions (*e.g.* growth rate and availability of the C/energy source). LAB used for lactic acid production are used to be classified as homofermentative (*Lactococcus, Enterococcus, Streptococcus* and some lactobacilli) as their hexose metabolism under non-limiting conditions is entirely via Embden-Meyerhof pathway to pyruvate which is then used to regenerate the reducing power (NADH) in the lactate dehydrogenase (LDH) catalyzed reaction to lactic acid. However at slow growth rate and low glycolytic flux mixed acid fermentation may take place and acetic acid, formic acid and ethanol are formed in addition to lactic acid [12]. The key enzyme in this metabolic shift *e.g.* in *L. lactis* is claimed to be pyruvate-formate lyase (PFL) [13]. There are two types of LDH for both enantiomers D-LDH and L-LDH. In addition some species have a racemase enzyme catalyzing the reaction between the two enantiomers. Thus enantiomerically pure lactic acid is produced by species with only one type of LDH and no racemase. A comprehensive list of different LAB strains used in lactic acid production is available elsewhere [1].

Biotechnical production of lactic acid may be based on several alternative micro-organisms. In addition to lactic acid bacteria filamentous fungi (*e.g. Rhizopus* spp.), other gram-positive bacteria (*e.g. Bacillus coagulans*) and metabolically engineered yeasts have been used also in industrial scale. The advantage of fungi is that they are active at and tolerate low medium pH. Low pH reduces significantly the consumption of neutralizing agent ($Ca(OH)_2$) in the fermentation stage and subsequent formation of gypsum ($CaSO_4$) in the product recovery stage. The advantage of filamentous fungi, *Bacillus* spp. and yeasts compared to lactic acid bacteria is their simple nutrient requirement in the fermentation medium. Filamentous fungi and *Bacillus* spp. are better suited to lignocellulosic fermentation raw materials as they are in general able to utilize pentose sugars in addition to hexoses. Anaerobic fermentation is generally speaking more feasible and this favors yeasts and lactic acid bacteria. When optimized the technical parameters such as product yield, R_P and final product concentration are quite similar for each of these production organisms.

In the wide literature on lactic acid production two examples based on other than lactic acid bacteria should be taken up. The first of them presents results with a thermotolerant *B. coagulans* strain [14]. High lactic acid $Y_{P/S}$ on both glucose and xylose (96 % and 88 %, respectively) were achieved at reasonable R_P (2.5 g/lh) and product concentration (100 g/l). Exceptionally high levels of lactic acid (200 g/l) were produced in fed-batch fermentation. Yeasts have been metabolically engineered aiming at lactic acid production since 1990's [15]. A recent article reported on metabolic engineering of *Candida utilis* having pyruvate decarboxylase deleted and a bovine L-lactate dehydrogenase expressed under the *pdc* promoter resulting in the production of lactic acid with high yield from glucose (95 %) and reasonable R_P (4.9 g/lh) ending up with lactic acid concentration of 103.3 g/l and more than 99.9 % enantiomeric purity [16].

Heterofermentative LAB (*Leuconostoc, Weissella* and some lactobacilli such as *Lb. brevis*) utilize both hexose and pentose sugars via phosphoketolase pathway (PKP). Several LAB possess the genes for PPP as well. The different pathways are presented in Fig. 1. Heterofermentative LAB may be applied for the production of side products such as polyols (mannitol, erythritol) and ethanol or acetic acid. This is only feasible if the markets for the side products are comparable to those of lactic acid and the production more than covers the added down-stream processing costs.

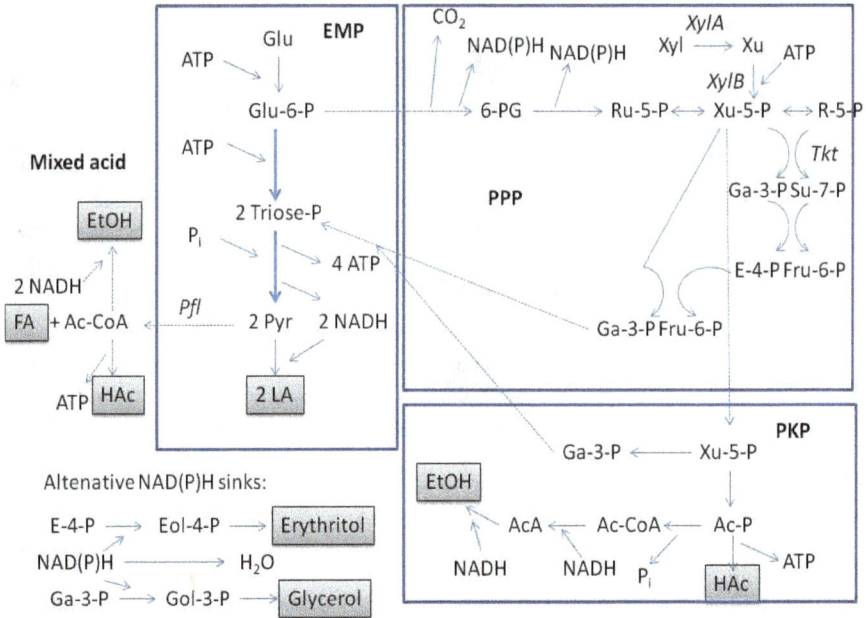

Figure 1. The main metabolic pathways in LAB. EMP: Embden-Meyerhof-Parnas pathway. PPP: pentose phosphate pathway. PKP: phosphoketolase pathway. Glu: glucose. LA: lactic acid. HAc: acetic acid. FA: formic acid. EtOH: ethanol. -P: energy-rich phosphate group. Pi: inorganic phosphate. Xyl: xylose. Xu: xylulose. 6-PG: 6-phosphogluconate. Ru: ribulose. R: ribose. Ga: glyceraldehyde. E: erythrose. Su: seduheptulose. Fru: fructose. Ac-CoA: acetyl-coenzymeA. Pyr: pyruvate. AcA: acetaldehyde. Ac-P: acetyl-phosphate. Pfl: pyruvate-formate lyase. XylA: xylose isomerase. XylB: xylulokinase. All carbohydrates are in D-form. Various metabolic end-products are presented with the dark background.

2. Future raw materials for production of lactic acid by LAB

The carboxylate platform is comprised of biological and chemical pathways that can be used in order to convert waste to bioproducts, such as lactic acid [17]. Lactic acid is a relatively cheap product, and one of the major challenges in its large-scale fermentative production is the cost of the raw material. This is the situation even in case of so called low-cost substrates [18]. Therefore, development of processes that utilize cheap raw materials at minimal costs have been under extensive studies. These substrates can be roughly classified as starch-based non-processed biomasses, lignocellulosic non-processed biomasses, and waste or side stream feedstocks. The former are nowadays generally considered as non-ideal feedstocks due to ethical reasons, and therefore they are not discussed in this review. Extensive reviews including starch-based feedstocks are available elsewhere [19]. With respect to future applications, the most likely raw materials for the lactic acid production are industrial side-streams and lignocellulosic biomasses. Recent advances in case of both raw material groups are discussed in the following.

As in other bioconversion processes, also in lactic acid production the focus of research has turned towards the use of lignocellulosic feedstocks. The major driving forces are fossil fuel deprivation and general paradigm change to bioeconomy, and the abundancy of lignocellulose materials. Generally, the effective utilization of lignocellulosic biomass for biochemical processes is limited due to seasonal availability, scattered distributions and high logistics cost [20]. The fermentation of lignocellulosic biomasses can also be hampered by inefficient pretreatment, high enzyme costs and end-product inhibition, formation of unwanted by-products under metabolism of pentoses, and carbon catabolite repression caused by the heterogeneous substrates. These challenges are further discussed in a recent review [8].

Paper industry residues and recycled paper products include various possible feedstocks for lactic acid production, which are together with agroindustrial residues discussed in a recent review [21]. Due to economical and ecological reasons, an intensive research interest is currently devoted to complex industrial by-products. In this field the advances presented during the past five years include the utilization of cellulosic biosludges from a Kraft pulp mill [22,23], and recycled paper sludge [24]. In both cases a nutrient supplement has increased the lactic acid productivity. LAB could be used for the bioconversion of hemicellulose fractions, e.g. from alfafa processing [25] to lactic acid. Direct conversion of xylan to lactic acid by LAB is already possible by use of genetically modified strains [26].

Food industry residues comprise a large variety of different biomasses and sludges that can be roughly categorized to agricultural wastes and food production wastes. Since the use of agricultural residues for lactic acid production is summarized in a recent review [21], it is not futher discussed here. Food production residues have been tested for bioconversion applications for ages, and the variety of used materials is large. Whey and other dairy industry residues are the prominent raw materials with respect to lactic acid production, reviewed in e.g. [27,28]. Whey retains about 55% of total milk nutrients, from which approximately 70% consists of lactose [29]. Availability of the lactose carbohydrate reservoir and the presence of other essential nutrients, such as proteins and phosphates, for the growth of microorganisms make whey and other cheese-making residues potent raw materials for the production of biochemicals.

Other quite often referred raw materials include brewery residues, especially spent grain [30], and winery wastes [31-35]. Additionally, there are various other proposed food industry residues that could fit to the lactic acid fermentation. The recently proposed include e.g. apple pomace [36], canned pineapple syrup [37], cashew apple juice [38], Jerusalem artichoke tubers [39], macaroni milk and rice-green pea-salad refectory wastes

[40], rice residues [41], sap from palmyra and oil palms [42], and spent coffee grounds [43]. Despite of the large variety of the raw materials, the main conclusions of these studies are that the optimization and control of pH and temperature is critical for the process, and that the supplementation of low-cost substrate with *e.g.* inorganic salts and yeast extract is necessary or at least improves the productivity remarkably. In a recent study the use of mixed cultures of *Lb. casei*, *Lb. helveticus*, and *S. thermophilus* was observed to reduce the demand of supplements compared to single strain cultures [44].

The required supplements and their concentrations depend on the low-cost substrate. Drawbacks of complex supplements are their cost and extensive down-stream processing required for the purification of lactic acid from fermentation broth, especially in applications requiring high purity. Therefore, the optimization of supplement concentration is essential. Although yeast extract is often considered superior to other supplements in terms of efficiency, its major drawback is the relatively high cost, and therefore substitutive supplements have been suggested. Equal productivities may be achieved via use of cheaper alternatives, such as inorganic phosphates [45], and microbial lysates [46-47]. It is notable that the use of lysates in combination to *e.g.* whey proteins could cause unwanted proteolytic activity. Other options for the increased productivity include *e.g.* the addition of manganese, which is a constituent of lactate hydrogenase [48], whey protein hydrolyzate [49], malt combing nuts [49], corn steep liquor [50], fish hydrolyzates and other fishery by-products [51-53], hydrolyzed spent cells [54] or red lentil flour [55]. It is likely that this is one of the future trends in lactic acid production, *i.e.* fermentation media are optimized from mixtures of different low-cost raw materials in order to avoid the use of expensive complex supplements.

The modern biorefineries are looking into oceans in order to find new abundant and less land- or water-using biomasses for the production of commodities. Among the plenty marine biomasses, brown seaweed and especially species *Laminaria japonica*, a common food in Japan, has been recognized as a potential raw material for the production of platform chemicals. *L. japonica* is interesting due to its high carbohydrate content and fast growth. Production of lactic acid from *L. japonica* hydrolyzates was reported in a recent study [56]. Another potential raw material for bioconversion is shrimp shell waste, which is produced in vast amounts as a by-product of food industry. It has been reported that the production of lactic acid can be combined to the recovery of biopolymer chitin, a precursor for largely applied chitosan [57-58]. Since the recovery of chitin is traditionally done via chemical processing, the integrated process offers both economical and ecological advantage. Similar to the previous examples of other food industry residues, also the marine food processing industry generates various different side streams, such as fish waste and shells that could perhaps be combined in biochemical production.

3. Novel LAB strains

Metabolic engineering in general is applied when *e.g.* $Y_{P/S}$, R_P, substrate flux through a desired pathway in growth phases or resting cells are aimed at. Metabolic engineering studies aiming at increased flux in glycolysis to lactic acid in LAB are fairly scarce. That may be explained by the fact that the metabolism of LAB is already tuned for efficient lactic acid production.

Some of these studies are listed in a review on metabolic engineering for lactic acid production [59]. The overexpression of L-LDH in *Lb. plantarum* can result 13-fold increase in LDH activity, and still show no effect on lactic acid production [60]. It has also been shown by overexpression that glyceraldehyde-3-P dehydrogenase (GAPDH) is not limiting the glycolytic flux either in growing or resting cells of a *L. lactis* strain [61]. Metabolic flux and control analysis (MFA and MCA) combined with the estimation of the kinetic parameters of the enzymes of a pathway are indeed needed in systematic and systemic approach to study and optimize also such seemingly simple - there is always growth and maintenance functions involved as well - metabolic pathway as that from glucose to lactic acid in LAB. An excellent view on topic is available in a review [62], which includes several references also for LAB (*e.g.* [63-66]).

More straightforward work on lactic acid production has been performed to achieve high enantiomeric purity by expressing and deleting respective genes for LDH. There are several examples of these as discussed in the recent review [59], such as construction of two different strains of *Lb. helveticus* for optically pure L-lactic acid production [67]. These strains differed from each other at the level of L-LDH activity (53 and 93 % higher than the wild type strain). Lactic acid production in a fermentation batch was equal to that of the wild type strain. However, at low pH when the growth and production are uncoupled, the strain with higher activity produced 20 % more lactic acid compared to construct with the lower activity.

Another straightforward target for the construction of genetically modified strains is widening of the raw materials for the production of lactic acid especially to lignocellulosic biomass-based materials. There are no reports on work to produce cellulolytic enzymes in LAB. Instead several groups have tried to produce xylanase in LAB [26]. This is however focused on heterofermentative LAB as they are naturally able to utilize pentoses and especially *Lb. brevis* as it has been shown to have endogenous beta-xylosidase activity [68]. Another approach is based on *L. lactis* IO-1 strain being able to metabolize xylose both via PKP and PPP [69]. PPP provides a homolactic fermentation route for pentoses. As the molecular biology tools or protocols for this strain were not available, another strain of *L. lactis* was used as the host. *XylRAB* genes from IO-1 strain were expressed in the host. *XylA* and *XylB* encode genes for xylose isomerase and xylulokinase, respectively. *XylR* is a putative transcriptional activator of the *XylAB* operon. In addition the gene for phosphoketolase was disrupted. Such a strain construct had homolactic fermentation for xylose. The rate of xylose fermentation was further improved by overexpressing the gene for transketolase, one of the enzymes in PPP. Almost theoretical $Y_{P/S}$ of lactic acid (1.58 vs. 1.67 mol/mol xylose) was achieved with lactic acid concentration of 50,1 g/l. Acetic acid concentration was as low as 0.3 g/l. Enantiomeric purity was very high (99.6 %). Similar approach has been applied for the production of D-lactic acid from xylose and L-arabinose [70.71].

Typical LAB fermentations are run at minimum pH of 5 – 5.5, which is much higher than the pKa-value of lactic acid (*i.e.* 3.8). Thus more than 90 % of the product exists as lactate. This is a major cost factor in the product recovery stage as well as the cause of high salt burden and/or gypsum formation. The tolerance to acid and low pH is difficult to explain at genetic level and thus hardly be affected by metabolic engineering methods on specific genes. A successful approach to engineer LAB strains for lower fermentation pH has been genome shuffling. *E.g.* populations from nitrosoguanidine (NTG) mutations and low pH acclimatization in chemostat cultivation have been used for the shuffling [72]. The resultant population grew at pH 3.8 and lowered pH by lactic acid formation down to 3.5. This is a promising result even though the population was not used with realistic sugar concentrations. Similar approach has been reported aiming at improving acid tolerance as well as R_P and glucose tolerance, respectively, with *Lb. rhamnosus* [73·74]. NTG and UV irradiation were used for mutagenesis and lethal mutants were fused from protoplasts. The best strain of [73] lowered pH down to 3.25 and increased average R_P by 60 % compared to the wild type strain. However, average R_P was still moderately low (ca. 1 g/lh). Final lactic acid concentration and $Y_{P/S}$ from glucose were 84 g/l and 82 %, respectively. In [74] higher $Y_{P/S}$ (> 95 %) and R_P (ca. 3.6 g/lh) were reached with the best strains on industrially relevant fermentation medium with 150 g/l glucose. The $Y_{P/S}$ from 200 g/l glucose was still 90 %, but the average R_P decreased to 2 g/lh. In a recent study *Lb. casei* mutants induced by NTG were screened in high glucose concentration (360 g/l) [75]. A mutant strain with highest osmotic tolerance produced 198.2 g/l lactic acid from 210 g/l glucose with increased R_P (5.5 g/lh).

4. Novel process technologies

From fermentor design point of view lactic acid production by LAB is quite simple and conventional as the process requires no gassing, gas exchange or gas mass transfer. When the production strain and fermentation conditions are optimized for lactic acid production there is no or little formation of side products (metabolites, cellular mass, exopolysaccharides). Thus *e.g.* the rheology of the fermentation broth is Newtonian and very close to that of water. Power consumption is mainly for the sake of homogeneity and reduction of gradients of pH-controlling agents. The biggest challenges for process technology are to minimize osmotic effects by substrates and the product, to reach high R_P and to minimize the costs and waste formation in the product recovery stage.

Typical fermentation approaches other than simple batch include repeated batch and fed-batch fermentation and continuous fermentation with cell-recycle as solutions with free cells and the use of immobilized cells in different reactor types (fixed or fluidized bed). A novel fed-batch strategy was developed recently by combining pH-control and substrate feeding [76]. The rationale behind the strategy was the linear relationship between the consumption amounts of alkali and that of substrate. Thus these two components were mixed together in the feeding liquid. This resulted in higher efficiency compared to batch fermentation, but the efficiency parameters were not especially high if compared with data from several other reports. By far the most studied method to increase the R_P and/or separate cell growth from product formation is based on the immobilization of the cells. These have also been reviewed [27]. Several immobilization methods have been applied including entrapment within gels such as alginate [77·78], modified alginate [79·80], or pectate [81], adsorption on granulated DEAE-cellulose [82] or porous glass [83], and biofilm formation on solid supports [84-85]. Solid incompressible supports and carrier materials such as granulated cellulose and porous glass may be applied in any scale and in various reactor designs while gels as compressible materials suit less well for larger scale especially in fixed-bed column reactors.

Immobilized cells may be utilized in various fermentation modes and reactor designs such as repeated batch or fed-batch, continuous fermentation with cell retention or recycle, in continuous stirred tank reactors (CSTR), fixed-bed or fluidized-bed reactors. High R_P (19-22 g/lh) have been achieved in a two-stage process with immobilized cells [86]. A special arrangement consisting of a CSTR for pH-control and substrate feeding and a fixed-bed reactor with immobilized cells was used in a concept with intermittent refreshing of the cells in a patent [87]. Short residence time within the column was possible because of the incompressible nature of the carrier material. Chemically pure product was achieved by using a production medium with few nutrients. Once the productivity decreased below a threshold value based on the consumption of alkali the cells were refreshed with nutrients. Incompressible carriers for cell adsorption have obvious advantages. However, new solutions to secure cell adherence on the carrier are required. This would facilitate efficient use of fluidized-bed reactors with minimal pressure losses in the reactor. Biological means for cell adherence may be one solution which could offer a further advantage to selectively keep the productive cells in the reactor.

Another approach to increase R_P is high cell-density fermentation with free cells recycled by membrane separation technique. This has been in use in industrial scale for lactic acid production already in 1980's in Finland. Several academic reports on this approach have been since published demonstrating very high R_P of 26 g/lh [88], 31.5 g/lh [89] and up to 57 g/lh [90].

It should be kept in mind that R_P is affected by the concentration of lactic acid. Thus not all published figures are comparable. Product inhibition may be diminished by in-situ recovery of the product. Electrodialysis [91-92], nanofiltration [93] and ion-exchange [94-95] have thus been coupled with the fermentation system.

Conventional lactic acid recovery from fermentation broth consists of cell and other solids separation, lactic acid precipitation as calcium lactate and precipitate recovery, acidification of the precipitate by sulfuric acid and the separation of the gypsum precipitate formed. The amount of gypsum is usually higher than the amount of lactic acid produced. Lately NatureWorks has reported to have reduced the formation of gypsum significantly. Probably this has been achieved by performing the fermentation at lower pH *e.g.* by using metabolically engineered yeast for the production of lactic acid. The amount of gypsum can be avoided by using electrodialysis for the acidification and separation of the acid and alkali formed with bipolar membranes [96]. The alkali formed may be recycled back to the fermentation. Electrodialysis has been considered too expensive technology for lactic acid recovery [97]. However, specific energy consumption of only 0.25 kWh/ kg lactic acid is presented [96]. Nanofiltration has been used as a pretreatment method to remove Mg- and Ca- and sulfate-ions and color before electrodialysis increasing significantly the capacity in electrodialysis [98]. Alternative techniques for lactic acid recovery are extraction [99] and use of ion-exchange [100-101], neither of which is a proper solution to the salt burden.

5. Conclusions

Lactic acid production in LAB has both cell mass and growth dependent portions. Typically LAB require several nutrient components for their growth increasing the fermentation and down-stream processing costs. Down-stream processing is especially important in the production of lactic acid for PLA. As R_P is the a major investment factor affecting costs, the minimization of medium and product purification costs should be accompanied by methods increasing cell mass concentration without excess growth. For this several different strategies have been applied so far mainly in academia (cell immobilization, cell-recycling and cell-retention). As history shows some of these could be applicable in industrial production as well, however pilot and demonstration plant studies and some risk-taking are required.

The main C/energy source spectrum available for LAB has been widened significantly. Reports of new possible substrates are frequently published, and the utilization of industrial side streams is a growing trend. Into this direction major successes have also been achieved with metabolic engineering providing strains for efficient production of lactic acid from pentoses as well, which is to promote sustainable use of renewables.

In an ideal fermentation process product inhibition should be minimized so that high R_P would be achieved even at high lactic acid concentrations resulting in feasible average productivities. For this purpose both acclimatization and mutagenesis has been applied successfully. However, it has to be considered how far can we go in respect to fermentation pH and lactic acid concentration. There are already remarkable alternatives to LAB with naturally better properties in this sense. Some success has been achieved with in-situ product recovery, but also these procedures lack experiences in any larger scale.

Conventional lactic acid production process with LAB is accompanied with the formation of large amounts of gypsum in the product recovery stage. Fermentation at lower pH diminishes this amount, but does not prevent its formation. Electrodialysis has been considered too expensive technique for the recovery of such cheap, bulk products as lactic acid. However, recent reports claim promising results with this technology. Forecasted figures for lactic acid market show up to one million tons per year. The growth would come mainly from the growth of PLA as a biodegradable polymer based on renewable raw materials. Economies of scale should decrease the production costs, but new technical approaches are also needed to reach these figures.

Author details

Sanna Taskila* and Heikki Ojamo
*University of Oulu, Faculty of Technology, Department of Process and
Environmental Engineering, Bioprocess Engineering Laboratory, Oulu, Finland*

---

*Corresponding Author

Abbreviations

η % - Efficiency, i.e. the ratio of YP/S to the maximum theoretical value
D-LDH - D-lactate dehydrogenase
LAB – lactic acid bacteria
L-LDH - L-lactate dehydrogenase
NTG – Nitrosoguanidine
R_P - Volumetric productivity g/l*h
SSF – Simultaneous saccharification and fermentation
PLA – poly lactic acid
PPP - Pentose phosphate pathway
PKP - Phosphoketolase pathway
$Y_{P/S}$ – Yield of lactic acid per substrate consumed g/g
$Y_{P/X}$ – Yield of lactic acid per cell mass g/g

6. References

[1] Wee YJ, Kim JN, Ryu HW. Biotechnological production of lactic acid and its recent applications. Food Technology and Biotechnology 2006;44 163-172.

[2] Datta R, Tsai SP, Bonsignore P, Moon SH, Frank JR. Technological and Economic-Potential of Poly(Lactic Acid) and Lactic-Acid Derivatives. Fems Microbiology Reviews 1995;16 221-231.

[3] Fukushima K, Sogo K, Miura S, Kimura Y. Production of D-lactic acid by bacterial fermentation of rice starch. Macromolecular Bioscience 2004;4 1021-1027.

[4] Jem JK, van der Pol JF, de Vos S. Microbial lactic acid, its polymer poly(lactic acid), and their industrial applications. Microbiology Monographs 2010;14 323-346.

[5] Vink ETH, Davies S, Kolstad JJ. The eco-profile for current Ingeo® polylactide production. Industrial Biotechnology 2010;6 212-224.

[6] Shah AA, Hasan F, Hameed A, Ahmed S. Biological degradation of plastics: A comprehensive review. Biotechnology Advances 2008;26 246-265.

[7] Nampoothiri KM, Nair NR, John RP. An overview of the recent developments in polylactide (PLA) research. Bioresource Technology 2010;101 8493-8501.

[8] Abdel-Rahman MA, Tashiro Y, Sonomoto K. Lactic acid production from lignocellulose-derived sugars using lactic acid bacteria: Overview and limits. Journal of Biotechnology 2011;156 286-301.

[9] Carr FJ, Chill D, Maida N. The lactic acid bacteria: A literature survey. Critical Reviews in Microbiology 2002;28 281-370.

[10] Murphy MG, Condon S. Correlation of Oxygen Utilization and Hydrogen-Peroxide Accumulation with Oxygen Induced Enzymes in Lactobacillus-Plantarum Cultures. Archives of Microbiology 1984;138 44-48.

[11] Sedewitz B, Schleifer KH, Gotz F. Physiological role of pyruvate oxidase in the aerobic metabolism of *Lactobacillus plantarum*. Journal of Bacteriology 1984;160 462-465.

[12] Zaunmuller T, Eichert M, Richter H, Unden G. Variations in the energy metabolism of biotechnologically relevant heterofermentative lactic acid bacteria during growth on sugars and organic acids. Applied Microbiology and Biotechnology 2006;72 421-429.

[13] Melchiorsen CR, Jokumsen KV, Villadsen J, Israelsen H, Arnau J. The level of pyruvate-formate lyase controls the shift from homolactic to mixed-acid product formation in *Lactococcus lactis*. Applied Microbiology and Biotechnology 2002;58 338-344.

[14] Ou MS, Ingram LO, Shanmugam KT. l(+)-Lactic acid production from non-food carbohydrates by thermotolerant *Bacillus coagulans*. Journal of Industrial Microbiology & Biotechnology 2011;38 599-605.

[15] Porro D, Brambilla L, Ranzi BM, Martegani E, Alberghina L. Development of Metabolically Engineered Saccharomyces-Cerevisiae Cells for the Production of Lactic-Acid. Biotechnology Progress 1995;11 294-298.

[16] Ikushima S, Fujii T, Kobayashi O, Yoshida S, Yoshida A. Genetic Engineering of Candida utilis Yeast for Efficient Production of L-Lactic Acid. Bioscience Biotechnology and Biochemistry 2009;73 1818-1824.

[17] Agler MT, Wrenn BA, Zinder SH, Angenent LT. Waste to bioproduct conversion with undefined mixed cultures: the carboxylate platform. Trends in Biotechnology 2011;29 70-78.

[18] Yadav AK, Chaudhari AB, Kothari RM. Bioconversion of renewable resources into lactic acid: an industrial view. Critical Reviews in Biotechnology 2011;31 1-19.

[19] John RP, Nampoothiri KM, Pandey A. Fermentative production of lactic acid from biomass: an overview on process developments and future perspectives. Applied Microbiology and Biotechnology 2007;74 524-534.

[20] Lin Y, Tanaka S. Ethanol fermentation from biomass resources: current state and prospects. Applied Microbiology and Biotechnology 2006;69 627-642.

[21] Alonso JL, Dominguez H, Garrote G, Gonzalez-Munoz MJ, Gullon B, Moure A, Santos V, Vila C, Yanez R. Biorefinery processes for the integral valorization of agroindustrial and forestal wastes. Cyta-Journal of Food 2011;9 282-289.

[22] Romani A, Yanez R, Garrote G, Alonso JL. SSF production of lactic acid from cellulosic biosludges. Bioresource Technology 2008;99 4247-4254.

[23] Romani A, Yanez R, Garrote G, Alonso JL, Parajo JC. Sugar production from cellulosic biosludges generated in a water treatment plant of a Kraft pulp mill. Biochemical Engineering Journal 2007;37 319-327.

[24] Marques S, Santos JAL, Girio FM, Roseiro JC. Lactic acid production from recycled paper sludge by simultaneous saccharification and fermentation. Biochemical Engineering Journal 2008;41 210-216.

[25] Sreenath HK, Moldes AB, Koegel RG, Straub RJ. Lactic acid production by simultaneous saccharification and fermentation of alfalfa fiber. Journal of Bioscience and Bioengineering 2001;92 518-523.

[26] Hu CY, Chi DJ, Chen SS, Chen YC. The direct conversion of xylan to lactic acid by *Lactobacillus brevis* transformed with a xylanase gene. Green Chemistry 2011;13 1729-1734.

[27] Kosseva MR, Panesar PS, Kaur G, Kennedy JF. Use of immobilised biocatalysts in the processing of cheese whey. International Journal of Biological Macromolecules 2009;45 437-447.

[28] Panesar PS, Kennedy JF, Gandhi DN, Bunko K. Bioutilisation of whey for lactic acid production. Food Chemistry 2007;105 1-14.

[29] Jelen P, Whey processing. In: Encyclopedia of dairy sciences. H.Roginski, J.W.Fuquay, P.F.Fox, (Eds.), Academic Press: London, 2003, pp. 2739-2751.

[30] Aliyu S, Bala M. Brewer's spent grain: A review of its potentials and applications. African Journal of Biotechnology 2011;10 324-331.

[31] Devesa-Rey R, Vecino X, Varela-Alende JL, Barral MT, Cruz JM, Moldes AB. Valorization of winery waste vs. the costs of not recycling. Waste Management 2011;31 2327-2335.

[32] Bustos G, Moldes AB, Cruz JM, Dominguez JM. Production of lactic acid from vine-trimming wastes and viticulture lees using a simultaneous saccharification fermentation method. Journal of the Science of Food and Agriculture 2005;85 466-472.

[33] Bustos G, Moldes AB, Cruz JM, Dominguez JM. Production of fermentable media from vine-trimming wastes and bioconversion into lactic acid by *Lactobacillus pentosus*. Journal of the Science of Food and Agriculture 2004;84 2105-2112.

[34] Bustos G, Moldes AB, Cruz JM, Dominguez JM. Formulation of low-cost fermentative media for lactic acid production with *Lactobacillus rhamnosus* using vinification lees as nutrients. Journal of Agricultural and Food Chemistry 2004;52 801-808.

[35] Bustos G, de la Torre N, Moldes AB, Cruz JM, Dominguez JM. Revalorization of hemicellulosic trimming vine shoots hydrolyzates trough continuous production of lactic acid and biosurfactants by L-pentosus. Journal of Food Engineering 2007;78 405-412.

[36] Gullon B, Yanez R, Alonso JL, Parajo JC. L-lactic acid. production from apple pomace by sequential hydrolysis and fermentation. Bioresource Technology 2008;99 308-319.

[37] Nakanishi K, Ueno T, Sato S, Yoshi S. L-Lactic Acid Production from Canned Pineapple Syrup by Rapid Sucrose Catabolizing *Lactobacillus paracasei* NRIC 0765. Food Science and Technology Research 2010;16 239-246.

[38] Honorato TL, Rabelo MC, Goncalves LRB, Pinto GAS, Rodrigues S. Fermentation of cashew apple juice to produce high added value products. World Journal of Microbiology & Biotechnology 2007;23 1409-1415.

[39] Ge XY, Qian H, Zhang WG. Enhancement of L-Lactic Acid Production in *Lactobacillus casei* from Jerusalem Artichoke Tubers by Kinetic Optimization and Citrate Metabolism. Journal of Microbiology and Biotechnology 2010;20 101-109.

[40] Omay D, Guvenilir Y. Lactic Acid Fermentation from Refectory Waste. Ekoloji 2011;20 42-50.

[41] Lu ZD, Lu MB, He F, Yu LJ. An economical approach for D-lactic acid production utilizing unpolished rice from aging paddy as major nutrient source. Bioresource Technology 2009;100 2026-2031.

[42] Chooklin S, Kaewsichan L, Kaewsichan L. Potential use of *Lactobacillus casei* TISTR 1500 for the bioconversion from palmyra sap and oil palm sap to lactic acid. Electronic Journal of Biotechnology 2011;14 .

[43] Mussatto SI, Machado EMS, Martins S, Teixeira JA. Production, Composition, and Application of Coffee and Its Industrial Residues. Food and Bioprocess Technology 2011;4 661-672.

[44] Secchi N, Giunta D, Pretti L, Garcia MR, Roggio T, Mannazzu I, Catzeddu P. Bioconversion of ovine scotta into lactic acid with pure and mixed cultures of lactic acid bacteria. Journal of Industrial Microbiology & Biotechnology 2012;39 175-181.

[45] Amrane A. Effect of inorganic phosphate on lactate production by Lactobacillus *helveticus* grown on supplemented whey permeate. Journal of Chemical Technology and Biotechnology 2000;75 223-228.

[46] Amrane A. Evaluation of lactic acid bacteria autolysate for the supplementation of lactic acid bacteria fermentation. World Journal of Microbiology & Biotechnology 2000;16 207-209.

[47] Coelho LF, de Lima CJB, Bernardo MP, Contiero J. d(-)-Lactic Acid Production by *Leuconostoc mesenteroides* B512 Using Different Carbon and Nitrogen Sources. Applied Biochemistry and Biotechnology 2011;164 1160-1171.

[48] Fitzpatrick JJ, Ahrens M, Smith S. Effect of manganese on Lactobacillus casei fermentation to produce lactic acid from whey permeate. Process Biochemistry 2001;36 671-675.

[49] Fitzpatrick JJ, O'Keeffe U. Influence of whey protein hydrolysate addition to whey permeate batch fermentations for producing lactic acid. Process Biochemistry 2001;37 183-186.

[50] Kim HO, Wee YJ, Kim JN, Yun JS, Ryu HW. Production of lactic acid from cheese whey by batch and repeated batch cultures of *Lactobacillus* sp RKY2. Applied Biochemistry and Biotechnology 2006;131 694-704.

[51] Gao MT, Hirata M, Toorisaka E, Hano T. Acid-hydrolysis of fish wastes for lactic acid fermentation. Bioresource Technology 2006;97 2414-2420.

[52] Beaulieu L, Desbiens M, Thibodeau J, Thibault S. Pelagic fish hydrolysates as peptones for bacterial culture media. Canadian Journal of Microbiology 2009;55 1240-1249.

[53] Vazquez JA, Montemayor MI, Fraguas J, Murado MA. High production of hyaluronic and lactic acids by *Streptococcus zooepidemicus* in fed-batch culture using commercial and marine peptones from fishing by-products. Biochemical Engineering Journal 2009;44 125-130.

[54] Gao MT, Hirata M, Toorisaka E, Hano T. Study on acid-hydrolysis of spent cells for lactic acid fermentation. Biochemical Engineering Journal 2006;28 87-91.

[55] Altaf M, Naveena BJ, Reddy G. Use of inexpensive nitrogen sources and starch for L(+) lactic acid production in anaerobic submerged fermentation. Bioresource Technology 2007;98 498-503.

[56] Jang S, Shirai Y, Uchida M, Wakisaka M. Production of L(+)-Lactic Acid from Mixed Acid and Alkali Hydrolysate of Brown Seaweed. Food Science and Technology Research 2011;17 155-160.

[57] Adour L, Arbia W, Amrane A, Mameri N. Combined use of waste materials - recovery of chitin from shrimp shells by lactic acid fermentation supplemented with date juice waste or glucose. Journal of Chemical Technology and Biotechnology 2008;83 1664-1669.

[58] Healy M, Green A, Healy A. Bioprocessing of marine crustacean shell waste. Acta Biotechnologica 2003;23 151-160.

[59] Singh SK, Ahmed SU, Pandey A. Metabolic engineering approaches for lactic acid production. Process Biochemistry 2006;41 991-1000.

[60] Ferain T, Garmyn D, Bernard N, Hols P, Delcour J. Lactobacillus-Plantarum Ldhl Gene - Overexpression and Deletion. Journal of Bacteriology 1994;176 596-601.

[61] Solem C, Koebmann BJ, Jensen PR. Glyceraldehyde-3-phosphate dehydrogenase has no control over glycolytic flux in *Lactococcus lactis* MG1363. Journal of Bacteriology 2003;185 1564-1571.

[62] Teusink B, Smid EJ. Modelling strategies for the industrial exploitation of lactic acid bacteria. Nature Reviews Microbiology 2006;4 46-56.

[63] Bai DM, Zhao XM, Li XG, Xu SM. Strain improvement and metabolic flux analysis in the wild-type and a mutant *Lactobacillus lactis* strain for L(+)-lactic acid production. Biotechnology and Bioengineering 2004;88 681-689.

[64] Neves AR, Ramos A, Nunes MC, Kleerebezem M, Hugenholtz J, de Vos WM, Almeida J, Santos H. In vivo nuclear magnetic resonance studies of glycolytic kinetics in *Lactococcus lactis*. Biotechnology and Bioengineering 1999;64 200-212.

[65] Even S, Lindley ND, Cocaign-Bousquet M. Transcriptional, translational and metabolic regulation of glycolysis in *Lactococcus lactis* subsp. cremoris MG 1363 grown in continuous acidic cultures. Microbiology-Sgm 2003;149 1935-1944.

[66] Cocaign-Bousquet M, Even S, Lindley ND, Loubiere P. Anaerobic sugar catabolism in *Lactococcus lactis*: genetic regulation and enzyme control over pathway flux. Applied Microbiology and Biotechnology 2002;60 24-32.

[67] Kylä-Nikkilä K, Hujanen M, Leisola M, Palva A. Metabolic engineering of Lactobacillus helveticus CNRZ32 for production of pure L-(+)-lactic acid. Applied and Environmental Microbiology 2000;66 3835-3841.

[68] Garde A, Jonsson G, Schmidt AS, Ahring BK. Lactic acid production from wheat straw hemicellulose hydrolysate by *Lactobacillus pentosus* and *Lactobacillus brevis*. Bioresource Technology 2002;81 217-223.

[69] Shinkawa S, Okano K, Yoshida S, Tanaka T, Ogino C, Fukuda H, Kondo A. Improved homo l-lactic acid fermentation from xylose by abolishment of the phosphoketolase pathway and enhancement of the pentose phosphate pathway in genetically modified xylose-assimilating *Lactococcus lactis*. Applied Microbiology and Biotechnology 2011;91 1537-1544.

[70] Okano K, Yoshida S, Yamada R, Tanaka T, Ogino C, Fukuda H, Kondo A. Improved Production of Homo-D-Lactic Acid via Xylose Fermentation by Introduction of Xylose Assimilation Genes and Redirection of the Phosphoketolase Pathway to the Pentose Phosphate Pathway in L-Lactate Dehydrogenase Gene-Deficient *Lactobacillus plantarum*. Applied and Environmental Microbiology 2009;75 7858-7861.

[71] Okano K, Yoshida S, Tanaka T, Ogino C, Fukuda H, Kondo A. Homo-D-Lactic Acid Fermentation from Arabinose by Redirection of the Phosphoketolase Pathway to the Pentose Phosphate Pathway in L-Lactate Dehydrogenase Gene-Deficient *Lactobacillus plantarum*. Applied and Environmental Microbiology 2009;75 5175-5178.

[72] Patnaik R, Louie S, Gavrilovic V, Perry K, Stemmer WPC, Ryan CM, del Cardayre S. Genome shuffling of *Lactobacillus* for improved acid tolerance. Nature Biotechnology 2002;20 707-712.

[73] Wang YH, Li Y, Pei XL, Yu L, Feng Y. Genome-shuffling improved acid tolerance and L-lactic acid volumetric productivity in *Lactobacillus rhamnosus*. Journal of Biotechnology 2007;129 510-515.

[74] Yu L, Pei X, Lei T, Wang Y, Feng Y. Genome shuffling enhanced L-lactic acid production by improving glucose tolerance of *Lactobacillus rhamnosus*. Journal of Biotechnology 2008;134 154-159.

[75] Ge XY, Yuan JA, Qin H, Zhang WG. Improvement of L-lactic acid production by osmotic-tolerant mutant of *Lactobacillus casei* at high temperature. Applied Microbiology and Biotechnology 2011;89 73-78.

[76] Zhang Y, Cong W, Shi SY. Application of a pH Feedback-Controlled Substrate Feeding Method in Lactic Acid Production. Applied Biochemistry and Biotechnology 2010;162 2149-2156.

[77] Idris A, Suzana W. Effect of sodium alginate concentration, bead diameter, initial pH and temperature on lactic acid production from pineapple waste using immobilized *Lactobacillus delbrueckii*. Process Biochemistry 2006;41 1117-1123.

[78] Shen XL, Xia LM. Lactic acid production from cellulosic waste by immobilized cells of *Lactobacillus delbrueckii*. World Journal of Microbiology & Biotechnology 2006;22 1109-1114.

[79] Rao CS, Prakasham RS, Rao AB, Yadav JS. Production of L(+) lactic acid by *Lactobacillus delbrueckii* immobilized in functionalized alginate matrices. World Journal of Microbiology & Biotechnology 2008;24 1411-1415.

[80] Rao CS, Prakasham RS, Rao AB, Yadav JS. Functionalized alginate as immobilization matrix in enantioselective L (+) lactic acid production by *Lactobacillus delbrucekii*. Applied Biochemistry and Biotechnology 2008;149 219-228.

[81] Panesar PS, Kennedy JF, Knill CJ, Kosseva MR. Applicability of pectate-entrapped *Lactobacillus casei* cells for L(+) lactic acid production from whey. Applied Microbiology and Biotechnology 2007;74 35-42.

[82] Lommi, H., Swinkels, W., Viljava, T., and Hammond, R. Bioreactor with immobilized lactic acid bacteria and the use thereof. WO patent 94/12614. 1994.

[83] Senthuran A, Senthuran V, Mattiasson B, Kaul R. Lactic acid fermentation in a recycle batch reactor using immobilized *Lactobacillus casei*. Biotechnology and Bioengineering 1997;55 841-853.

[84] Dagher SF, Ragout AL, Sineriz F, Bruno-Barcena JM. Cell Immobilization for Production of Lactic Acid: Biofilms Do It Naturally. Advances in Applied Microbiology, Vol 71 2010;71 113-148.

[85] Qureshi N, Annous BA, Ezeji TC, Karcher P, Maddox IS. Biofilm reactors for industrial bioconversion processes: employing potential of enhanced reaction rates. Microbial Cell Factories 2005;4 .

[86] Schepers AW, Thibault J, Lacroix C. Continuous lactic acid production in whey permeate/yeast extract medium with immobilized *Lactobacillus helveticus* in a two-stage process: Model and experiments. Enzyme and Microbial Technology 2006;38 324-337.

[87] Viljava, T. and Koivikko, H. Method for preparing pure lactic acid. US patent 5,932,455. 1999.

[88] Richter K, Nottelmann S. An empiric steady state model of lactate production in continuous fermentation with total cell retention. Engineering in Life Sciences 2004;4 426-432.

[89] Xu GQ, Chu J, Wang YH, Zhuang YP, Zhang SL, Peng HQ. Development of a continuous cell-recycle fermentation system for production of lactic acid by *Lactobacillus paracasei*. Process Biochemistry 2006;41 2458-2463.

[90] Kwon S, Yoo IK, Lee WG, Chang HN, Chang YK. High-rate continuous production of lactic acid by *Lactobacillus rhamnosus* in a two-stage membrane cell-recycle bioreactor. Biotechnology and Bioengineering 2001;73 25-34.

[91] Gao MT, Hirata M, Koide M, Takanashi H, Hano T. Production of L-lactic acid by electrodialysis fermentation (EDF). Process Biochemistry 2004;39 1903-1907.

[92] Min-tian G, Koide M, Gotou R, Takanashi H, Hirata M, Hano T. Development of a continuous electrodialysis fermentation system for production of lactic acid by *Lactobacillus rhamnosus*. Process Biochemistry 2005;40 1033-1036.

[93] Jeantet R, Maubois JL, Boyaval P. Semicontinuous production of lactic acid in a bioreactor coupled with nanofiltration membranes. Enzyme and Microbial Technology 1996;19 614-619.

[94] Monteagudo JM, Aldavero M. Production of L-lactic acid by *Lactobacillus delbrueckii* in chemostat culture using an ion exchange resins system. Journal of Chemical Technology and Biotechnology 1999;74 627-634.

[95] Senthuran A, Senthuran V, Hatti-Kaul R, Mattiasson B. Lactate production in an integrated process configuration: reducing cell adsorption by shielding of adsorbent. Applied Microbiology and Biotechnology 2004;65 658-663.

[96] Wee YJ, Yun JS, Lee YY, Zeng AP, Ryu HW. Recovery of lactic acid by repeated batch electrodialysis and lactic acid production using electrodialysis wastewater. Journal of Bioscience and Bioengineering 2005;99 104-108.

[97] Akerberg C, Zacchi G. An economic evaluation of the fermentative production of lactic acid from wheat flour. Bioresource Technology 2000;75 119-126.

[98] Bouchoux A, Roux-de Balmann H, Lutin F. Investigation of nanofiltration as a purification step for lactic acid production processes based on conventional and bipolar electrodialysis operations. Separation and Purification Technology 2006;52 266-273.

[99] Yankov D, Molinier J, Kyuchoukov G, Albet J, Malmary G. Improvement of the lactic acid extraction. Extraction from aqueous solutions and simulated fermentation broth by means of mixed extractant and TOA, partially loaded with HCl. Chemical and Biochemical Engineering Quarterly 2005;19 17-24.

[100] Gullon B, Alonso JL, Parajo JC. Ion-Exchange Processing of Fermentation Media Containing Lactic Acid and Oligomeric Saccharides. Industrial & Engineering Chemistry Research 2010;49 3741-3750.

[101] Moldes AB, Alonso JL, Parajo JC. Recovery of lactic acid from simultaneous saccharification and fermentation media using anion exchange resins. Bioprocess and Biosystems Engineering 2003;25 357-363.

Chapter 10

Application of Amylolytic *Lactobacillus fermentum* 04BBA19 in Fermentation for Simultaneous Production of Thermostable α-Amylase and Lactic Acid

Bertrand Tatsinkou Fossi and Frédéric Tavea

Additional information is available at the end of the chapter

1. Introduction

Lactic acid bacteria (LAB) have diverse applications for both animals and humans. Food, pharmaceutical and chemical industries rely on these microorganisms to produce fermented beverage, foods and other important compounds of industrial interests. In recent years the industrial relevance of lactic acid bacteria is on an increasing trend because of the application of lactic acid as chemical for the production of biodegradable plastics [1]. Typical LAB are Gram-positive, non-sporing, catalase-negative, devoid of cytochromes, anaerobic but aerotolerant cocci or rods that are acid-tolerant and produce lactic acid as the major end product during sugar fermentation [2]. Although most LAB are unable to degrade starch because of the lack of the amylolytic activity, a few exhibit this activity and are qualified as amylolytic lactic acid bacteria (ALAB) which are able to decompose starchy material through the amylases production during the fermentation processes [3]. Regarding the importance and availability of starchy biomass in the world, amylases and lactic acid production from starch appear as two potential industrial applications of ALAB. Amylases play important role in degradation of starch and are produced in bulk from microorganisms and represent about 25 to 33% of the world enzyme market [4]. The spectrum of amylases application has widened in many fields, such as clinical, medical and analytical chemistry as well as in the textile, food, fermentation, paper, distillery and brewing industries [4]. The advantages of using thermostable amylases in industrial processes include the decreased risk of contamination, cost of external cooling and increased diffusion rate [4]. Several thermostable α-amylases have been purified from *Bacillus* sp. and the factors influencing their thermostability have been investigated [5]. However, no study has yet dealt with thermostable amylase from lactic acid bacteria (LAB). The use of thermostable amylases from *Lactobacillus* is of advantage as they are generally non-pathogenic. On the other hand, the major end product of LAB fermentation, lactate, has applications as a preservative, acidulant and flavouring agent in the food industry, because of the tartness provided by lactate and also because lactate is generally regarded as safe (GRAS) [6].

Thus a thermostable amylase producing lactic acid bacterium would be a potential candidate for food industries and especially for the making of high density gruel from starchy raw material as corn or wheat [7]. This would require a good knowledge of the conditions required to optimally produce amylase and lactic acid of good quality. The present study deals with the co-production of thermostable α-amylase and lactic acid from a LAB, *Lactobacillus fermentum* 04BBA19, isolated from a starchy waste of a soil sample from the western region of Cameroon.

2. Background and significance

2.1. Amylolytic lactic acid bacteria

Amylolytic lactic acid bacteria (ALAB) have been reported from different tropical starchy fermented foods, made especially from roots as cassava and sweet potato or grains as maize sorghum and rice. Strains of *Lactobacillus plantarum* have been isolated from African cassava-based fermented products [8], *L. plantarum* A6 (LMG 18053) have been isolated from retted cassava in the Congo [9] and *Lactobacillus manihotivorans* OND32 have been isolated from cassava sour starch fermentations in Colombia [10]. Amylolytic strains of *Lactobacillus fermentum* were isolated for the first time from Benin maize sourdough (ogi and mawè) by Agati et al. [11]. Sanni et al. [12] described amylolytic strains of *L. plantarum* and *L. fermentum* strains in various Nigerian traditional amylaceous fermented foods. ALAB are generally screened in fermented amylaceous foods. Owing to their relatively high starch content, starchy biomass appears as an important eco-niche for the screening and isolation of ALAB, which can be industrially applied to convert starch into mono- and disaccharides for lactic acid fermentation. The composition of the microbiota and in particular the occurrence of ALAB is determined by the way the raw material is processed [13]. Most ALAB isolated belong to the *Lactobacillus* genus, however few studies reported the existence of amylolytic activity in some strains of *Bifidobacterium* isolated from the human large intestinal tract [14, 15]. The distribution of amylolytic microorganisms in the human large intestinal tract has been investigated in various individuals of different ages using anaerobic cultures techniques. So far, twenty one amylolytic bifidobacteria have been isolated from adult faeces and tested for rice fermentation [16].

Owing to the ability of their α-amylases to partially hydrolyze raw starch, ALAB can ferment different types of amylaceous raw material, such as corn [17], potato [18], or cassava [19] and different starchy substrates [20, 21, 8]. Amylolytic LAB utilize starchy biomass and convert it into lactic acid in a single step fermentation. ALAB are mainly used in food fermentation, they are involved in cereal based fermented foods such as European sour rye bread, Asian salt bread, sour porridges, dumplings and non-alcoholic beverage production.

Few of them are used for production of lactic acid in single step fermentation of starch [1].
The common method to produce lactic acid from starchy biomass involves the pretreatment
for gelatinisation and hydrolysis (liquefaction and saccharification). The liquefaction of the
starch is carried out at high temperatures of 90–130 °C for 15 min followed by enzymatic
saccharification to glucose and subsequent conversion of glucose to lactic acid by
fermentation [22, 1]. This two-step process involving consecutive enzymatic hydrolysis and
fermentation makes it economically unattractive. The bioconversion of carbohydrate
materials to lactic acid can be made much more effective by coupling the enzymatic
hydrolysis of carbohydrate substrates and microbial fermentation of the derived glucose
into a single step. This has been successfully employed for lactic acid production from raw
starch materials with many representative bacteria including *Lactobacillus* and *Lactococcus*
species [23, 20, 24, 21].

Because at industrial scale, the use of glucose addition is an expensive alternative, there is
interest in the use of a cheaper source of carbon, such as starch, the most abundantly
available raw material on earth next to cellulose. This, in combination with amylolytic lactic
acid bacteria may help to decrease the cost of the overall fermentation process. Amylolytic
lactic acid bacteria can convert the starch directly into lactic acid. Development of
production strains which ferment starch to lactic acid in a single step is necessary to make
the process economical. Very few bacteria have been reported so far for direct fermentation
of starch to lactic acid [1, 25, 26] Approximately 3.5 billion tonnes of agricultural residues are
produced per annum in the world [27]. The use of a specific carbohydrate feedstock
depends on its price, availability, and purity. Although agro-industrial residues are rich in
carbohydrates, their utilization is limited [27]. Different food/agro-industrial products or
residues form the cheaper alternatives to refined sugars as substrates for lactic acid
production. Sucrose-containing materials such as molasses are commonly exploited raw
materials for lactic acid production. Starch produced from various plant products is a
potentially interesting raw material based on cost and availability. Laboratory-scale
fermentations have been reported for lactic acid production from starch by *Lactobacillus
amylophilus* GV6 [20], *L. amylophilus* B4437 [28], *Lactobacillus amylovorus* [29, 23, *Lactococcus
lactis* combined with *Aspergillus awamorii* [30] and *Rhizopus arrhizus* [31]. *L. amylophilus*
NRRL B4437 [32], *L. amylovorus* [17] and *L. amylophilus* GV6 are exceptions that have been
described to actively ferment starch to lactic acid and this may lead to alternative process of
industrial lactic acid production [23, 20]. To make the process cost effective in terms of
substrate, various groups have worked on acid/enzyme hydrolysis of starchy substrates
followed by *Lactobacillus* fermentation or simultaneous saccharification and fermentation by
co-culture/mixed culture fermentations. It has been reported that starch is used as substrate
in two steps [1].

2.2. Thermostable amylases

Amylases are among the most important enzymes and are of great significance in present-
day biotechnology. Although they can be derived from several sources, such as plants,
animals and microorganisms; enzymes from microbial sources generally meet industrial

demands. The spectrum of amylase application has widened in many other fields, such as clinical, medical and analytical chemistries, as well as their widespread application in starch saccharification and in the textile, food, brewing and distilling industries. Thermostability is one of the main features of many enzymes sold for bulk industrial usage. Thermostable α-amylases are of interest because of their potential industrial applications. They have extensive commercial applications in starch liquefaction, brewing, sizing in textile industries, paper and detergent manufacturing processes. [33,34, 35]. The advantages of using thermostable amylases in industrial processes include the decreased risk of contamination and cost of external cooling, a better solubility of substrates, a lower viscosity allowing accelerated mixing and pumping [36]. Several thermostable α-amylase have been purified from *Bacillus sp.* and the factors influencing their thermostability have been investigated, but the thermostability of amylases from lactic acid bacteria have attracted very few scientific attention. *Lactobacillus amylovorus, Lactobacillus plantarum, Lactobacillus manihotivorans*, and *Lactobacillus fermentum* are some of the lactic acid bacteria exhibiting amylolytic activity which have been studied [37, 10, 38, 5, 39, 40]. However, most of α-amylase from these bacteria presented weak thermostability compared to those of genus *Bacillus*. Owing to the important acidification of fermenting medium by most lactic acid bacteria, the production of thermostable amylase by a lactic acid bacterium under submerged or solid-state fermentation can help to reduce the risk of contamination caused by undesirable micro-organisms during the process [41, 42]. Another advantage is the non-pathogen character of the genus *Lactobacillus* that allows their utilization in food fermentation processes.

2.3. Lactic acid

Lactic acid a water soluble and highly hygroscopic aliphatic acid is present in humans, animals and microorganisms. It is the first biotechnologically produced multi-functional versatile organic acid having wide range of applications [1], namely as a preservative in many food products. It can be produce by LAB trough fermentation or synthetically from lactonitrile [43]. It is non-volatile, odorless organic acid and is classified as GRAS (Generally Recognized As Safe) for use as a general purpose food additive. The lactic acid consumption market is dominated by the food and beverage sector since 1982 [1]. More than 50% of lactic acid produced is used as emulsifying agent in bakery products [44]. It is used as acidulant/flavoring/pH buffering agent or inhibitor of bacterial spoilage in a wide variety of processed foods, such as candy, breads and bakery products, soft drinks, soups, sherbets, dairy products, beer, jams and jellies, mayonnaise, and processed eggs, often in conjunction with other acidulants. Lactic acid or its salts are used in the disinfection and packaging of carcasses, particularly those of poultry and fish, where the addition of aqueous solutions during processing increased shelf life and reduced microbial spoilage. The esters of calcium and sodium salts of lactate with longer chain fatty acids have been used as very good dough conditioners and emulsifiers in bakery products. The water retaining capacity of lactic acid makes it suitable for use as moisturizer in cosmetic formulations. Ethyl lactate is the active ingredient in many anti-acne preparations. The natural occurrence of lactic acid in human body makes it very useful as an active ingredient in cosmetics [45]. Lactic acid has long been used in pharmaceutical formulations, mainly in topical ointments, lotions, and parenteral solutions. It also finds applications in the preparation of biodegradable polymers for medical uses such as surgical sutures, prostheses and controlled drug delivery systems [45]. Because of ever-increasing amount of plastic wastes worldwide, considerable research and development efforts have been devoted towards making a single-use, biodegradable substitute of conventional thermoplastics. Biodegradable polymers are classified as a family of polymers that will degrade completely – either into the corresponding monomers or into products, which are otherwise part of nature – through metabolic action of living organisms. The demand for lactic acid has been increasing considerably, owing to the promising applications of its polymer, the polylactic acid (PLA), as an environment-friendly alternative to plastics derived from petrochemicals. PLA has received considerable attention as the precursor for the synthesis of biodegradable plastic [46]. The lactic acid polymers have potentially large markets, as they many advantage like biodegradability, thermo plasticity, high strength etc., have potentially large markets. The substitution of existing synthetic polymers by biodegradable ones would also significantly alleviate waste disposal problems. As the physical properties of PLA depend on the isomeric composition of lactic acid, the production of optically pure lactic acid is essential for polymerization. L-Polylactic acid has a melting point of 175–178 °C and slow degradation time. L-Polylactide is a semicrystalline polymer exhibiting high tensile strength and low elongation with high modulus suitable for medical products in orthopedic fixation (pins, rods, ligaments etc.), cardiovascular applications (stents, grafts etc.), dental applications, intestinal applications, and sutures [45].

3. Materials and methods

3.1. Samples

Twenty-eight samples of soils were collected from main geographic zones of Cameroon in four localities: (Ngaoundere, Yaounde, Bafoussam and Mbouda) at the factories where starchy wastes are frequently submitted to natural fermentation. Four kinds of factories were investigated: "gari" factories, corn and cassava mills, cassava plantation after harvesting and treatment of tubers and flour markets. At the site where degradation of starchy material was remarkable and visible, one to five grams of soils were collected and transferred to polyethylene aseptic bag, the factory age recorded; the samples were finally transported to the laboratory and analyzed in the same week.

3.2. Screening of thermostable amylases and lactic acid producing bacteria

The starch degrading amylolytic lactic acid bacterial strains were isolated from different samples of soil. Amylolytic micro-organisms were firstly enriched by introduction of 1 g of soil sample in 100 ml Erlenmeyer flasks containing 50 ml of enrichment liquid medium, composed of (gram per litre): 5 g soluble starch, 5 g peptone, 5 g yeast extract 0.5 g $MgSO_4.7H_2O$, 0.01g $FeSO_4.7H_2O$, 0.01 g NaCl. Enrichment of thermostable amylases producing bacteria was carried out by heating Erlenmeyer flasks at 90°C for 5 min followed by incubation in an alternative shaker at 37-40°C and speed of 150 oscillations per minute for 24 h. Amylases producing bacteria strains were screened on agar plate, containing (gram per liter): 10 g soluble starch, 5 g peptone, 5 g yeast extract, 0.5 g $MgSO_4.7H_2O$, 0.01g $FeSO_4.7H_2O$, 0.01 g NaCl, 15 g agar. Incubation at 37-40°C was carried out for 48 h, after which the plates were stained with lugol solution (Gram iodine solution: 0.1% I_2 and 1% KI). The colonies with the largest halo forming zone were pre-selected and tested for Gram staining and catalase activity.

Preliminary tests were carried out to determine the heat stability of the amylase of each isolate as we described previously [47, 48]. The gas production from glucose, growth at different temperature (10, 40, 45 °C) as well as the ability to grow in different concentration of NaCl was determined as described by Schillinger and Lucke [49] and Dykes et al [50]. The isolates which were Gram positive and catalase negative, non-motile and producing heat stable amylase and lactic acid were finally selected and identified using API 50 CH test kit (bioMerieux, France). The APILAB PLUS database identification was used to interpret the results.

3.3. Microbial growth, amylase and lactic acid production

In order to study microbial growth, amylase and lactic acid production, the microorganism was propagated at 40°C for 70 h in 50 ml of a basal medium containing: soluble starch, 1% (w/v); yeast extract, 0.5 % (w/v) placed in 100 ml Erlenmeyer flask with shaking at 150 oscillations per minute in an alternative shaker (Kotterman, Germany). The initial pH of the medium was adjusted to 6.5 using 0.1 M HCl. After removal of cells by centrifugation (8000xg, 30 min, 4°C) in centrifugator (Heraeus, Germany), the supernatant was considered as the crude enzyme solution and was also used for lactic acid evaluation.

3.4. Optimisation of raw starch degrading thermostable amylase and lactic acid production

The amylase and lactic acid production was optimized by studying the effect of cultural and environmental variables (carbohydrate and nitrogen sources, metal salts and surfactants) individually and simultaneously. The effect of carbohydrate sources was studied by replacing soluble starch in basal medium with different sugars, gelatinized and raw natural crude starch sources (glucose, fructose, maltose, amylose, amylopectine, cassava, corn, rice tapioca, and sorghum flours at final concentration of 1% (w/v)). Nitrogen sources were tested by replacing yeast extract with various nitrogen sources (peptone, tryptone, beef extract, soyabean meal, ammonium sulphate, and urea at final concentration of 1.5% (w/v)). The effect of metal salts was studied by adding individually various metal salts ($CaCl_2.2H_2O$, $MgSO_4.7H_2O$, $FeSO_4.7H_2O$, $FeCl_3$, NaCl at concentration of 0.1% (w/v)). Similarly the effect of surfactants was studied by supplementing the culture medium with Tween 80 and Tween 40 at concentration of 1.5% (v/v).

All media containing gelatinized starch sources were autoclaved at 121 °C for 20 min, while for the media containing raw starch flour, starch powder was sterilized by washing in ethanol and added to sterile nutrient broth.

3.5. Partial enzyme purification

The culture supernatant was supplemented with solid ammonium sulphate to 65% (w/v) final concentration, with mechanical stirring at 4°C. The suspension was retained for 1 h at 4 °C, and centrifuged at 8000 g for 30 min at the same temperature. The resultant supernatant was brought to 70 % w/v ammonium sulphate saturation at 4°C. 50-70% (w/v) ammonium sulphate precipitate was recovered, dissolved in 0.1 M phosphate buffer and dialysed using Spectra/PorR, VWR 2003 dialysis membrane overnight against the same buffer at 4°C and used as partial purified enzyme solution.

3.6. Effect of temperature and pH on activity and stability

The optimal temperature for amylase activity was determined by assaying activity between 30 and 100°C for 30 min in 50 mM phosphate buffer. Measurement of optimum pH for amylase activity was carried out under the assay conditions for pH range of 3.0-10.0, using 50mM of three buffer solutions: Tris-HCl (pH 3.0), Na_2HPO_4-Citrate (pH 4.0 – 6.0), and Glycine-NaOH (pH 7.0-10.0).

The temperature stability was determined by incubating the partial purified enzyme solution in water bath for temperature range of 30-100°C for 30, 60, 90, 120, 180 min and then cooled with tap water. The remaining α-amylase activity was measured and expressed as the percentage of the activity of untreated control taken as 100%. The first order inactivation rate constants, k_i were calculated from the equation: $lnA = lnA_0 - k_i t$, where A_0 is the initial value of amylase activity and A the value of amylase activity after a time t (min).

For the determination of pH stability, the enzyme was incubated in a water bath at 60 °C at varying pH value for 30 min. The residual activity was detected under the same conditions and expressed as the percentage of the activity of untreated control taken as 100%.

3.7. Effect of metal salts and chelating agent

The effect of metal salts and EDTA on amylase activity was determined by adding 0.05 to 0.1% (w/v) of metal salts ($CaCl_2.2H_2O$, $MgSO_4.7H_2O$, $FeSO_4.7H_2O$, NaCl, $FeCl_3$, $CuSO_4.5H_2O$) and EDTA to the standard assay. The effect of metal salts and chelating agent on amylase activity were evaluated by pre-incubating the enzyme in the presence of effectors for 30 min at 60°C. The remaining amylase activity was determined and expressed as the percentage of the activity of untreated control taken as 100%.

3.8. Analytical methods

Cell growth was evaluated by reading the absorbance of culture medium at 600 nm using a Secoman spectrophotometer and numeration of total colony forming unit by 10-fold serial dilution of fermented broth and pour plating on MRS-starch agar (De Man Rogosa and Sharpe medium in which glucose has been replaced by soluble starch (Prolabo-Merck Eurolab, France)). In order to evaluate the capacity of microorganism to acidify the culture medium, the pH of the fermented broth was measured using an electronic pH meter (Mettler Seven S20, Japan)

The amylolytic power of Lactic acid bacteria was determined using the method of wells by inoculation of 10 µl of microbial strain in 4 mm depth micro-wells on the surface of MRS-starch agar plate. The starch hydrolysis halo was revealed after 48 h of incubation using iodine solution. The amylolytic power was defined as the average diameter (mm) of hydrolysis halo provoked by a strain after its inoculation in micro-well on MRS-starch agar plate for 48 h incubation at optimum temperature of growth for three assays

The activity of amylase both in crude and purified extracts was assayed by iodine method. In a typical run, 5 ml of 1% soluble starch solution and 2 ml of 0.1M phosphate buffer (pH 6.0) were mixed and maintained at a desired temperature for 10 min, then 0.5 ml of appropriately diluted enzyme solution was added. After 30 min the enzyme reaction was stopped by rapidly adding 1ml of 1M HCl into the reaction mixture. For the determination of residual starch, 1 ml of the reaction mixture was added to 2.4 ml of diluted iodine solution and its optical density was read at 620 nm using a spectrophotometer (Secoman). One unit of amylase activity (U) was defined as the amount of enzyme able to hydrolyse 1 g of soluble starch during 60 min under the experimental condition. The lactic acid was determined according to Kimberley and Taylor [51]. The nature of amylase (endo-acting or exo-acting) was determined according to Ceralpha method (Megazyme) which uses a blocked maltoheptaoside as substratre [57].

The affinity of the enzyme preparation from selected LAB toward raw cassava starch was studied by incubating 0.2 g of raw cassava flour with 1ml of the enzyme solution at 60 °C for 15 min. After centrifugation, the α-amylase activity of the supernatant was measured and the adsorption percentage was calculated as follows: $Adsorption\,(\%) = \dfrac{A-B}{A} \times 100$, A is the original α-amylase activity and B is the α-amylase activity in the supernatant after adsorption on raw potato starch granules.

For the determination of raw starch digestibility, raw cassava was used and the reaction mixture containing 100 U of α-amylase preparation from the selected LAB and 100 mg of raw cassava starch in a final volume of 10ml dispensed in 100ml Erlenmeyer flasks were incubated in alternative water bath shaker at 60°C and 150 oscillations per min. After a time interval of 6 h, the reducing sugars liberated in the reaction mixtures were determined by dinitrosalicyclic acid method [58].

Light microscopy was used for the examination of the effect of enzyme on raw starch granules using Olympus microscope BH-2.

4. Results and discussion

4.1. Biochemical properties of amylolytic LAB isolated

From the 28 samples of soil collected from different localities of Cameroon, 90 amylolytic isolates were screened but only 9 isolates (04BBA15, 04BBA19, 05BBA22, 05BBA23,

14BYA42, 20BBA60, 17BNG51, 23BYA21, 26BMB81) presented very high amylolytic power (≥15mm) and were qualified as amylase overproducing isolates. The amylolytic power was defined as the average diameter (mm) of starch hydrolysis halo (Fig.1.) provoked by a strain after its inoculation in micro-well on MRS-starch agar plate for 48 h incubation at optimum temperature of growth for three assays. The amylolytic power is an expression of the capacity of an isolate to degrade starch during the culture. Among the amylase overproducing isolates, two (04BBA19, 26BMB81) were aero-anaerobic non spore forming, gram positive and catalase negative bacteria; this characteristic is proper to lactic acid bacteria. Microscopic observation showed rod cells. Biochemical characteristics of these isolates were carried out using API 50 CH kit bioMerieux system, the results are summarized in Table 1, the isolates were tested for their possibility to ferment 50 carbohydrates, and this fermentation profile was use for their numerical identification. According to their biochemical profile, 04BBA19 and 26BMB81 were respectively identified as *Lactobacillus fermentum* and *Lactobacillus plantarum*. The strain 04BBA19 (*Lactobacillus fermentum*) presented a very high amylolytic power, as it was able to cause a starch hydrolysis halo of 45 ±1.5mm on MRS-starch agar plate after 48 h of incubation at 40 °C; consequently it was selected for further studies. The preliminary test of thermostability carried out on its crude extract amylase showed that it produce a very high thermostable enzyme and it was selected for an application on simultaneous production of thermostable amylase and lactic acid from starchy material.

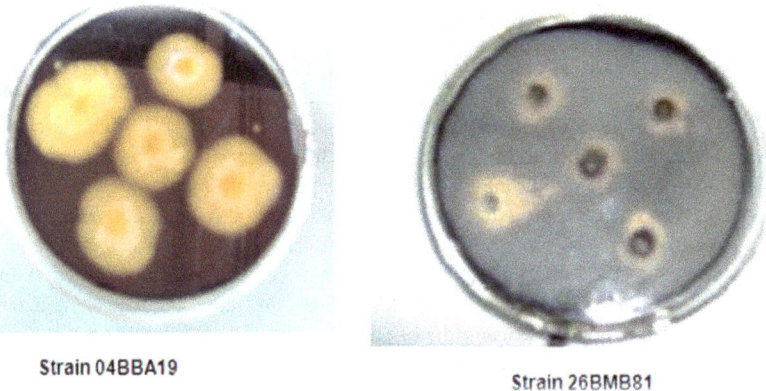

Strain 04BBA19

Strain 26BMB81

Figure 1. Plate assays for detection of amylase activity of lactic acid bacteria (04BBA19, 26BMB81) on MRS-starch agar plate medium. The diameter of hydrolysis halo was revealed by flooding the plates with Iodine solution (0.1% I₂+1% KI) after 48 h of culture at 40°C

Test number	1	2	3	4	5	6	7	8	9	10	11	12	13	14	15	16	17	18	19	20	21	22	23	24	25	26	27	28	29
Strains cod	Heterofermentative	Gaz production	Optimum temperature of growth	Growth at 10°C	Dextran	Ammonia from Arginine	Nitrate reduction	Glycerol	Erythritol	D-arabinose	L-arabinose	Ribose	D-xylose	L-xylose	Adonitol	ß methyl-D-Xyloside	Galactose	Glucose	Fructose	Mannose	Sorbose	Rhamnose	Dulcitol	Inositol	Mannitol	Sorbitol	α-Methyl-D-mannoside	α-Methyl-D-glucoside	N-Acetyl-Glucosamine
04BBA19	+	+	45°C	-	-	+	-	-	-	-	-	+	-	-	-	-	+	+	+	+	-	-	-	nc	-	-	-	-	-
26BMB81	-	+	40°C	-	-	-	-	-	-	-	-	+	+	-	-	-	+	+	+	+	-	-	-	-	+	+	-	+	+

Test number	30	31	32	33	34	35	36	37	38	39	40	41	42	43	44	45	46	47	48	49	50	51	52	53	54	55	56	Identification species (API 50CHL)
Strains cod	Amygdalin	Arbutin	Aesculin	Salicin	Cellobiose	Maltose	Lactose	Melibiose	Sucrose	Trehalose	Inulin	Melezitose	Raffinose	Starch	Glycogen	Xylitol	ß Gentiobiose	D-turanose	D-lyxose	D-tagatose	D-fucose	L-fucose	D-arabitol	L-arabitol	Gluconate	2-gh-Gluconate	5-Keto-Gluconate	
04BBA19	-	-	nc	-	-	+	+	+	+	-	-	-	+	+	-	-	-	-	-	-	-	-	-	-	+	-	-	Lactobacillus fermentum
26BMB81	+	+	+	+	+	+	+	+	+	+	nc	+	+	+	+	-	+	+	-	-	-	-	-	-	-	+	-	Lactobacillus plantarum

+, positive reaction; -negative reaction, nc, non -conclusive

Table 1. Biochemical characteristics of amylases overproducing *Lactobacillus* isolated from soils

4.2. Amylase and lactic acid production

In the presence of starch as carbon source at 40°C, *L. fermentum* 04BBA19 strain grew, exhibited amylolytic activity and produced lactic acid in the culture medium. The amylase production pattern in *L. fermentum* 04BBA19 (Fig.2) indicates that the induction of amylase took place during the lag phase (after 10 h of incubation) in the presence of starch. The level of amylase production increased significantly during the exponential phase of growth. Lactic acid production became visible around 15 h after incubation and also increased considerably during the exponential phase of growth. Cell growth, amylase and lactic acid production reached maxima values at the same time (40 h of fermentation). The values of those maxima were 1.1×10^9 cfu/ml, 107.3 ± 0.5 U/ml, 8.7 ± 0.5 g/l for cell growth, amylase activity, and lactic acid production respectively. Such coincidence shows that amylase production by *L. fermentum* 04BBA19 was tightly linked to cell growth. These results are in agreement with the report of Goyal et al. [53], Liu and Xu [54] on the relationship between pattern of cell growth and amylase production. The decline of cell growth and amylase production after the peak occurred around 50 h of incubation and could be attributed to the rise of lactic acid concentration in fermented broth [6] or to the rise of protease levels [55]. The acidification was also expressed by the decrease of initial pH of culture broth (Fig. 2). The initial pH of culture broth declined significantly and reached a value of 3.0 around 50 h of incubation and then remained constant.

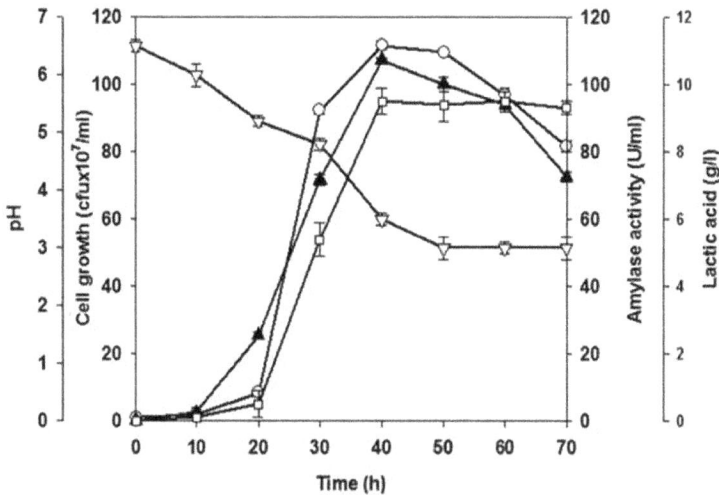

Figure 2. Time course of growth (○), pH (▽), α-amylase (▲) and lactic acid (□) production by *L. fermentum* 04BBA19 in 1% (w/v) soluble starch medium at 40°C, pH 6.0. The data shown are averages of triplicates assays within 10% of the mean value.

The study of cell growth and amylase production as a function of temperature (Fig. 3a) showed that *L. fermentum* 04BBA19 exhibited maximal growth and amylase activity at 45°C, confirming thus the strong relationships between cell growth and amylase production. On the other hand the maximum value of lactic acid was produced at the same temperature. Many other investigators reported that maximum amylase production occurred at the optimum growth temperature [56, 53]. These results are contrary to the findings of Chandra et al.[57] who studied the growth and amylase production of *Bacillus licheniformis* CUM 305. They have observed that this microorganism grew very well at 30°C, but did not produce α-amylase at that temperature. In addition, Saito and Yamamoto [58] found α-amylase production at 50°C and cell growth at a temperature lower than 45°C for another strain of *B. licheniformis*.

The amylase and lactic acid production by *L. fermentum* 04BBA19 was influenced significantly by initial pH of culture broth (Fig. 3b). Maximum amylase and lactic acid production was achieved for pH range of 4.0-6.5. These results could be explained by the fact that pH generally act by inducing morphological change in microorganism which facilitate enzyme production [59].

4.3. Optimisation of amylase and lactic acid production

Amylase production is known to be induced by a variety of carbohydrate, nitrogen compounds and minerals [60, 61]. In order to achieve high enzyme yield, efforts are made to develop a suitable medium for proper growth and maximum secretion of enzyme, using an adequate combination of carbohydrates, nitrogen and minerals [53, 62].

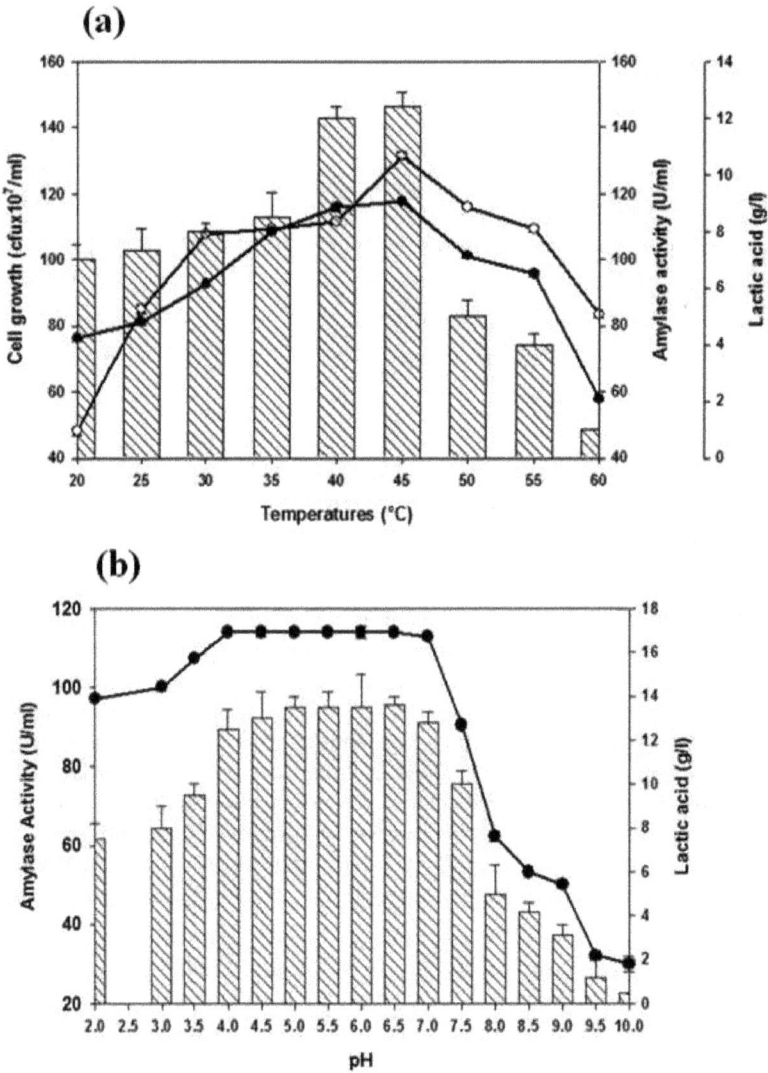

Figure 3. (a) Effect of temperature on microbial growth (O), α-amylase (●) and lactic acid (▨) production. (b) Effect of initial pH of culture broth on α-amylase (●) and lactic acid (▨) production. The data shown are averages of triplicate assays within 10% of the mean value.

From the use of different carbohydrate sources in the present study, soluble starch proved to be the best inducer of amylase production (Table 1). In the presence of soluble starch at concentration of 1% (w/v), the enzyme yield reached 107.0±1.2 U/ml after 48 hours of fermentation, while in the presence of raw cassava starch at the same concentration, the enzyme yield was 67.1±0.5 U/ml. These results are in agreement with the reports of Cherry et al. [63], Saxena et al. [4] who reported maximum amylase production when starch was used as carbohydrate source. In the presence of glucose and fructose, amylase production was almost nil; and that was a proof that glucose and fructose repressed amylase synthesis by *L. fermentum* 04BBA19. This observation is in agreement with the reports of Theodoro and Martin [64] showing that synthesis of carbohydrate degrading enzymes in some microbial species leads to catabolic repression by substrate such as glucose and fructose. Similar results were observed by Halsetine et al. [65] for the production of amylase by the hyperthemophilic archeon *Sulfolobus solfataricus*. According to them, glucose prevented α-amylase gene expression and not only secretion of performed enzyme. Since amylase yield is higher with amylose (92.3 U/ml) as carbohydrate source than with amylopectin (50.1 U/ml), the *L. fermentum* 04BBA19 amylase is more efficient for hydrolysis of alpha-1,4 linkages than those of alpha-1,6. The amylase production increased with the soluble starch concentration (Fig. 4), reaching a maximum (180.5 ± 0.3 U/ml) at the concentration range of 8-16 % (w/v). These optimum starch concentrations for amylase production by *L. fermentum* 04BBA19 are higher than that observed for amylase production in *Bacillus* sp. PN5 reported by Saxena et al. [4]. This microorganism presented an optimum soluble starch concentration of 0.6% (w/v) for amylase production. The lactic acid production also increased with the soluble starch concentration, the optimum starch concentration for lactic acid production was achieved at the same range of concentration for amylase production.

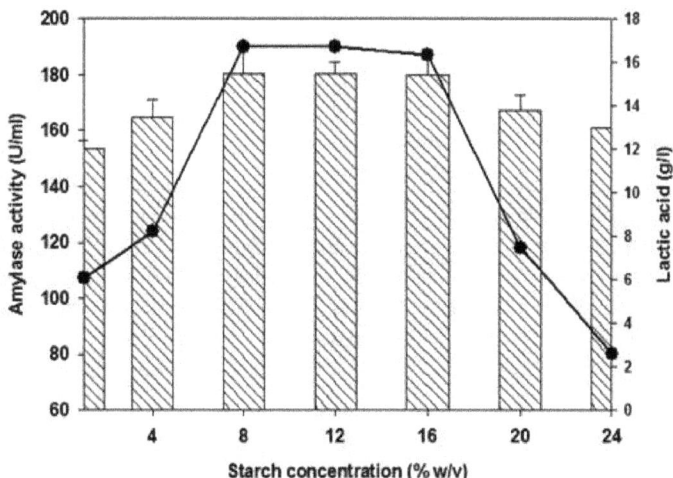

Figure 4. Effect of starch concentration on α-amylase (●) and lactic acid production (▨) by *L. fermentum* 04BBA19. The data shown are averages of triplicate assays with SD within 10% of mean value

Among the various gelatinized starchy sources tested, corn and sorghum flour were found to be the most suitable for α-amylase and lactic acid production by *L. fermentum* 04BBA19 while for the raw starchy sources tested, potato starch was most suitable (Table 2). On the other hand the level of lactic acid was more important when corn and sorghum flours were used. The good production of α-amylase and lactic acid when these starchy flours are used is based on their composition; they also contain proteins and vitamins which are required by lactic acid bacteria for their growth, enzymes and acids production [66].

Among nitrogen sources used in the present study, soya bean meal and yeast extract showed significant effect on α-amylase and lactic acid production. Soya bean meal, rich in protein is a potential nutrient for lactic acid fermentation. Similar results were obtained by several authors. Goyal et al. [53] reported that soybean meal presented a positive effect and was the best nitrogen source for raw starch digesting thermostable α-amylase production by the *Bacillus* sp I-3 strain. The yeast extract was also reported to be a potential nutrient for lactic acid fermentation, since it contains vitamins, amino acids [66]. Though all nitrogen sources are positively influencing enzyme production by *L. fermentum* 04BBA19, an inverse behaviour has been observed with other bacterial strains, for instance, Tanyildizi et al. [67] reported zero effect of yeast extract on amylase production by *Bacillus* sp.

All metal salts tested in this study increased amylase and lactic acid production by *L. fermentum* 04BBA19, except $CuSO_4.5H_2O$ that acted as inhibitor. The inhibition of amylase production by $CuSO4.5H2O$ was also reported by Wu et al. [68] for the *Bacillus* sp CRP strain. Copper ion acted as poisonous compound for this strain and consequently inhibited amylase synthesis. The effect of $CaCl_2.2H_2O$ was the most important, and was in agreement with the observation of Gangadharan et al. [61] who described the rise of amylase production by *B. amyloliquefaciens* when $CaCl_2.2H_2O$ was supplemented to the culture medium. The supplementation of metal ions has been reported to provide good growth and also influence higher enzyme production. Most α-amylases are metalloenzymes and in most cases, Ca^{2+} ions are required for maintaining the spatial conformation of the enzyme, thus play an important role in enzyme stability [61].

From the surfactants tested in this study, Tween-80 appeared to be the best surfactant sources for amylase production by *L. fermentum* 04BBA19. Similar results were obtained by Reddy et al. [69]. These authors reported that the supplementation of culture medium with Tween-80 resulted in a marked increase in the yields of thermostable β-amylase and pullullanase by *C. thermosulfurogenes* SV2, and that the stimulation of enzyme production was greater when the surfactants were added after 18 h of incubation of culture. Beside stimulation, the surfactants caused and increased secretion of the enzymes into extracellular fluid [59].

From various environmental factors tested for α-amylase and lactic acid production by *L. fermentum* 04BBA19, it has been observed that all factors that increase amylase synthesis also positively affect lactic acid production. The optimization of the basal medium by supplementation of all carbohydrate, nitrogen, mineral and surfactant sources (excepted $CuSO_4.5H_2O_4$) in culture medium resulted to a significant improvement of enzyme and lactic acid yield. In the optimized medium, amylase activity and lactic acid content reached 732.4±0.4 U/ml and 53.2±0.4 g/l respectively.

Parameters	Enzyme yield (U/ml)	Lactic acid(g/l)
Carbohydrate sources (1% w/v)		
Glucose	$0.1\pm0.0^{d*}$	14.3 ± 0.5^a
Fructose	0.2 ± 0.0^d	12.1 ± 0.5^b
Maltose	0.1 ± 0.0^d	12.8 ± 0.4^b
Amylose	92.3 ± 0.1^b	12.2 ± 0.1^b
Amylopectin	50.1 ± 0.5^c	10.3 ± 0.5^c
Soluble starch	107.3 ± 0.5^a	8.7 ± 0.5^d
Nitrogen sources (1.5% w/v)		
Yeast Extract	107.3 ± 0.5^b	8.7 ± 0.5^b
Beef extract	92.4 ± 0.5^c	7.3 ± 0.2^b
Peptone	88.3 ± 1.7^d	7.1 ± 0.3^b
Tryptone	75.3 ± 0.5^e	6.5 ± 0.5^b
Soya bean meal	397.3 ± 0.4^a	29.2 ± 0.4^a
Ammonium sulphate	95.4 ± 1.5^c	$7.2.\pm0.8^b$
Urea	76.3 ± 0.3^e	5.3 ± 0.6^c
Minerals (0.1% w/v)		
$CaCl_2. 2H_2O$	412.1 ± 0.6^a	33.2 ± 0.1^a
$MgSO_4. 7H_2O$	315.1 ± 0.4^b	31.2 ± 0.5^a
$FeSO4. 7H_2O$	237.3 ± 0.7^c	20.2 ± 0.4^b
NaCl	315.2 ± 0.9^b	22.1 ± 0.6^b
$CuSO_4.5H_2O$	12.2 ± 0.6^e	3.2 ± 0.3^d
Surfactants (1.5% w/v)		
Tween-40	209.5 ± 0.1^b	27.3 ± 0.4^b
Tween-80	215.1 ± 0.3^a	35.2 ± 0.3^a
Gelatinized starchy sources (1 %w/v)		
Corn flour	303.5 ± 0.2^a	36.3 ± 0.6^a
Cassava flour	182.3 ± 0.4^c	24.2 ± 0.8^d
Sorghum flour	305.8 ± 0.7^a	35.2 ± 0.1^a
Rice flour	187.3 ± 0.8^c	30.1 ± 0.5^b
Tapioca flour	237.4 ± 0.6^b	$27.2\pm0.7c$
Raw starchy sources		
Cassava starch	67.1 ± 0.5^c	21.3 ± 0.4^a
Potato starch	87.2 ± 0.5^a	23.4 ± 0.1^a
Cocoyam starch	78.6 ± 0.2^b	22.7 ± 0.4^a
Media		
Basal medium	107.5 ± 0.3	8.7 ± 0.5^b
Optimized medium	732.3 ± 0.4	53.2 ± 0.4^a

Table 2. Effect of different parameters on α-amylase and lactic acid production by *L. fermentum* 04BBA19 in submerged state fermentation at 45 °C and initial pH 6.5.

The basal medium contained soluble starch, 1% (w/v); yeast extract, 0.5 % (w/v); while the optimized medium contained all parameters without $CuSO_4.5H_2O$. The data shown are averages of triplicate assays with SD within 10% of mean value. For each group of parameters (Carbohydrate, Nitrogen, Mineral, Starchy sources, media), means with different superscripts within columns are significantly different ($p<0.05$).

4.4. Enzyme properties

The amylase produced by *L. fermentum* 04BBA19 showed high affinity toward cassava raw starch granules with 80% adsorption and brought about 79% hydrolysis of 1% (w/v) suspension of raw cassava starch. On the other hand, the enzyme was able to hydrolyze blocked p-nitro phenyl methyl heptaoside, releasing a yellow compound (p-nitro phenol) with maximum absorption at 530 nm. This result was a proof that amylase from *L. fermentum* 04BBA19 is an endo acting amylase (α-amylase), since the blocked p-nitro phenyl methyl heptaoside is known to be hydrolysed only by endo-acting amylases [38].

The enzyme exhibited maximum activity at 60-70°C and maintained 100% of its initial activity at 80°C for 30 min of heat treatment (Fig 5-a). When the enzyme was treated for the same time (30 min.) at 90°C and 100°C, the remaining activities were 90 and 87% respectively. These results showed the thermophilic character and very high thermostability of α-amylase from *L. fermentum* 04BBA19. In general, most of lactic acid bacteria do not produce amylases. However, this property have been observed in some genera of lactic acid bacteria, especially in *L. plantarum* and *L. amylovorus* [37], *L. manihotivorans* [70, 13], *L. fermentum* OGI E1 [38]. But amylases produced by these strains are not thermostable. Traditionally high thermostable and thermophiles amylases are found in *Bacillus* and *Thermococcus* genera as: *B. amyloliquefaciens* [71]; *B. licheniformis* [72]; *B. stearothermophilus* [73]; *B. subtilis* and *T. aggreganes* [74], *T. profundus* [75], *Bacillus* sp PN5 [4], *B. cohnii* US147 [35], *Chromohalobacter* sp. TVSP 101 [76].

Fig. 6 shows the thermostability pattern of α-amylase from *L. fermentum* 04BBA19 at 80°C, 90°C and 100°C when the time of heat treatment is beyond 30 min. Table 3 presents the thermal inactivation rate constant (k_i) and half-life (T) at these temperatures. The half-life of this enzyme is higher than that of α-amylase from *B. licheniformis*: 120 min at 70°C [76]. The thermal stability was considerably improved by addition of 0.1% (w/v) $CaCl_2.2H_2O$. Goyal et al. [53] obtained a half-life value of 3.5 h at 80°C with α-amylase from *Bacillus* sp.I-3 in the presence of 0.1 % (w/v) calcium chloride, while under the same conditions; α-amylase from *L. fermentum* 04BBA19 displayed a half-life of 6.1 h.

Due to its high thermostability, α-amylase from *L. fermentum* 04BBA19 could be highly competitive in industrial bioconversion reactions, as compared to α-amylase from *Bacillus*. In addition, this competitiveness is enhanced by the fact that lactobacilli, due to their non-pathogen character, are easily used in food industry [6].

The *L. fermentum* 04BBA19 α-amylase is active and stable in pH range of 4.0 – 7.0 (Fig. 5-b), which is the pH range of many foods. In this respect, this amylase could be used in starch hydrolysis, brewing and baking.

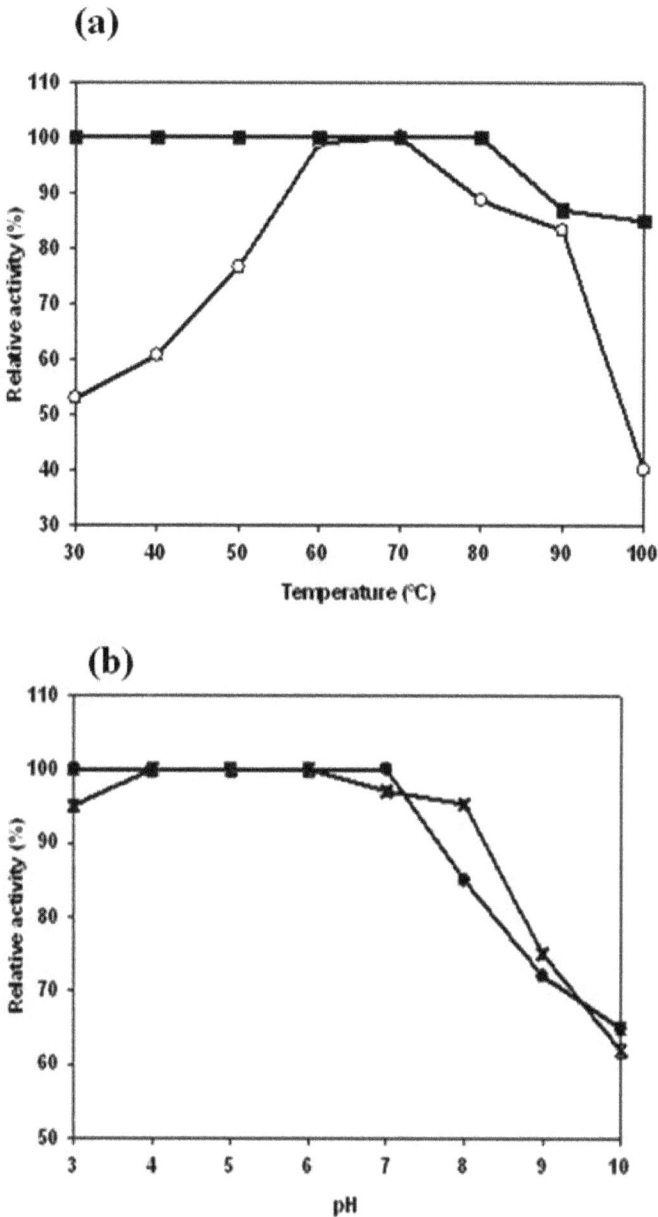

Figure 5. (a) Effect of temperature on activity (○) and stability (■) of α-amylase from *L. fermentum* 04BBA19. (b) Effect of pH on activity (✕) and stability (■) of α-amylase from *L. fermentum* 04BBA19. The data shown are averages of triplicate assays within 10% of the mean value.

The metal salts generally act on activity of enzyme through their ions. The enzyme activity was highly improved by Ca^{2+}, while Fe^{2+}, Fe^{3+}, Na^+ and Mg^{2+} had less significant effect. On the contrary Cu^{2+} and EDTA acted as inhibitors (Fig. 7). The behaviour of the enzyme towards metal ions, particularly calcium, indicates its metalloenzyme nature, which is confirmed by the action of EDTA.

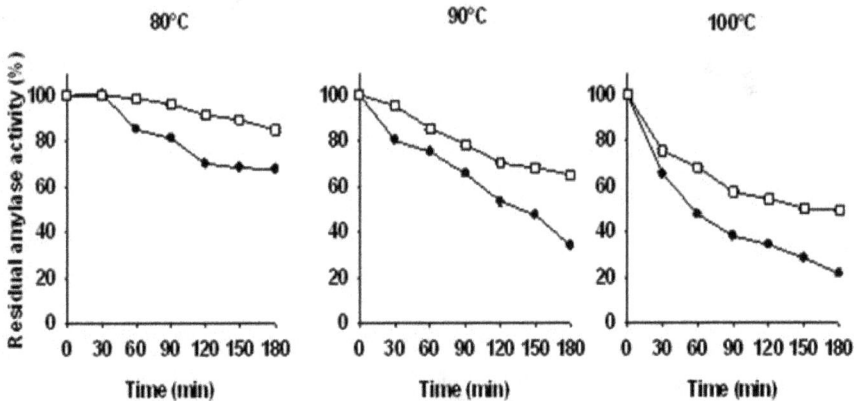

Figure 6. Thermostability pattern of α-amylase from *L. fermentum* 04BBA19, at 80, 90, 100°C without CaCl₂.2H₂O (●) and with 0.1% (w/v) CaCl₂.2H₂O (□). The enzyme was pre-incubated at optimum pH, for 30, 60, 90, 120 and 180 min at temperatures (80, 90 and 100°C). The remaining activity was determined incubating the enzyme at optimum temperature, 60°C for 30 min. The data shown are averages of triplicate assays with SD within 10% of mean value.

	Temperatures					
	80 °C		90°C		100°C	
CaCl₂.2H₂O (% w/v)	k_i (10^{-3}.min^{-1})	T (min)	k_i (10^{-3}.min^{-1})	T (min)	k_i.(10^{-3}.min^{-1})	T (min)
0	3.4	204.0	5.6	123.8	7.9	87.7
0.1	1.9	364.8	2.6	266.6	3.8	182.4

Table 3. Inactivation rate constant (k_i) and half-live (T) of amylase from *L. fermentum* 04BBA19 at 80, 90 and 100°C in the absence and the presence of 0.1% (w/v) CaCl₂.2H₂O.

The main ALAB that have been isolated for the past decade are summarized in Table 4. No study has dealt with the thermostability of their amylases, except the case reported by Aguilar et al. (2000) concerning the properties of the extracellular amylase produced by *L. manihotivorans* LMG 18010ᵀ. This strain produced an amylase with a moderate themostability exhibiting maximum activity at 55°C.

The strain *L. fermentum* 04BBA19 appears as the first ALAB producing highly thermostable amylase. The potential industrial application of this strain could be the bioconversion of inexpensive raw material as starch into lactic acid in single step process. On the other hand this strain and its amylase are potential candidates for food industries (making of high density gruels, baking, brewing) and for the production of biodegradable plastic from starchy raw material.

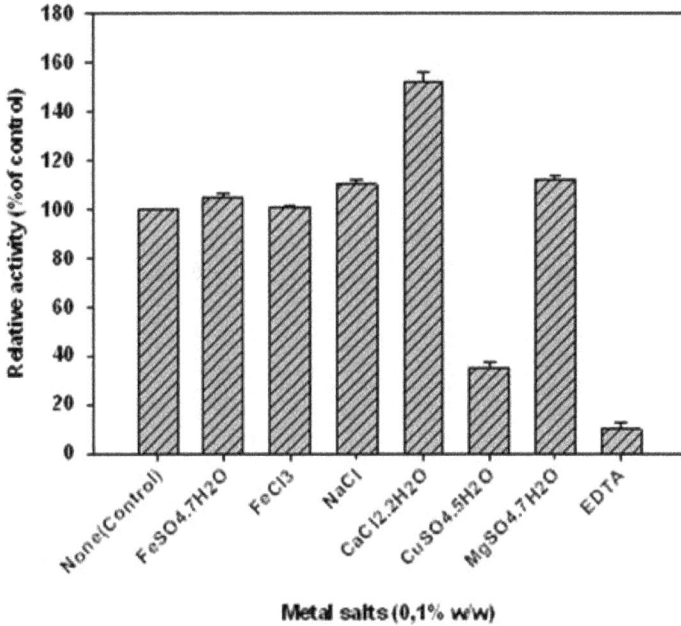

Figure 7. Effect of metal salts and EDTA on the activity of α-amylase from *L. fermentum* 04BBA19. The data shown are averages of triplicate assays with SD within 10% of mean value.

Bacteria	Strains	References
L. fermentum	04BBA19	[47, 48]
L. manihotivorans	LMG18010^T	[69]
L. fermentum	Ogi E1	[11]
L. fermentum	MW2	[11]
L. fermentum	K9	[12]
L. acidophilus	L9	[78]
L. amylovorus	ATCC33622	[23]
L. amylovorus	B-4542	[29]
L. amylovorus		[17]
L. manihotivorans	OND32T	[10, 13]
L. manihotivorans		[13]
L. manihotivorans	LMG 18011	[79]
L. acidophilus		[78]

L. plantarum	A6	[80, 9]
L. plantarum	LMG18053	[9]
L. plantarum	NCIM 2084	[81]
Streptococus. bovis	148	[82]
Lactobacillus sp	LEM 220,	[85]
Lactobacillus sp	LEM 207	[85]
Leuconostoc sp		[86]
Leuconostoc	St3-28	[80]
S. macedonicus		[87]
L. amylolyticus		[88]
L. amylophilus	JCIM 1125	[84]
L. amylophilus	B 4437	[28, 32]
L. amylophilus	GV6	[20]
Bifidobacterium adolescentis	Int57	[15]
B. adolescentis	ZS8	[16]

Table 4. The main amylolytic lactic bacteria strains isolated during the past two decade

5. Conclusion

L. fermentum 04BBA19 which is a soil isolate produced very high thermostable α-amylase. This is the first study dealing with high thermostable amylase from a lactic acid bacterium. According to its properties, this enzyme is a good candidate for starch hydrolysis at high temperature. An economical process could be attained through the use of this enzyme at the liquefaction stage at high temperatures.

On the other hand the fact that thermostable amylase and lactic acid production can be combined in single fermentation step would not only provide a way to make gruels with high energy density, but also improve its safety, since lactic acid bacteria fermentation is an efficient way to inhibit food-borne pathogens.

Owing to the importance of this finding, further studies will focus on the development of an accurate method for preparing high energy density complementary food using local starchy sources and the *L. fermentum* 04BBA19 strain.

Author details

Bertrand Tatsinkou Fossi[*]
Department of Microbiology and Parasitology, University of Buea, Buea, Cameroon

Frédéric Tavea
Department of Biochemistry, University of Douala, Douala, Cameroon

[*] Corresponding Author

Acknowledgement

We gratefully acknowledge the assistance of the Ministry of Higher Education and Brewing society "Société Anonyme des Brasseries du Cameroun" (SABC) who supported this work through Research-Development Grant Programme.

6. References

[1] Reddy G, Altaf M, Naveeana BJ, Ventkateshwar M., Vijay Kumar E (2008) Amylolytic bacterial fermentation. A review. Biotechnology Advanced 26: 22-34.

[2] Axelsson L (2004) Lactic acid bacteria: classification and physiology. In: Salminen S, von Wright A, Ouwehand A, editors. Lactic acid bacteria: microbiological and functional aspects. 3rd rev. and exp. ed.New York: Marcel Dekker, Inc.; 2004. p. 1-66.

[3] Asoodeh A, Chamani J, Lagziana M (2010) A novel thermostable, acidophilic α-amylase from a new thermophilic "Bacillus sp. Ferdowsicous" isolated from Ferdows hot mineral spring in Iran: Purification and biochemical characterization. Inter. J. Biol. Macromol. 46: 289-297.

[4] Saxena RK, Dutt K, Argawal L, Nayyar P (2007) A highly thermostable and alkaline amylase from a Bacillus sp PN5. Biores. Technol. 98: 260-265.

[5] Haki GD, Rakshit, SK (2003) Developments in industrially important thermostable enzymes: a review. Biores. Technol. 89: 17-34.

[6] Singh SK, Ahmed SU, Pandey A (2006) Metabolic engineering approaches for lactic acid production. Process Biochem. 41: 991-1000.

[7] Nguyen TT, Loiseau G, Icard-Vernière C, Rochette I, Trèche S, Guyot, JP (2007) Effect of fermentation by amylolytic lactic acid bacteria in process combinations, on characteristics of rice/soybean slurries: A new method for preparing high energy density complementary foods for young children. Food Chem. 100: 623-631.

[8] Songré-Ouattara, LT, Mouquet-Rivier C., Icard-Vernière C, Humblot C, Diawara B, Guyot JP (2008) Enzyme activities of lactic acid bacteria from a pearl millet fermented gruel (ben-saalga) of functional interest in nutrition. Int. J. of Food Microbiol. 128: 395–400.

[9] Giraud E, Lelong B, Raimbault M (1991) Influence of pH and initial lactate concentration on the growth of Lactobacillus plantarum. Appl Microbiol Biotechnol 36: 96–9.

[10] Morlon-Guyot J, Guyot JP, Pot B, Jacobe de Haut I, Raimbault M (1998). Lactobacillus manihotivorans sp. Nov., a new starch-hydrolysing lactic acid bacterium isolated during cassava sour starch fermentation. Int. J. Syst. Bacteriol. 48: 1101-1109.

[11] Agati VJP, Guyot J, Morlon-Guyot P, Talamond, Hounhouigan DJ (1998). Isolation and characterization of new amylolytic strains of Lactobacillus fermentum from fermented maize doughs (mawe and ogi) from Benin. J Appl Microbiol 85: 512–20.

[12] Sanni A, Morlon-Guyot J, Guyot JP. New efficient amylase-producing strains of Lactobacillus plantarum and L. fermentum isolated from different Nigerian traditional fermented foods. Int J Food Microbiol 2002;72:53–62.

[13] Guyot JP, Calderon M, Morlon-Guyot JP (2000). Effect of pH control on lactic acid fermentation of starch by *Lactobacillus manihotivorans* LMG 180010T. J. Appl. Microbiol. 88: 176-182.

[14] Ji, GE., Han HK, Yun SW, Rhim SL (1992). Isolation of amylolytic *Bifidobacterium* sp. Int57 . J. Microbiol. Biotechnol. 2: 85–91.

[15] Lee SK, Kim, YB, Ji, GE, (1997). Purification of amylase secreted from *Bifidobacterium adolescentis*. J. Appl. Microbiol. 83: 267–272.

[16] Lee JH, Lee SK,. Parka KH, In K,. Jib GE (1999). Fermentation of rice using amylolytic *Bifidobacterium*. Int J Food Microbiol 50: 155–161.

[17] Nakamura LK (1981). *Lactobacillus amylovorus*, a new starch-hydrolyzing species from cattle waste-corn fermentations. Int J Syst Bacteriol: 31:56–63.

[18] Chatterjee M, Chakrabarty SL, Chattopadhyay BD, Mandal RK (1997) Production of lactic acid by direct fermentation of starchy wastes by an amylase-producing Lactobacillus. Biotechnol Lett: 19:873–4.

[19] Giraud E, Champailler A, Raimbault M (1994) Degradation of raw starch by a wild amylolytic strain of *Lactobacillus plantarum*. Appl Environ Microbiol 60: 4319–23.

[20] Vishnu C, Seenayya G, Reddy G. Direct fermentation of various pure and crude starchy substrates to L(+) lactic acid using *Lactobacillus amylophilus* GV6.World J Microbiol Biotechnol 2002;18:429–33.

[21] Naveena BJ, Altaf Md, Bhadrayya K, Madhavendra SS, Reddy G (2005) Direct fermentation of starch to L(+) lactic acid in SSF by *Lactobacillus amylophilus* GV6 using wheat bran as support and substrate—medium optimization using RSM. Process Biochem 40: 681–90.

[22] Anuradha R, Suresh AK, Venkatesh KV (1999) Simultaneous saccharification and fermentation of starch to lactic acid. Process Biochem 35: 367–75.

[23] Zhang DX, Cheryan M (1994) Starch to lactic acid in a continuous membrane reactor. Process Biochem 29: 145–50.

[24] Naveena BJ, Vishnu C, Altaf Md, Reddy G (2003) Wheat bran an inexpensive substrate for production of lactic acid in solid state fermentation by *Lactobacillus amylophilus* GV6-optimization of fermentation conditions. J Sci Ind Res62: 453–6.

[25] Goel MK. Biotechnology: an overview; 1994 (http://www.rpi.edu/dept/chem-eng/Biotech-Environ/goel.html).

[26] Altaf M, Venkateshwar M, Srijana M, Reddy G (2007) An economic approach for L-(+) lactic acid fermentation by *Lactobacillus amylophilus* GV6 using inexpensive carbon and nitrogen sources. J Appl Microbiol;103:372–80.

[27] Pandey A, Soccol CR, Rodriguez-Leon JA, Nigam P (2001). Solid state fermentation in biotechnology: fundamentals and applications. New Delhi: Asiatech Publishers

[28] Mercier P, Yerushalami L, Rouleau D, Dochania D (1992). Kinetics of lactic acid fermentations on glucose and corn by *Lactobacillus amylophilus*. J Chem Technol Biotechnol 55:111–21.

[29] Cheng P, Muller RE, Jaeger S, Bajpai R, Jannotti EL (1991) Lactic acid production from enzyme thinned cornstarch using *Lactobacillus amylovorus*. J Ind Microbiol 7: 27–34

[30] Kurusava H, Ishikawa H, Tanaka H. L-lactic acid production from starch by co-immobilized mixed culture system of *Aspergilus awamori* and *Streptococcus lactis*. Biotechnol Bioeng 1988;31: 183–7.

[31] Kristoficova L, Rosenberg M, Vlnova A, Sajbidor J, Cetrik M. Selection of Rhizopus strains for L (+) lactic acid and gammalinolenic acid production. Folia Microbiol 1991;36:451–5.

[32] Nakamura LK, Crowell CD (1979). *Lactobacillus amylophilus*, a new starch hydrolyzing species from swine waste-corn fermentation. Dev Ind Microbiol 20: 531– 40.

[33] Prasana VA (2005). Amylases and their applications. Afr. J Biotechnol 4(13): 1525-1529.

[34] Sivaramakrishnan S, Gangadharan D, Madhavan K (2006) α-Amylase from microbial sources-An overview on recent developments. Food technol. Biotechnol. 44(2): 173-184.

[35] Ghorbel RE, Maktouf S, Massoud EB, Bejar S, Chaabouni SE (2009) New Thermostable Amylase from *Bacillus cohnii* US147 with a Broad pH Applicability. Appl. Biochem. Biotechnol. 157: 50–60.

[36] Lin LL, Chyau CC, Hsu WH (1998) Production and properties of a raw starch degrading amylase from thermophilic and alkaliphilic Bacillus sp TS-23. Biotechnol Appl Biochem, 28 (1): 61-68.

[37] Giraud E, Cunny G (1997) Molecular characterization of the α-amylase genes of *Lactobacillus plantarum* A6 and *Lactobacillus amylovorus* reveals an unusual 3′ end structure with direct tandem repeats and suggest a common evolutionnary origin. Gene 198: 149-157.

[38] Talamond, P, Desseaux V, Moreau Y, Santimone M, Marchis-mouren G (2002) Isolation, characterization and inhibition by acarbose of the α-amylase from *Lactobacillus fermentum* : comparison with *Lb. manihotivorans* and *Lb. plantarum*. Comparative Biochem. and Physiol. 133: 351-360.

[39] Qui and Yao (2007) Lactic acid production, rice straw. Bioresources 2(3): 419-429.

[40] Sawadogo-Lingani H, Diawara B.Traoré AS, Jakobsen M (2008) Technological properties of *Lactobacillus fermentum* involved in the processing of dolo and pito, West African sorghum beers, for the selection of starter cultures. J.Appl Microbiol, 104(3): 873-882.

[41] Champagne C (1998). La production des ferments dans l'industrie laitière Ed. Agriculture et Agroalimentaire, Canada,pp : 32-145.

[42] Edelman S, Westerlund-Wikström B, Leskelä S, Kettunen H, Rautonen N, Apajalahti J., Korhonen TK (2002) In vitro adhesion specificity of indigenous *lactobacilli* within the avian intestinal tract. *Appl Environ Microbiol* 68: 5155-5159.

[43] Hofvendahl K, Hahn-Hagerdal B (1997) L-lactic acid production from whole wheat flour hydrolysate using strains of *Lactobacilli* and *Lactococci*. Enzyme Microbiol Technol 20: 301-307.

[44] Datta RS, Sai PT, Patric B, Moon SH, Frank JR (1993). Technological and economic potential of polylactic acid and lactic acid derivatives. International congress on chemicals from biotechnology, Hannover, Germany p. 1–8.

[45] Wee YJ, Kim JN, Ryu HW (2006). Biotechnological production of lactic acid and its recent applications. Food Technol Biotechnol 44: 163–72.

[46] Senthuran A, Senthuran V, Mattiasson B, Kaul R (1997). Lactic acid fermentation in a recycle batch reactor using immobilized *Lactobacillus casei*. Biotechnol Bioeng 55:843–53.

[47] Tatsinkou, FB, Tavea F, Jiwoua C, Ndjouenkeu R (2009) Screening of thermostable amylase producing bacteria and yeasts strains from some cameroonian soils. Afr. J. Microbiol. Res. 3(9): 504-514.

[48] Tatsinkou, FB, Tavea F, Jiwoua C, Ndjouenkeu R (2011) Simultaneous production of raw starch degrading highly thermostable a-amylase and lactic acid by *Lactobacillus fermentum* 04BBA19. Afr. J. Biotechol. 10(34): 6564-6574.

[49] Schillinger U, LuKe F.K (1987). Identification of lactobacilli from meat and meat products. Food Microbiology 4, 199– 208.

[50] Dykes GA, Britz TJ, von Holy A (1994). Numerical taxonomy and identification of lactic acid bacteria from spoiled, vacuum packaged Vienna sausages. Journal of Applied Bacteriology 76, 246– 252.

[51] Kimberley AC, Taylor, C (1996) A simple colorimetric assay for muramic acid and lactic acid. Appl. Biochem. Biotechnol. 56: 49-58.

[52] Miller GL (1959) Use of dinitrosalicylic acid reagent for determination of reducing sugar. Anal. Chem. 31: 426-429.

[53] Goyal N, Gupta JK, Soni. SK (2005) A novel raw starch digesting thermostable α-amylase from *Bacillus* sp. I-3 and its use in the direct hydrolysis of raw potato starch. Enzyme and Microb. Technol. 37: 723-734.

[54] Liu DX, Xu Y (2008) A novel raw starch digesting α-amylase from a newly isolated *Bacillus* sp YX-1: purification and characterization. Biores. Technol. 99: 4315-4320.

[55] Hiller P, Wase D, Emery, AN (1996) Production of α-amylase by *Bacillus amyloliquefaciens* in batch and continuous culture using a defined synthetic medium. Biotechnol. Lett. 18 : 795-799.

[56] Burhan A, Nisa U, Gökhan, C, Ômer C, Ashabil, Osman, G (2003) Enzymatic properties of a novel thermostable, thermophilic, alkaline amylase and chelator resistant amylase from an alkaliphilic *Bacillus* sp isolate ANT-6. Process Biochem. 38: 1397-1403.

[57] Chandra AK, Medda S, Bhadra, AK (1980) Production of extracellular thermostable α-amylase by *Bacillus licheniformis*. J. Ferment. Technol. 58: 1-10.

[58] Saito N, Yamamoto K (1975) Regulatory factors affecting α-amylase production in *Bacillus licheniformis*. J. Bacteriol. 121: 848-856.

[59] Gupta R, Gigras P, Mohapatra H, Goswami VK, Chauhan, B (2003) Microbial α-amylases: a biotechnological perspective. Process Biochem. 38: 1599-1616.

[60] Muralikrishna G, Nirmala M, (2005) Cereal α-amylase –an overview. Carbohydrate polymer 60: 193-173.

[61] Gangadharan D, Sivaramakrishna, K, Nampoothiri KM, Sukumaran RK, Pandey A (2008) Response surface methodology for the optimisation of alpha amylase production by *Bacillus amyloliquefaciens*. Biores. Technol. 99: 4597-4602.

[62] Sodhi HK, Sharma K, Gupta, JK, Soni SK (2005) Production of a thermostable -amylase from *Bacillus* sp PS-7 by solid state fermentation and its synergistic use in the hydrolysis of malt starch for alcohol production. Process Biochem. 40: 525-534.

[63] Cherry HM, Hussain T, Anwar M.N (2004) Extracellular glucoamylase from isolate of *Aspergillus fumigates*. Pak. J. Biol. Sci. 7(11): 1988-1992.

[64] Theodoro CE, Martin ML (2000) Culture conditions for the production of thermostable amylase by *Bacillus* sp. Brazilian J. Microbiol. 31: 298-302.

[65] Halsetine C, Rolfsmeier M, Blum P (1996) The glucose effect and regulation of α-amylase synthesis in the hyperthermophilic archeon Sulfolobus solfataricus . J.Bacteriol. 178: 945-950.

[66] Gao L, Yang H, Wang X, Huang Z, Ishii M, Igarashi Y, Cui Z (2008) Raw straw fermentation using lactic acid bacteria. Biores.Technol. 99: 2742-2748.

[67] Tanyildizi, MS, Ozer D, Ebiol, M (2005) Optimisation of α-amylase production by *Bacillus* sp using response surface methodology. Processs Biochem. 40, 2291-2296.

[68] Wu WX, Mabinadji J, Tatsinkou FB (1999) Effect of culture conditions on the production of an extracellular thermostable alpha-amylase from an isolate of *Bacillus* sp. J. Zhejiang Univ Agric Life Sci 25: 404-408.

[69] Reddy PR, Reddy G Seenayya G (1999). Enhance production of thermostable β-amylase and pullulanase in the presence of surfactant by *Clostridium thermosulfurogenes* SV2. Process Biochem. 34: 87-92.

[70] Aguilar G, Morlon-Guyot J, Trejo-Aguilar, B, Guyot, JP (2000) Purification and Characterization of an extracellular alpha-amylase produced by *Lactobacillus manihotivorans* LMG 18010T, an amylolytic lactic bacterium. Enzyme Microb. Technol. 27: 406-413.

[71] Underkoffer L (1976) Microbial enzyme In Miler B. Litsky W. (Eds). Industrial Microbiology. Mc Graw-Hill, New York.

[72] Viara NI, Elena PD, Elka IE (1992) Purification and characterisation of a thermostable alpha amylase from *Bacillus licheniformis*. J. of Bacteriol. 28: 277-289.

[73] Vihinen M, Mantsala P (1990). Characterization of a thermostable *Bacillus stearothermophilus* alpha-amylase. Biotechnol. Appl. Biochem. 12: 427-435.

[74] Canganella F, Andrade C, Antranikian G (1994) Characterization of amylolytic and pullulytic enzymes from thermophilic archea and from a new *Ferividobacterium* species. Appl. Microb. Biotechnol. 42 : 239-245.

[75] Kwak Y, Akeba T, Kudo T (1998) Purification and characterization of enzyme from hyperthermophilic archeon *Thermococcus profundus*, which hydrolyses both α-1-4 and α-1-6 glucosidic linkages. J. Ferment. Bioeng, 86: 363-367.

[76] Prakash B, Vidyagar M, Madhukumar MS, Muralikrishna G, Sreeramulu K (2009). Production, purification and characterization of two extrememely halotolerant thermostable, and alkaline stable α- amylases from *Chromohalobacter* sp TVSP 101. Process Biochem. 44: 210-215.

[77] Bayramoglu G Yilmaz M Arica Y (2003) Immobilization of a thermostable a-amylase onto reactive membranes: kinetics characterization and application to continuous starch hydrolysis. Food chem. 57: 83-90.

[78] Lee Hs, SeG, Carter S (2001) Amylolytic cultures of *Lactobacillus acidophilus*: potential probiotics to improve dietary starch utilization. J food Sci. 66:2.

[79] Ohkouchi Y, Inoue Y (2006) Direct production of L(+)-lactic acid from starch and food wastes using *Lactobacillus manihotivorans* LMG18011. Bioresour. Technol 97:1554–62.

[80] Mette Hedegaard Thomsen, Guyot JP, Kiel P. Batch fermentations on synthetic mixed sugar and starch medium with amylolytic lactic acid bacteria. Appl Microbiol Biotechnol 2007;74: 540–6.

[81] Krishnan S, Bhattacharya S, Karanth NG (1998) Media optimization for production of lactic acid by *Lactobacillus plantarum* NCIM 2084 using response surface methodology. Food Biotechnol 12: 105–21.

[82] Junya Narita, Nakahara S, Fukuda H, Kondo A (2004) Efficient production of L-(+)-lactic acid from raw starch by *Streptococcus bovis* 148. J Biosci Bioeng 97: 423–5.

[83] Yumoto I, Ikeda K (1995) Direct fermentation of starch to L(+)-lactic acid using *Lactobacillus amylophilus*. Biotechnol Lett 17: 543–6.

[84] Champ MO, Szylit P, Raimbault M, Abdelker N (1983) Amylase production by three Lactobacillus strains isolated from chicken crop. J Appl Bacteriol 55:487–93.

[85] Lindgren S, Refai O (1984) Amylolytic lactic acid bacteria in fish silage. J Appl Bacteriol 57: 221–8.

[86] Diaz-Ruiz G, Guyot JP, Ruiz-Teran F, Morlon-Guyot J, Wacher C (2003) Microbial and physiological characterization of weakly amylolytic but fast-growing lactic acid bacteria: a functional role in supporting microbial diversity in pozol, a Mexican fermented maize beverage. Appl Environ Microbiol 69:4367–74.

[87] Bohak I, Back W, Richter L, Ehrmann M, Ludwing W, Schleifer KH (1998) *Lactobacillus amylolyticus* sp. nov., isolated from beer malt and beer wort. Syst Appl Microbiol : 21:360–4.

Chapter 11

Lactic Acid Bacteria as Source of Functional Ingredients

Panagiota Florou-Paneri, Efterpi Christaki and Eleftherios Bonos

Additional information is available at the end of the chapter

1. Introduction

Lactic acid bacteria (LAB) are widespread microorganisms which can be found in any environment rich mainly in carbohydrates, such as plants, fermented foods and the mucosal surfaces of humans, terrestrial and marine animals. In the human and animal bodies, LAB are part of the normal microbiota or microflora, the ecosystem that naturally inhabits the gastrointestinal and genitourinary tracts, which is comprised by a large number of different bacterial species with a diverse amount of strains [1,2].

Phylogenetically the LAB belong to the *Clostridium* branch of Gram positive bacteria. They are non-sporing, aero tolerant anaerobes that lack catalase and respiratory chain, with a DNA base composition of less than 53 mol% G+C [3,4]. According to their morphology LAB are divided to robs and cocci and according to the mode of glucose fermentation to homofermentative and heterofermentative. The homofermentative LAB convert carbohydrates to lactic acid as the only or major end-product, while the heterofermentative produce lactic acid and additional products such as ethanol, acetic acid and carbon dioxide [5,6]. Thus, the main metabolism of LAB is the degradation of different carbohydrates and related compounds by producing primarily lactic acid and energy. Although many genera of bacteria produce lactic acid as primary or secondary fermentation products, typical lactic acid bacteria are those of the Lactobacillales order, including the following genera: *Lactobacillus, Carnobacterium, Lactococcus, Streptococcus, Enterococcus, Vagococcus, Leuconostoc, Oenococcus, Pediococcus, Tetragonococcus, Aerococcus* and *Weissella* [7].

Many strains of LAB are among the most important groups of microorganisms used in the food and feed industries, although some of the genus Pediococcus cause deterioration of foods, which results in their spoilage [4]. LAB have been used in food preservation and for the modification of the organoleptic characteristics of foods, for example flavors and texture [2]. Various strains of LAB [8] can be found in dairy products (yoghurt, cheese), fermented meats (salami), fermented vegetables (olives, sauerkraut), sourdough bread, etc [9]. The European Food Safety Authority (EFSA) has stated that several LAB strains can be considered to have "Qualified Presumption of Safety" QPS-status [9].

Moreover, nowadays, LAB play an important role in the industry for the synthesis of chemicals, pharmaceuticals, or other useful products (Figure 1). Also, the biotechnological production of lactic acid has recently reported that offers a solution to the environmental pollution by the petrochemical industry [10].

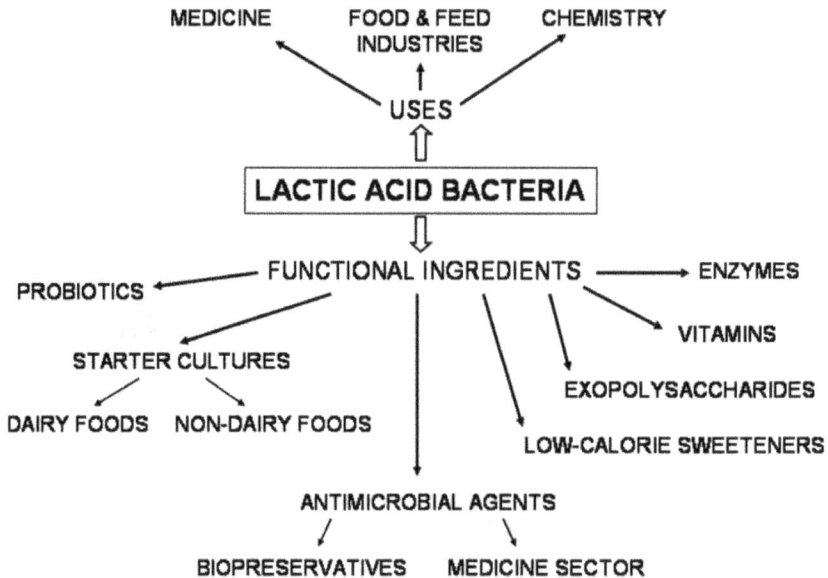

Figure 1. Uses and Functional Ingredients of Lactic Acid Bacteria

This chapter will discuss recent applications of LAB as source of probiotics, starter cultures, antimicrobial agents, vitamins, enzymes and exopolysaccharides, especially those that can satisfy the increasing consumer's demands for natural products and functional foods in relation with human health.

2. Lactic acid bacteria as source of probiotics

Etymologically the term probiotics is derived from the Greek "probios" which means "for life". In 1974 Parker [11] defined as probiotics "organisms and substances which contribute to intestinal microbial balance". Fuller in 1989 [12] defined as probiotic "a live microbial feed supplement which beneficially affects the host animal by improving its intestinal microbial balance". Later the Food and Agriculture Organization / World Health Organization defined probiotic bacteria as "live microorganisms which when administered in adequate amounts confer a health benefit on the host" [13]. Since probiotics can colonize the gastrointestinal tract and exert their beneficial effect long term, without requiring continuous medical intervention, they have been used for a century to treat a variety of mucosal surface infections (gut, vagina), but their use decreased after the appearance of antibiotics. However, today, probiotics are considered as an alternative solution to antibiotics due to the increasing spread of antibiotic resistance and the need for treatment cost reduction [14].

Microorganisms considered as commercial probiotics are mainly of the Lactobacillus genus with over one hundred species recognized, for example: *L. acidophilus, L. rhamnosus, L. reuteri, L. casei, L. plantarum, L. bulgaricus, L. delbrueckii, L. helveticus* [15-17]. Lactobacilli are Generally Recognized As Safe (GRAS) organisms [18,19].

Probiotic bacteria are very sensitive to many environmental stresses, such as acidity, oxygen and temperature [20,21] and they must fulfill some functional and physiological aspects such as [21,22]: a) Adherence to the intestinal epithelium and colonization of the lumen of the tract. b) Ability to stabilize the intestinal microbiota. c) Counteracting the action of harmful microorganisms. d) Production of antimicrobial substances. e) Stimulation of the immune response.

There are subcategories of the general term probiotic [23,24] which are: a) Probiotic drugs: intended to cue, treat and prevent disease. b) Probiotic foods: food ingredients and dietary supplements. c) Direct-fed microbials: probiotics for animal use. d) Designer probiotics: genetically modified. Generally, foods containing probiotic bacteria fall in the category of functional foods [25].

2.1. Mechanism of action of probiotics

Probiotics have multiple and diverse effects on the host. The main mechanisms of action of probiotic bacteria by which they improve mucosal defenses of the gastrointestinal tract include:

a. Antimicrobial activity: The probiotics block the colonization of pathogenic bacteria by decreasing luminal pH, inhibiting bacterial invasion and adhesion to epithelial cells and producing antimicrobial compounds such as bacteriocins and defensins, organic acids and hydrogen peroxide. The interaction of LAB with the mucosal epithelial cells of the gastrointestinal tract and the lymphoid cells in the gut enhance the gut immune response against ingested pathogens [26,27].
b. Enhancement of mucosal barrier function against ingested pathogens: It is achieved with the increasing mucus production through modulation of cytoskeletal and tight junctional protein phosphorylation. The probiotic bacteria compete with pathogenic bacteria for epithelial binding sites, inhibiting the colonization of strains like *Salmonella* and *E. coli* [28,29]. Probiotic bacteria interact with the epithelial cells of the gut, either directly (via cell compounds like DNA, lipoteichoic acids and cell-surface polysaccharides) or indirectly (through production of bioactive metabolites) [30]. The enhancement of mucosal barrier function may be an important mechanism by which probiotics benefit the host in various diseases such as Type 1 diabetes [31,32].

c. Immunomodulation: Specific strains of probiotics might influence the innate and the acquired immune system, thus playing an important role in human diseases. Probiotic bacteria may affect the epithelial cells, the dendritic cells, the monocytes / macrophages and the various types of lymphocytes (Natural killer cells, T-cells and T-cell redistribution) directly or secondarily [33,34]. This action of probiotics could be important for the elimination of neoplastic host cells [22]. Moreover, the effects of probiotics on B-lymphocytes and antibody production resulted in an increase in IgA secretion and the enhancement of response to vaccination [34]. Recently, it was also reported that probiotics can have positive effects on the respiratory system by preventing and reducing the severity of respiratory infections, because of an increase of IgA in the bronchial mucosa [35].

2.2. Probiotics and health

Functional properties of probiotics have been demonstrated for various therapeutic applications. Nevertheless, the health benefits provided by probiotics are strain-specific, therefore no probiotic strain will have all proposes benefits, not even strains of the same species [36]. Among the LAB probiotic strains *L. rhamnosus* GG and *L. casei Shirota* have the strongest human health efficacy in the management of lactose intolerance, rotaviral diarrhea and antibiotic associated diarrhea [17]. An optimal single oral dose, based on detection of the bacteria in human feces is 10^9 bacterial colony forming units (CFU) [37], while in other reports 10^6-10^7 CFU / g of food are considered adequate amounts [13,17,38].

Moreover, in animal nutrition dietary probiotics or direct fed microbials, term which is preferred in the USA, are able to help the maintenance of a healthy intestinal microflora. This microflora may serve to improve performance and health status of the animals, but also to suppress food born pathogens such as *Salmonella* and *Campylobacter*. These conditions are necessary for the production of safe meat and meat products [39]. For instance, the gastrointestinal tract of broilers can be colonized by ingested probiotic bacteria from the first days of their life, which results in shorter period for the achievement of microflora stability. Also, in another example the dietary inclusion of probiotics in young calves' milk replacers may improve their growth performance [40].

Some of the beneficial effects of probiotics are well established as shown in Table 1.

Lactic Acid Bacteria	Effects on human health	References
Lactobacillus rhamnosus GG	May shorten the course of rotavirus causing diarrhea. Helps to alleviate the symptoms of ulcerative colitis and atopic dermatitis.	41-45
Lactobacillus casei	Reduces the severity and duration of diarrhea. It can stimulate the immune system of the gut and alleviates the symptoms of Crohn's disease	17, 43, 46

Lactic Acid Bacteria	Effects on human health	References
Lactobacillus acidophilus	Secretes lactic acid which reduces the pH of the gut and inhibits the development of pathogens (*Salmonella spp, E. coli*). Reduces blood cholesterol.	17, 46, 47
Lactobacillus johnsonii	Effective in inhibition of *H. pylori* and against inflammation	17, 36
Lactobacillus plantarum	Produces short-chain fatty acids that block the generation of carcinogenic agents by reducing enzyme activities	17, 36
Lactobacillus fermentum	Effective in restoration of a normal microflora. Effective against bacterial vaginosis flora	48
Lactobacillus reuteri	Reduces the duration of diarrhea	49
Enterococcus faecium	Can reduce blood cholesterol leading to decreased blood pressure	50-52

Table 1. Lactic acid bacteria derived probiotics and human health

LAB derived probiotics have potential health benefits in the following situations:

1. Diarrheal diseases:
   a. Infective diarrhea. The most studied gastrointestinal condition treated by probiotics is acute infectious diarrhea in infants. Children represent a main target of studies due to the importance of limiting the spread of diseases and decreasing the need of antibiotics (Aureli et al. 2011). Clinical trials with LAB derived probiotics (*L. rhamnosus* GG; *L. reuteri*; *L. casei*; *L. delbrueckii subsp. Bulgaricus*) support the efficacy of these probiotics in preventing diarrhea [14,49,53], due to their direct or indirect interaction with the enterotoxins [54].
   b. Antibiotic associated diarrhea. A variety of probiotic bacteria, mainly lactobacilli have been used in the treatment and prevention of antibiotic associated diarrhea [55,56]. In a recent study *L. acidophilus* and *L. casei* seemed to be effective in reducing the risk of development of diarrhea [57]. Nevertheless, the results obtained were from pilot studies, so further investigation is needed to evaluate the efficacy of probiotics on such disorders.
   c. Clostridium difficile associated diarrhea. *C. difficile* is an opportunistic pathogen often responsible for diarrhea in vulnerable people. *L. rhamnosus* GG, has demonstrated positive effects on treated patients [14].
   d. Travelers diarrhea. Probiotics with Lactobacilli did not seem to be effective on such diarrhea, which is caused by bacteria, in particular enterotoxigenic E. coli [14,53].
   e. Radiation induced diarrhea. Although there is little research on this subject, probiotics seem to be promising in decreasing radiation diarrhea [53].

2.   Inflammatory bowel disease

LAB may affect positively the intestinal mobility and relieve constipation, possibly through a reduction of the intestinal pH [58].

   a.   Pouchitis. It is a chronic inflammation of the ileal pouch. The treatment with probiotics such as *L. rhamnosus* GG and *L. acidophilus* reduced the risk of pouchitis due to decreased mucosal inflammation [59,60].
   b.   Crohn's disease. This disease can involve the whole gastrointestinal tract and is characterized by inflammatory processes occurring deeper in the tissues. Among other typical treatments *L. rhamnosus* is used aiming at decreasing the rate and the severity of disease after surgery [61].
   c.   Ulcerative colitis. It is an acute or chronic disease only affecting the large bowel. LAB probiotics (*L. acidophilus*) provide some promising initial indications [53].

3.   Irritable bowel syndrome

This term is used to describe a heterogenous group of gastrointestinal symptoms, like diarrhea, constipation, bloating and abdominal pain. *L. plantarum* strain 299V and *E. faecium* PR88 could be effective treatments against this syndrome [62,63].

4.   Prevention of colon cancer

The anticarcinogenic effect of probiotiocs may be attributable to a combination of mechanisms like the induction of pro- or anti-inflammatory and secretary responses that could inhibit carcinogenesis [22]. In vitro studies with lactobacillus strains have shown anti-mutagenic activities. However, there is no evidence yet that probiotics can protect against the development of colon cancer in humans [64,65]. Although, it is hypothesized that the strains tested may have anti-carcinogenic effects by reducing the activity of the enzyme β-glucuronidase.

5.   Helicobacter pylori

It is a common chronic bacterial infection in humans, which causes many problems, such as chronic gastritis, septic ulcers and gastric cancer. Probiotics *like L. salivarius, L. casei Shirota* and *L. acidophilus* appear to be promising in inhibiting the growth of H. pylori in vitro [66,67]. Moreover, *L. johnsonii* was also shown effective to inhibit *H. pylori* [68].

6.   Lactose intolerance

It is the most common disorder of the intestinal carbohydrate digestion. In both adults and children it has been shown that probiotics can improve the lactose digestion by reducing the intolerance symptoms and slowing orocecal transit [69,70].

7.   Blood cholesterol

Recently it has been suggested that some strains of probiotic bacteria, *like L. acidophilus, L. plantarum* and *Enterococcus faecium* could significantly reduce blood cholesterol and increase resistance of low density lipoprotein oxidation, leading to decrease of blood pressure [50-52].

8.   Other disorders

The majority of probiotics use has focused on diseases related to the gut, but there are studies that evaluated probiotics, in allergic conditions, including atopic dermatitis, rhinitis, bacterial vaginosis and food allergies [53,71].

a.   Atopic Dermatitis. It is the most common of the chronic skin disorders, known as eczema. Investigations have shown that probiotics like *L. rhamnosus* GG, can prevent or reduce the symptoms [42]. Even eczema can be prevented if mothers ingest probiotics during pregnancy and neonatals ingest them during the first 6 months of their life [53].
b.   Bacterial Vaginosis. Probiotics (*L. rhamnosus* GR-1 and *L. fermentum* RC-14) are considered to have theraupetic benefits in vaginosis. Probably this is due to the large numbers of lactobacilli in the healthy vaginal microflora [72,73].
c.   Other ailments. Probiotics such as *Lactococcus*, *Pediococcus* and *Leuconostoc* can prevent or limit mycotoxinogenic mould growth [74-78]. Moreover, LAB according to their bacterial strain could bind aflatoxin B₁ both in vivo and in vitro [79]. It was reported in studies that *L. paracasei* ST11 reduced body and abdominal fat [80]. These probiotic bacteria seemed to have an anti-obese action. Probably intestinal bacteria may regulate body weight by affecting the host's metabolic neuroendocrine and immune functions [80]. Additionally, probiotics may have anticariogenic effects, preventing and treating dental caries and generally be effective in the oral cavity and the treatment of periodontal disease [81,82].

3. Lactic acid bacteria as source of starter cultures

3.1. Starter cultures and functional starter cultures in fermentation of foods

LAB for a long time have been applied as starter cultures in fermented foods and beverages, because they can improve nutritional, organoleptic, technological and shelf-life characteristics [83,84]. LAB initiate rapid and adequate acidification in the raw materials, through the production of various organic acids from carbohydrates. Lactic acid is the most abundant, followed by acetic acid, whilst LAB can also produce ethanol, bacteriocins, aroma compounds, exopolysaccharides and some enzymes [85]. Earlier the production of fermented foods and beverages was obtained on a spontaneous fermentation, due to the microflora naturally present on the raw materials. Later on, the direct addition of selected starter cultures to the food matrix was preferred by the food industry. The advantages were the high degree of control over the fermentation process and the standardization of the final product [84].

As starter culture can be defined a microbial preparation of a large number of one or more microorganisms which is introduced to a raw material aiming to produce a fermented food by accelerating and steering its fermentation process [86,87].

The industries of fermented foods mainly utilize commercial starter cultures for the direct inoculation to the food matrix, which are available as frozen and freeze dried concentrates or lyophilized preparations [88].

Recently the use of functional starter cultures in food and beverage fermentation is being explored. These cultures have at least one functional property, contributing in the improvement of the fermentation process, enhancing the quality and of the end safety product and conferring health benefits [84]. Nevertheless, the selection of starter cultures must also eliminate undesirable side effects like the formation of D-lactic acid or a racemate of lactic acid (DL) or the formation of biogenic amines [84,89].

3.2. Probiotics as functional starter cultures

A category of successful starter cultures are LAB produced probiotic cultures. Firstly, Metchnikoff [90] discovered the beneficial effects of LAB on human health, through the consumption of yoghurts and fermented milks. Currently probiotic cultures are used for a number of products such as yoghurt, yoghurt drinks, infant formulas, dietary supplements, etc [91]. Yoghurt is manufactured using *Streptococcus thermophilus* and *Lactobaccilus delbrueckii subsp. bulcaricus* as starter cultures [17].

A manufacture in order to choose any probiotic microbial strain to be used as starter culture or better as a blend with a traditionally used starter culture (co-culture), must check the following aspects [92]: 1) The ability of the probiotics to grow in a medium to increase the cells counts. 2) The robustness of the organism to withstand the freezing and drying stages of preparation. 3) The tolerance to acidity of the gastric acid and the bile salts during their passage in the gastrointestinal tract. Thus, the probiotic strains must be stable in order to claim the health benefits [92].

LAB are used as starter cultures either in dairy or non-dairy products (Table 2).

Genus	Application in dairy foods	Application in non-dairy foods
Lactobacillus spp	Cultured dairy products, cheese, yoghurt, kefir	Sausage, sourdough bread, fermented vegetables
Lactococcus spp	Cheese, butter milk sour cream, cultured dairy products	-
Leuconostoc spp	Cheese, cultured dairy products, sour cream, buttermilk	Fermented vegetables
Streptococcus thermophilus	Cheese, yoghurt	-
Pediococcus	-	Sausage, fermented vegetables
Tetragenococcus	-	Soy sauce
Oenococcus	-	Wine

Table 2. Lactic acid bacteria used as starter cultures in fermented foods

3.3. Functional starter cultures in fermented dairy products

Traditionally, LAB have been used in the fermentation of dairy products, as a simple and safe way of preserving such foods. The main species of LAB that can potentially be used as probiotic cultures in dairy products belong to the *Lactobacillus spp* (*L. acidophilus*, *L. lactis*, *L. casei*, *L. plantarum*, *L. rhamnosus*, *L. reuteri*, *L. delbrueckii subsp. bulgaricus*) or to the *Enterococcus spp.* (*E. faecalis*, *E. faecium*) [17,92].

Dairy products are considered as ideal vehicles for delivering probiotics to the human gut. Yoghurt is considered the most important, followed by cultured buttermilk, kefir, cheeses, ice-cream [17,22,92] or frozen desserts like chocolate mousse [93]. Moreover proteolytic strains of LAB produced probiotics are used to release bioactive peptites called angiotensin I-converting enzyme inhibitors, which are examined for their hypotensive role [94]. Furthermore interaction between probiotics and starter cultures are possible, either as synergism (e.g. yoghurt) or antagonism (e.g. bacteriocins which exhibit antibiotic properties) [20].

A minimum viable LAB count of 10^6 CFU/g in fermented dairy food is recommended for the claimed health benefits [95].

3.4. Functional starter cultures in fermented non-dairy products

3.4.1. Fermented meat and meat products

The preservation of meat and meat products by fermentation has been used from ancient times and it was based mainly on natural meat microorganisms. Recently, researchers begun to develop starter cultures for meat products, in order to ensure standard quality for the fermentation process [87]. In 1995 the first LAB meat starter culture used by Niven et al. [96] in the USA was a pure culture of *Pediococcus cerevisiae*. Essential requirements of meat LAB starter cultures are the immediate and rapid production of organic acids at the start of the fermentation, which will result in a pH below 5.1 [97]. Therefore, the original characteristics of the foods are changed, resulting in enhancement of the final products [98].

As commercial meat LAB starter cultures the species more used belong to the Lactobacillus and Pediococcus strains, which can be isolated from dry sausages [97], sauerkraut [20], or smoked salmon [99]. Strains of the above LAB were found to have the best survival activity under acidic conditions and high levels of bile salts [98]. The role of the starter culture as aforementioned is for the safety of foods by inactivating pathogens and spoilage microorganisms via the acid and bacteriosin production. Therefore, the production of biogenic amines is inhibited and microbial growth is suppressed, without the use of antibiotics [97].

Several studies have reported that LAB from meat and meat products can have antibiotic resistance [100]. Thus, before using novel starter cultures or probiotic cultures it is important to check that they do not contain transferable resistance genes [97]. In addition, the selection of LAB starter cultures for sausage production must not have amino decarboxylase activity.

Otherwise, biogenic amines will be produced in foods, such as histamine, tryptamine, tyramine, cadaverine, putrescine and phenylethylamine which have toxic effects [101].

3.4.2. Fermented vegetables

LAB fermentation of vegetables can be achieved due to the presence of carbohydrates. Usually fermented vegetable juices are produced from cabbage, red beet, carrot, celery and tomato [18,102]. Also, LAB play an important role in pickles and table olives fermentation, affecting the final flavour and shelf-life [103,104].

3.4.3. Starter cultures in silages

Ensiling is a traditional method of preserving forages and is widely used all over the world. It is based on natural fermentation, where LAB ferment water-soluble carbohydrates into organic acids, mainly lactic acid or acetic and formic acids, under anaerobic conditions. Inoculation of LAB is often used as silage additive to enhance lactic acid fermentation [18]. This results in decreasing pH, inhibiting detrimental anaerobes and preserving the nutritional value and palatability of the forage [105,106].

Among the LAB genera frequently used are *Lactobacillus plantarum*, Enterococcus *faecium*, *Pediococcus acidilactici*, *Pediococcus pentoseceus* and *Lactobacillus acidophilus*, with usual rates 10^5-10^6 viable cells / g [107]. Feeding ruminants with silages that have been treated with LAB beyond improving their performance, it is believed to induce probiotic effects [108].

4. Lactic acid bacteria as source of antimicrobial agents

LAB derived probiotic bacteria display a wide range of antimicrobial activities. Some strains of LAB produce non specific antimicrobial substances (short chain fatty acids, hydrogen peroxide) while others produce toxins (bacteriosins, bacteriosin-like components) [109]. Short chain fatty acids (formic, acetic, propionic, butyric and lactic acids) which are produced during the anaerobic metabolism of carbohydrates, decrease the pH. It has been considered that these acids are responsible for the domination of mucosal ecosystems by LAB [110]. Also, hydrogen peroxide inhibits the growth of pathogens [111].

4.1. Bacteriocins

Most of bacteriocins originating by Gram positive bacteria are produced from LAB. They are proteins that have bacteriocidal activity against species closely related to the bacteriocin producing strains, which could be applied in food preservation and health care [112,113]. Traditionally bacteriocin production has been considered an important characteristic in the selection of probiotic strains, while nowadays it is considered that they may function within the gastrointestinal tract [114], perhaps as alternatives to antibiotics for medical and veterinary use [115]. Generally bacteriocins are cationic peptides which display hydrophobic or amphilitic properties and usually the bacterial membrane is the target for their action [116].

The majority of bacteriocin produced by LAB are distinguished from classical antibiotics because: a) They are ribosomally synthesized and have a relatively narrow killing spectrum. b) They can be divided into two main groups, produced by Gram-negative and Gram-positive bacteria [117,118].

Bacteriocins according to their structure and characteristics can be classified mainly in the following classes:

- Class I (lantibiotics), small peptides [119].
- Class II, small heat-stable proteins which are further divided into subclasses such as IIa (pediocin-like bacteriocins) and IIb (two peptite bacteriocins) [119].
- Class III (helveticin) [120].

Bacteriocins mainly produced by *Lactobacillus acidophilus* have strong antimicrobial capacity against various food pathogens [113]. Bacteriocins can act as bactericidal or bacteriostatic, a distinction which is strongly dependent on bacteriocin dose and degree of purification, physiological state of the indicator cells and experimental conditions such as incubation temperature, pH, presence of agents disrupting cell wall integrity, etc [121,122].

4.2. Traits of LAB derived bacteriocins

LAB derived bacteriocins are suitable to use as food preservatives due to their characteristics: a) protein nature – they are inactivated by proteases in the gastrointestinal tract. b) Non-toxic and generally non-immunogenic. c) Thermoresistant thus the antimicrobial activity remains after pasteurization and sterilization. d) Affect most of the Gram-positive bacteria. e) Genetic determinants generally located in plasmid facilitating genetic manipulation to increase the variety of natural peptides. f) Usually act on the bacterial cytoplasmic membrane having no cross resistance with antibiotics [122,123].

Some benefits of the use of bacteriocins as food preservatives are: a) extended shelf-life of foods. b) reduction of the risk of transmission of food born pathogenic bacteria. c) Amelioration of economic losses due to food spoilage. d) No addition of chemical preservatives. e) Decrease of the intensity of heat treatments resulting in better preservation of food nutrients and sensory properties of the food. f) Marketing of "novel" foods, less acidic, less salty and with higher water content [123].

Nicin (lantibiotic – class I) is the first bacteriosin produced by LAB (*Lactococcus lactis*) whilst today it is used in many countries as biopreservatives in foods [109]. Nicins have a dual mode of action: a) Binding to lipid II thus preventing correct cell wall synthesis and b) employing lipid II as a docking molecule to initiate a process of membrane insertion and pore formation which leads to rapid cell death [109]. Nisin-producing bacteria can be found in about 30% of human milk samples. This substance may protect mothers from mastitis and infants from toxication by pathogenic skin flora like Staphylococcus aureus [124]. Except from nicin, currently pediocin PA-1/AcH from several Pediococcus strains and enterocin AS-48 from Enterococcus faecalis are used as biopreservatives [119].

There are at least three ways in which bacteriocins can be incorporated into a food to ameliorate its safety: a) By using a purified or semi-purified bacteriocin preparation as food ingredient. b) By introducing an ingredient that has earlier been fermented with a bacteriocin producing strain. c) By using a bacteriocin-producing culture in fermented products to produce the bacteriocin in situ [125].

Additionally, bacteriocin production can contribute to the probiotic functionality of intestinal LAB, while in certain cases may be directly responsible for it, with respect to either beneficially modulating the gut microbiota or inhibiting some gastrointestinal pathogenic bacteria [30].

Consequently bacteriocins, derived from LAB can cover a broad field of applications, including the food industry and the medical sector, mainly in combination with other treatments to increase their effectiveness in humans and animals [126]. In the latter, bacteriocins can be used as growth promoters, instead of antibiotics, which have been banned in the European Union since 2006 [127].

5. Lactic acid bacteria as source of vitamins

Human life cannot exist without vitamins, because they are involved in essential functions e.g. cell metabolism and antioxidant activities. Humans cannot synthesize most of these vitamins, although it is well known that some intestinal bacteria like LAB can produce some vitamins (folate, vitamin B_{12} or cobalamin, vitamin K_2 or menaguino, riboflavin and thiamine) [128,129]. The gut microbiota has been recognized as a source of some water-soluble vitamins, while such vitamins have also been reported as results of the LAB fermentation in yogurt, cheeses and other fermented foods.

5.1. Folate

Folate is the term used to describe the folic acid derivatives, such as the folyl glutamates which are naturally present in foods and folic acid that is the chemically synthesized form of folate, commonly used for food fortification and nutritional supplements. Folate belongs to the B-group of vitamins and participates in many metabolic pathways like the biosynthesis of DNA and RNA and the inter-conversions of amino acids. Moreover, folate possesses antioxidant capacity that protects the genome by preventing free radical hack of DNA [130].

Dietary folate is essential for humans, since it cannot be synthesized by mammalian cells. Folate can be found in legumes, leafy greens, some fruits and vegetables, in liver and fermented dairy products [131], especially in yogurts, where it may be increased depending on the starter cultures used and the storage condition, to values above 200 μg / lt [132]. Epidemiological studies indicated that folate deficiency is associated with a variety of disorders like Alzheinmer's disease, coronary heart diseases, osteoporosis and increased risk of breast and colorectal cancer [130,133].

LAB having the ability to produce folate belong to the *Lactobacillus spp (L. lactis, L. plantarum, L. bulgaricus), Streptococcus spp.* and *Enterococcus spp.* Nevertheless, some lactobacilli strains (*L. gasseri, L. salivarius, L. acidophilus* and *L. johnsonii*) used as both starter cultures and probiotics, cannot synthesize folate due to their lacking in some genes involved in folate biosynthesis [130]. Furthermore, it has been reported that some starter cultures and probiotic lactobacillus strains in non-dairy foods utilize more folate than they produce [128,130]. For this reason nowadays the food industry focuses on the strategy to select and use folate producing probiotic strains, to produce fermented products with elevated amounts of "natural" folate concentrations, without increasing production cost, although increasing health benefits [130,133].

5.2. Vitamin B$_{12}$

Vitamin B$_{12}$ or cobalamin is required for the metabolism of fatty acids, amino acids, nucleic acids and carbohydrates [134]. Vitamin B$_{12}$ cannot be synthesized by mammals and must be obtained from exogenous sources like foods or the intestinal microbiota [128]. It has been reported that among the microorganisms some members of the *Lactobacillus spp* have the ability to produce this vitamin. In particular a probiotic strain of *L. reuteri* which exhibits hypocholesterolaemic activity in animals can produce B$_{12}$ [135].

Vitamin B$_{12}$ deficiency can cause various pathological disorders that affect the haematopoietic (pernicious anaemia), nervous and cardiovascular system. Furthermore, this deficiency in male animal models influenced the number of offspring which showed growth retardation and decrease in some blood parameters [136].

5.3. Vitamin K

Vitamin K is involved in blood clotting, tissue calcification, atherosclerotic plaque and bones and kidneys function [137]. Vitamin K is present as phylloquinone (Vitamin K$_1$) in green plants and as menaquinone (K$_2$) produced by some intestinal bacteria, like LAB and especially strains of the genera *Lactococcus*, *Lactobacillus*, *Enterococcus*, *Leuconostoc* and *Streptococcus* [128]. Vitamin K deficiency has been involved in some clinical disorders like intracranial hemorrhage in newborn infants and possible bone fracture resulting from osteoporosis [129]. LAB producing menoquinone could be useful to supplement vitamin K requirements in humans [138].

5.4. Riboflavin

Riboflavin or vitamin B$_2$ is necessary in cellular metabolism, being the precursor of coenzymes acting as hydrogen carriers in biological redox reactions [129]. Although, riboflavin is present in many foods such as dairy products, meat, eggs, green vegetables, its deficiency occurs with damages in the liver, skin and changes in the brain glucose metabolism [128,129], with symptoms like hyperaemia, sore throat, odema of oral and mucous membranes, cheilosis and glossitis [139].

Currently, riboflavin-producing LAB strains were isolated and used as a convenient biotechnological application for the preparation of bread (fermented sourdough) and pasta to enrich them with vitamin B$_2$ [140].

6. Lactic acid bacteria as source of enzymes

LAB possess an extensive collection of enzymes many of which have the potential to influence the composition and the processing, organoleptic properties and quality of foods and feeds. LAB release various enzymes into the gastrointestinal tract and exert potential synergistic effects on digestion and alleviate symptoms of intestinal malabsorption [141]. In other cases these organisms may serve as a source for the preparation of enzyme extracts that are able to function under the environmental conditions of fermentation [142]. The enzymatic activity has been studied mainly in LAB isolated from wine or other fermented foods like cheeses and yoghurt [143,144]. Species of *Lactococcus* and *Pediococcus* are the LAB most commonly associated with fermented foods [143]. The LAB produced enzymes and in particular amylases which are the most stable can be used in sourdough technology for the natural improvement of bread texture [145]. Moreover, LAB contribute to the aroma and flavor of fermented foods. Certain peptidases produced by *Lactococcus lactis subsp. cremoris* improved the sensory quality of cheese [146]. In addition, proteolysis and lipolysis may enhance the flavour of most varieties of cheese [147]. LAB strains isolated from a traditional Spanish Genestoso cheese were evaluated for the enzymatic activity and it was reported that dipeptidase activity of high level was found for *Lactococcus spp*, enterolytic activity was detected for *Enterococcus spp.*, while carboxypeptidase activity was very low or undetectable [147].

Also, enzymes play an important role in winemaking. Wine flavor and aroma apart from aromas originating in grapes and alcoholic fermentation, is derived mainly from the activity of the LAB, through the action of their enzymes. These bacteria grow in wine during malolactic fermentation, following alcoholic fermentation, while a broad range of secondary modifications improve the taste and flavor of wine [144].

7. Lactic acid bacteria as source of exopolysaccharides

7.1. Definition and classification of exopolysaccharides

A number of LAB can produce a variety of long chain sugar polymers, called exopolysaccharides (EPS) which are mainly employed for the production of fermented dairy products. They are synthesized either extracellularly from sucrose by glycansucrases or intracellularly by glycosyltransferases from sugar nucleotide precursors [148]. These EPS can be classified according to their chemical composition and biosynthesis mechanism as homopolysaccharides, consisting of a single type of monosaccharide and heteropolysaccharides consisting of repeating units of two or more types of monosaccharides, substituted monosaccharides and other units like phosphate, acetyl and glycerol [149-151].

Homopolysaccharides are further divided into fructans including levan and inuline-type and glucans including dextran, mutan, alteran and b-1, 3 glucan [152]. On the other hand, heteropolysaccharides demonstrate little structural similarity to one another. Their production is influenced by the bacterial growth, phase, medium composition (carbon and nitrogen source), pH and temperature [153]. They can be produced by *Lactococcus spp.* and *Lactobacillus spp.* and they play a crucial role in the food industry [151]. Homopolysaccharides can be introduced in sourdough products, influencing the structural quality and backing ability in bakery products, while heteropolysaccharides are used as food additives in dairy products [154]. EPS contribute to the organoleptic quality of the fermented foods, in texture, taste perception, mouth-feel and stability [153,155]. The above researchers reported that there is no information about the effects of bacterial EPS in non-dairy foods, such as meat products, sauerkraut and vinegar. Although EPS are tasteless, they prolong the time that the milk product spends in the mouth, enhancing its delicacy through an improved volatilization of the intrinsic flavors [153,155].

7.2. Applications of exopolysaccharides in the industry

In the last years, EPS derived from LAB have received increasing interest because of their GRAS status and their properties. EPS can improve the rheology of fermented foods (viscosity and elasticity) as natural biothickeners, emulsifiers, gelling agents and physical stabilizers to bind water and limit syneresis [153,156]. In particular commercial products like LAB dextran could be utilized apart from foods in gel filtration products, in the pharmaceutical industry, as blood volume expander and flow improver, in chemistry as paper and metal plating processes, in enhanced oil recovery and in chromographic media. Furthermore, levan can find use in the food industry as biothickener, while alteran as low-viscocity factor, extender, etc [128,151,157]. Additionally, EPS may produce oligosaccharides having prebiotic properties that could find important applications in functional foods [158]. The successful application of EPS in the manufacture of fermented milks is determined by the ability to bind water, interact with proteins and increase the viscosity of the milk serum phase [159]. Although, many LAB strains are able to produce EPS, their yield is low [149] and their industrial applications for the improvement of the properties of food products are limited [155].

7.3. Potential health benefits of exopolysaccharides

Apart from the technological benefits some EPS derived from LAB are claimed to have beneficial physiological effects on consumer's health. These benefits are detectable at very low concentrations [153]. The EPS due to their increased viscosity in foods may remain for longer time in the gastrointestinal tract and therefore be beneficial to the transient colonization by probiotic bacteria [153,160]. Another health benefit is the generation of short chain fatty acids by colonic microflora degradation in the gut. Several of these fatty acids are possibly involved in the prevention of colon cancer [153,159]. In addition, LAB synthesized EPS appear to have anti-tumor, anti-ulcer, immuno-modulating and cholesterol-lowering activity [155].

7.4. Factors limiting the use of exopolysaccharides

The EPS used in the industry represent only a small fraction of the biopolymers used. The reasons are their economical production which needs a global knowledge of their biosynthesis and an adapted bioprocess technology [151]. Moreover, large scale production of LAB derived EPS is low, since LAB being anaerobes, are relatively inefficient in converting energy from carbohydrates, compared to aerobes [149,161]. So this technological barrier must be overcome for cost effective production of EPS. Furthermore, the genetic instability of EPS production is a problem to industrial applications, resulting in loss or reduction of production or change in the composition of EPS [162].

Increasing the knowledge on EPS structure may lead to the production of the "designer" EPS, including the modification primary in structure by altering their physical properties, their function and their production levels. However, for such production of EPS legal approval and the acceptance by the consumers and the food industries are required [151,163]. However, if these biomolecules are to be developed commercially, they must be cost effective.

7.5. Negative effects of exopolysaccharides

In some circumstances EPS cause food spoilage. For instance during the fermentation of wine or cider the final products receive undesirable properties. The EPS synthesis is responsible for dental plaque that results in dental caries. Moreover, the accumulation of EPS cause many technical and hygienic problems in the cheese and milk industries [156,164].

8. Lactic acid bacteria as source of low-calorie sweeteners

Recently, low calorie sugars produced from LAB have attracted the interest of researchers, industries and consumers, since they can find application as vital food ingredients mainly in foods marketed as "diabetic foods", like sugar-free candies, cookies and chewing gums [165,166]. Manitol, sorbitol, xylitol, tagatose and thehalose are sweeteners produced by LAB. These substances are polyols, i.e. sugar alcohols and can be produced in food fermentation processes. They can be incorporated directly to foods or be produced in the food by LAB, leading to the production of foods containing such sweeteners [165]. *Leuconostoc* and *Lactobacillus spp.* seem to be the most promising producers of these sweeteners [167,168].

A number of health benefits have been attributed to these LAB produced low calories sweeteners, like low glycemic index, osmotic diuretics, weight control, antiplaque, prebiotic. These GRAS substances could be used especially by children, diabetic patients and weight watchers [165,166].

9. Conclusion

Lactic acid bacteria are very promising sources for novel products and applications, especially those that can satisfy the increasing consumer's demands for natural products and functional foods. They can be used in the diet of humans and animals, with particular role in their health status. Despite recent advances, the study of LAB and their functional ingredients is still an emerging field of research that has yet to realize its full potential.

Author details

Panagiota Florou-Paneri and Efterpi Christaki
Laboratory of Nutrition, Faculty of Veterinary Medicine, Aristotle University of Thessaloniki, Thessaloniki, Greece

Eleftherios Bonos
Animal Production, Faculty of Technology of Agronomics, Technological Educational Institute of Western Macedonia, Florina, Greece

10. References

[1] Aureli P, Capurso L, Castellazzi AM, Clerici M, Giovannini M, Morelli L, Poli A, Pregliasco F, Salvini F, Zuccotti (2011) Probiotics And Health: An Evidence-based Review. Pharmacol. res. 63: 366-376.

[2] Barinov A, Bolotin A, Langella P, Maguin E, Van De Guchte M (2011) Genomics Of The Genus *Lactobacillus*. In: Sonomoto K, Yokota A, editors. Lactic Acid Bacteria and Bifidobacteria: Current Progress in Advanced Research. Caister Academic Press, Portland, USA.

[3] Stiles ME, Holzapfel WH (1997) Lactic Acid Bacteria Of Foods And Their Current Taxonomy. Int. j. food Microb. 36: 1–29.

[4] Johnson-Green P (2002) Introduction To Food Biotechnology. CRC Press, Boca Raton.

[5] Halasz A (2009) Lactic Acid Bacteria. In: Lasztity R, editor. Food Quality and Standards (Vol. 3). EOLSS Publishers Co Ltd, UK.

[6] Jay J (2000) Modern Food Microbiology (6th edition). Aspen, Maryland

[7] Hutkins RW (2006) Microbiology and Technology of Fermented Foods. Blackwell Publishing, Iowa, USA. p. 24.

[8] Teitelbaum JE, Walker WA (2002) Nutritional Impact Of Pre- and Probiotics As Protective Gastrointestinal Organisms. Annu. rev. nutr. 22: 107-138.

[9] Korhonen J (2010) Antibiotic Resistance Of Lactic Acid Bacteria. Dissertations in Forestry and Natural Sciences, University of Eastern Finland.

[10] Hamdan AM, Sonomoto K (2011) Production Of Optically Pure Lactic Acid For Bioplastics. In: Sonomoto K, Yokota A, editors. Lactic Acid Bacteria and Bifidobacteria: Current Progress in Advanced Research. Caister Academic Press, Portland, USA.

[11] Parker RB (1974) Probiotics, The Other Half Of The Antibiotic Story. Animal nutr. health. 29: 4-8.

[12] Fuller R (1989) Probiotics In Man And Animals. J. appl. bact. 66: 365-378.

[13] FAO/WHO – Food and Agriculture Organization / World Health Organization (2001) Health And Nutritional Properties Of Probiotics In Food Including Powder Milk With Live Lactic Acid Bacteria, Report Of A Joint FAO/WHO Expert Consultation On Evaluation Of Health And Nutritional Properties Of Probiotics In Food Including Powder Milk With Live Lactic Acid Bacteria, Cordoba. Argentina. http://www.who.int/ foodsafety/ publications/ fs_management/ en/ probiotics.pdf. Accessed 2011 Dec 15.

[14] O'May GA, Macfarlane GT (2005) Health Claims Associated With Probiotics. In: Tamime AY, editor. Probiotic Dairy Products. Blackwell Publishing, Oxford, UK.

[15] Krishnakumar V, Gordon IR (2001) Probiotics: Challenges And Opportunities. Dairy ind. Int. 66: 36-40.

[16] Playne MJ, Bennet LE, Smithers GW (2003) Functional Dairy Foods And Ingredients. Aust. J. dairy technol. 58: 242-264.

[17] Shah NP (2007) Functional Cultures And Health Benefits. Int. dairy j. 17: 1262-1277.

[18] Avall-Jaaskelainen S, Palva A (2005) Lactobacillus Surface Layers And Their Applications. FEMS microbiol. rev. 29: 511-529.

[19] Choi SS, Kang BY, Chung MJ, Kim SD, Park SH, Kim JS, Kang CY and Ha NJ (2005) Safety Assessment of Potential Lactic Acid Bacteria *Bifidobacterium longum* SPM1205 Isolated from Healthy Koreans. J. microbiol. 43: 493-498.

[20] Heller KJ (2001) Probiotic Bacteria In Fermented Foods: Product Characteristics And Starter Organisms. Am. j. clin. nutr. 73(S): 374S-379S.

[21] Parvez S, Malik KA, Kang SA, Kim (2006) Probiotics And Their Fermented Food Products Are Beneficial For Health. J. applied microbiol. 100: 1171-1185.

[22] Soccol CR, Vandenberghe LPDS, Spier MR, Medeiros ABP, Yamaguishi CT, Lindnen JDD, Pandey A, Thomaz-Soccol V (2010) The Potential Of Probiotics: A Review. Food technol. biotechnol. 48: 413-434.

[23] Ahmed F (2003) Genetically Modified Probiotics In Foods. Trends biotechnol. 21: 491-497.

[24] Sanders ME (2009) How Do We Know When Something Called "Probiotic" Is Really A Probiotic? A Guideline For Consumers And Health Care Professionals. Functional Food Rev. 1: 3-12.

[25] Stanton C, Gardiner G, Meehan H, Collins K, Fitzgerald G, Lynch PB, Ross RP (2001) Market Potentials For Probiotics. Am. j. clin. nutr. 73(S): 476-483.

[26] Bourlioux P, Koletzko B, Guarner F, Braesco V (2003) The Intestine And Its Microflora Are Partners For The Protection Of The Host: Report In Danone Symposium The Intelligent Intestine. Am. j. clin. nutr. 78: 675-683.

[27] Mazahreh AS, Ershidat OTM (2009) The Benefits Of Lactic Acid Bacteria In Yoghurt On The Gastrointestinal Function And Health. Pakistan j. nutr. 8: 1404-1410.

[28] Sherman PM, Johnson-Henry KC, Yeung HP, Ngo PSC, Goulet J, Tompkins TA (2005) Probiotics Reduce Enterohemorrhagic Escherichia Coli O157:H7- And Enteropathogenic *E. Coli* O127:H6-Induced Changes In Polarized T84 Epithelial Cell Monolayers By Reducing Bacterial Adhesion And Cytoskeletal Rearrangements. Infect. immun. 73: 5183-5188.

[29] Lin CK, Tsai HC, Lin PP, Tsen HY, Tsai CC (2008) *Lactobacillus Acidophilus* LAP5 Able To Inhibit The *Salmonella Cholerasuis* Invasion To The Human Caco-2 Epithelial Cell. Anaerobe. 14: 251-255.

[30] O'Shea EF, Cotter PD, Stanton C, Ross RP, Hill C (2012) Production Of Bioactive Substances By Intestinal Bacteria As A Basis For Explaining Probiotic Mechanisms: Bacteriocins And Conjugated Linoleic Acid. Int. J. Food Microbiol. 152: 189-205.

[31] Watts T, Berti I, Sapone A, Gerarduzzi T, Not T, Zielke, Fasano A (2005) Role Of The Intestinal Tight Junction Modulator Zonulin In The Pathogenesis Of Type I Diabetes In BB Diabetic prone Rats. Proc. natl. acad. Sci. USA. 102: 2916 –2921.

[32] Meddings J (2008) The Significance Of The Gut Barrier In Disease. Gut. 57: 438–440.

[33] Walker WA (2008) Mechanisms Of Action Of Probiotics. Clinic. inf. dis. 46(S2): S87.

[34] Ng SC, Hart AL, Kamm MA, Stagg AJ, Knight SC (2009) Mechanisms Of Action Of Probiotics: Recent Advances. Inflamm. bowel dis. 15: 300-308.

[35] Perdigon G, Alvarez S, Medina M, Vintini E, Roux E (1999) Influence Of The Oral Administration Of Lactic Acid Bacteria On Iga Producing Cells Associated To Bronchus. Int. j. immunopathol. Pharmacol. 12: 97–102.

[36] Figueroa-Gonzalez I, Quijano G, Ramirez G, Cruz-Guerrero (2011) Probiotics And Prebiotics – Perspectives And Challenges. J. sci. food agric. 91: 1341-1348.

[37] Tannonck GW (2003) Probiotics: Time For A Dose Of Realism. Curr. i. intestinal microbiol. 4: 33-42.

[38] Vaz-Velho M, Todorov SD (2011) Potential Probiotic Evaluation OF Bacteriocin Producing *Lactobacillus Plantarum* ST16PA Isolated From Papaya (*Carica Papaya*). 12th ASEAN Food Conference, Bangkok, Thailand.

[39] Vila B, Esteve-Garcia E, Brufau J (2010) Probiotic Micro-organisms: 100 Years Of Innovation And Efficacy; Modes Of Action. World poult. Sci. 66: 369-380.

[40] Frizzo LS, Soto LP, Zbrun MV, Signorini ML, Bertozzi E, Sequeira G, Rodriguez Armesto R, Rosmini MR (2011) Effect Of Lactic Acid Bacteria And Lactose On Growth Performance And Intestinal Microbial Balance Of Artificially Reared Calves. Livestock sci. 140: 246-252.

[41] Thomas MR, Litin SC, Osmon DR, Corr AP, Weaver AL, Lohse CM (2001) Lack Of Effect Of *Lactobacillus* GG On Antibiotic-associated Diarrhea: A Randomized, Placebo-controlled Trial. Mayo Clinic Proc. 76: 883-889.

[42] Isolauri E, Arvola T, Sutas Y, Moilanen E, Salminen E (2000) Probiotics in the management of atopic eczema. Clin. Experim. Allergy 30: 1604-1610.

[43] Kaur IP, Chorpa K, Saini A (2002) Probiotics: Potential Pharmaceutical Applications. Eur. j. pharmaceut. sci. 15: 1-9.

[44] Kruis W, Fric P, Pokrotnieks J, Lukas M, Fixa B, Kascak M, Kamm MA, Weismueller J, Beglinder C, Stolte M, Wolff C, Schulze J (2004) Maintaining Remission Of Ulcerative Colitis With The Probiotic *Escherichia Coli Nissle* 1917 Is As Effective As With Standard Mesalazine. Gut 53: 1617-1623.

[45] Henker J, Muller S, Laass MW, Schreiner A, Schulze J (2008) Probiotic Echerichia Coli Nissle 1917(EcN) For Successful Remission Maintainance Of Ulcerative Colitis In Children and Adolescents: An Open-label Pilot Study. Z. Gastroenterol. 46: 874-875.

[46] Itsaranuwat P, Shal-haddad K, Robinson PK (2003) The Potential Therapeutic Benefits Of Consuming 'Health-Promoting' Fermented Dairy Products: A Brief Update. Int. j. dairy technol. 56: 203–210.

[47] Gill H, Prasad J (2008) Bioactive Components Of Milk: Probiotics, Immunomodulation, And Health benefits. In: Bosze Z. Advances in Experimental Medicine and Biology. Springer, New York, USA. Pp. 423-464.

[48] Reid G, Charbonneau D, Erb J, Kochanowski B, Beuerman D, Poehner R, Bruce AW (2003) Oral Use Of Lactobacillus Rhamnosus GR-1 And L. Fermentum RC-14 Significantly Alters Vaginal Flora: Randomized Placebo-Controlled Trial In 64 Healthy Women. FEMS immunol. med. microbiol. 35: 131-134.

[49] Shornikova AV, Casas IA, Mykkanen H, Salo E, Vesikari T (1997) Bacteriotherapy With *Lactobacillus Reuteri* In Rotavirus Gastroenteritis. Pediatric inf. dis. j. 16: 1103-1107.

[50] Goel AK, Dilbaghi N, Kamboj DV, Singh L (2006) Probiotics: Microbial Therapy For Health Modulation. Defence sci. j. 56: 513-529.

[51] Liong MT, Shah NP (2005) Roles Of Probiotics And Prebiotics On Cholesterol: The Hypothesized Mechanisms. Nutrafood 4: 45-57.

[52] Cavallini DC, Bedani R, Bomdespacho LQ, Vendramini RC, Rossi EA (2009) Effects Of Probiotic Bacteria, Isoflavoned And Simvastatin On Lipid Profile And Aherosclerosis In Cholesterol-fed Rabbits: A randomized Double-Blind Study. Lipids health dis. 8, Article 1.

[53] Harish K, Varghese T (2006) Probiotics In Humans – Evidence Based Review. Calicut med. j. 4: e3.

[54] Bomba A, Nemcova R, Mudronova D, Guba P (2002) The Possibilities Of Potentiating The Efficacy Of Probiotics. Trends food sci. technol. 13: 121-126.

[55] Cremonini F, Di Caro S, Nista EC, Bartolozzi F, Capelli G, Gasbarini G, Gasbarrini A (2002) Meta-Analysis: The Effect Of Probiotic Administration On Antibiotic-Associated Diarrhoea. Aliment. pharmacol. ther. 16: 1461-1467.

[56] Sazawal S, Hiremath G, Dhingra U, Malik P, Deb S, Black RE (2006) Efficacy Of Probiotics In Prevention Of Acute Diarrhoea: A Meta-Analysis Of Masked, Randomized, Placebo-Controlled Trials. Lancet infect. dis. 6: 374-382.

[57] Yoon SS, Sun J (2011) Probiotics, Nuclear Receptor Signaling, and Anti-Inflammatory Pathways. Gastroenterol. res. pract. Vol: 2011, Article ID: 971938.

[58] Mallett AK, Bearne CA, Rowland IR (1989) The Influence Of Incubation Ph On The Activity Of Rat And Human Gut Flora Enzymes. J. appl. bacteriol. 66: 433–437.

[59] Laake K, Bjorneklett A, Bakka A, MIdtvedt T, Norin K, Eide TJ (1999) Influence Of Fermented Milk On Clinical State, Faecal Bacterial Count And Biochemical Characteristics In Patient With Ileal-Pouch-Anal-Anastomosis. Microb. ecol. health dis. 11: 211-217.

[60] Gosselink MP, Schouten WR, Van Lieshout LM, Hop WC, Laman JD, Ruseler-Van Embden JG (2004) Delay Of The First Onset Of Pouchitis By Oral Intake Of The Probiotic Strain Lactobacillus Rhamnosus GG. Dis. Colon rectum. 47: 876-884.

[61] Prantera C, Scribano ML, Falasco G, Andreoli A, Luzi C (2002) Ineffectiveness Of Probiotics In Preventing Recurrence After Curative Resection For Crohn's Disease: A Randomized Controlled Trial With *Lactobacillus* GG. Gut 51: 405-409.

[62] Hunter J, Lee A, King T, Barratt M, Linggood M, Blades J (1996) *Enterococcus Faecium* Strain PR88 – An Effective Probiotic. Gut 38(S1), A62.

[63] Niedzielin K, Kordecki K, Kosik R (1998) New Possibility In The Treatment Of Irritable Bowel Syndrome: Probiotics As A Modification Of The Microflora Of The Colon. Gastroenterology 114: A402.

[64] Brady LJ, Gallaher DD, Busta FF (2000) The Role Of Probiotic Cultures In The Prevention Of Colon Cancer. The j. nutr. 130(2S): 410S-455S.

[65] Rafter J. (2003) Probiotics And Colon Cancer. Best Pract. Res. Clin. Gastroenterol. 17: 849-859.

[66] Hamilton-Miller JM (2003) The role of probiotics in the treatment and prevention of Helicobacter pylori infection. Int. j. antimicrob. agents. 22: 360-366.

[67] Cats A, Kulpers EJ, Bosschaert MA, Pot RG, Vandenbroucke-Gauls CM, Kusters JG (2003) Effect Of Frequent Consumption Of *Lactobacillus Casei*-Containing Milk Drink In Helicobacter *Pylori*-Colonised Subjects. Pharmacol. therap. 17: 429-435.

[68] Marteau P, De Vrese M, Cellier CJ, Schrezenmeir J (2001) Protection From Gastrointestinal Diseases With The Use Of Probiotics. Am. j. clin. nutr. 73: 430S–436S.

[69] Saltzman JR, Russell RM, Golner B, Barakat S, Dallal GE, Goldin BR. (1999) A Randomized Trial Of *Lactobacillus Acidophilus* BG2FO4 To Treat Lactose Intolerance. Am. j. clin. nutr. 69: 140–6.

[70] Roberfroid MB (2000) Prebiotics And Probiotics: Are They Functional Foods? Am. j. clin. nutr. 71(S): 1682S-1687S.

[71] Pessi T, Sutas Y, Hurme M, Isolauri E (2000) Interleukin-10 Generation In Atopic Children Following Oral *Lactobaccilus Rhamnosus* GG. Clin. experim. Allergy. 30: 1804-1808.

[72] Reid G, Bruce AW, Fraser N, Heinemann C, Owen J, Henning B (2001) Oral Probiotics Can Resolve Urogenital Infections. FEMS immunol. med. microbiol. 30: 49-52.

[73] De Vrese M, Schrezenmeis J (2002) Probiotics And Non-Intestinal Infectious Conditions. Br. j. nutr. 88(S1): S59-66.

[74] Suzuki I., Nomura M., Morichi T. (1991) Isolation Of Lactic Acid Bacteria Which Suppress Mold Growth And Show Antifungal Action. Milchwissenschaft, 46: 635–639.

[75] Florianowicz T (2001) Antifungal Activity Of Some Microorganisms Against Penicillium Expansum. Eur. food res. technol. 212: 282–286.

[76] Mandai V., Sen S.K., Mandai N.C. (2007) Detection, Isolation And Partial Characterization Of Antifungal Compound(S) Produced By Pediococcus Acidilactici LAB 5. Nat. prod. commun. 2: 671–674.

[77] Sathe SJ, Nawani NN, Dhakephalkar PK, Kapadnis BP (2007) Antifungal Lactic Acid Bacteria With Potential To Prolong Shelf-Life Of Fresh Vegetables. J. appl. microbiol. 103: 2622–2628.

[78] Gerez CL, Torino MI, Rollan G, Font de Valdez G (2009) Prevention Of Bread Mould Spoilage By Using Lactic Acid Bacteria With Antifungal Properties. Food Control. 20: 144–148.

[79] Kankaanpaa P, Tuomola E, El-Nezami H, Ahokas J, Salminen SJ (2000) Binding Of Aflatoxin B1 Alters The Adhesion Properties Of *Lactobacillus Rhamnosus* Strain GG In Caco-2 Model. J. food prot. 63: 412–414.

[80] Deshpande G, Rao S, Patole S (2011) Progress In The Field Of Probiotics: Year 2011. Curr. Opin. Gastroenterol. 27: 13-18.

[81] Chen F, Wang D (2010) Novel Technologies For The Prevention And Treatment Of Dental Caries: A Patent Survey. Expert. opin. Ther. Pat. 20: 681–694.

[82] Flichy-Fernandez AJ, Alegre-Diago T, Penarrocha-Oltra D, Penarrocha-Diago (2010) Probiotic Treatment In The Oral Cavity: An Update. Med. oral patol. oral cir. bucal. 15: e677-680.

[83] Wood BJB, Holzapfel WH (1995) The Genera of Lactic Acid Bacteria. London, Blackie Academic & Professional, UK.

[84] Leroy F, De Vuyst L (2004) Lactic Acid Bacteria As Functional Starter Cultures For The Food Fermentation Industry. Trends food sci. technol. 15: 67-78.

[85] De Vuyst L, Leroy F (2007) Bacteriocins From Lactic Acid Bacteria: Production, Purification, And Food Applications. J. mol. Microbiol. biotechnol. 13: 194-199.

[86] Ray B (1992) The Need For Food Biopreservation. In: Ray B, Daeschel M, editors. Food biopreservatives of microbial origin Boca Raton, Florida: CRC Press. pp. 1–23.

[87] Caplice E, Fitzgerald G F (1999) Food Fermentations: Role Of Microorganisms In Food Production And Preservation. Int.j food microbiol. 50: 131–149.

[88] Sandine WE (1996) Commercial Production Of Dairy Starter Cultures. In: Cogan TM, Accolas JP, editors. Dairy Starter Cultures. Wiley-VCH. New York, USA. pp. 191–206.

[89] Joosten HMLJ, Gaya P, Nunez, M (1995) Isolation Of Tyrosin Decarboxylaseless Mutants Of A Bacteriocin-Producing Enterococcus Faecalis Strain And Their Application In Cheese. J food prot. 58: 1222–1226.

[90] Metchnikoff II, Chalmer Mitchell P (1910) Nature Of Man Or Studies In Optimistic Philosophy. Kessinger Publishing, Whitefish, MT, USA.

[91] Hansen EB (2002) Commercial Bacterial Starter Cultures For Fermented Food Of The Future. Int. j. food microbiol. 78: 119-131.

[92] Tamine AY, Saarela M, Korslund Sondergaard A, Mistry VV, Shah NP (2005) Production And Maintenance Of Viability Of Probiotic Micro-organisms In Dairy Products. In: Tamine AY, editor. Probiotic Dairy Products. Blackwell Publishing Ltd, Oxford, UK. pp. 39-72.

[93] Aragon-Alegro LC, Alarcon-Alegro JH, Cardarelli HR, Chiu MC, Sadd SMI (2007) Potentially Probiotic And Symbiotic Chocolate Mousse. LWT-Food sci. technol. 40: 669–675.

[94] Conlin PR, Chow D, Miller ER, Svetkey LP, Lin PH, Harsha DW, Moore TJ, Sacks FM, Appel LJ (2000). The Effect Of Dietary Patterns On Blood Pressure Control In Hypertensive Patients: Results From The Dietary Approaches To Stop Hypertension (DASH) Trial. Am. j. hypertension 13: 949–955.

[95] Karma BKL, Emata OC, Barraquio VL (2007) Lactic Acid And Probiotic Bacteria From Fermented And Probiotic Dairy Products. Sci. Diliman. 19: 23_34.

[96] Erkkila S (2001) Bioprotective And Probiotic Meat Starter Cultures For The Fermentation Of Dry Sausages. Academic Dissertation, Department of Food Technology, University of Helsinki, Finland.

[97] Ammor MS, Mayo B (2007) Selection Criteria For Lactic Acid Bacteria To Be Used As Functional Starter Cultures In Dry Sausage Production: An Update. Meat sci. 76: 138-146.

[98] Zhang W, Xiao S, Samaraweera H, Lee EJ, Ahn DU (2010) Improving Functional Value Of Meat Products. Meat Sci. 86: 15-31.

[99] Todorov SD, Furtado DN, Saad SMI, Tome E, Franco BDGM (2011) Potential Beneficial Properties Of Bacteriocin-producing Lactic Acid Bacteria Isolated From Smoked Salmon. J. appl. microbiol. 110: 971-986.

[100] Gevers D, Danielsen M, Huys G, Swings J (2003) Molecular Characterization Of Tet(M) Genes In Lactobacillus Isolates From Different Types Of Fermented Dry Sausage. Appl. Environm. Microbiol. 69: 1270–1275.

[101] Suzzi G, Gardini F (2003) Biogenic Amines In Dry Fermented Sausages: A Review. Int.j. food microbiol. 88: 41–54.

[102] Buruleanu L, Nicolescu CL, Bratu MG, Manea I, Avram D (2010) Study regarding some metabolic features during lactic acid fermentation of vegetable juices. Romanian biotechnol. Letters. 15: 5177-5188.

[103] Fleming HP (1984) Development In Cucumber Fermentation. Solid State Fermentation Symposium, London, UK. pp. 241-252.

[104] Medina E, Gori C, Servili M, de Castro A, Romero C, Brenes M (2010) Main Variable Affecting The Lactic Acid Fermentation Of Table Olives. Int. j. food sci. tech. 45: 1291-1296.

[105] Weinberg ZG, Ashbell G, Hen Y, Azrieli A (1993) The Effect Of Applying Lactic Acid Bacteria At Ensiling On The Aerobic Stability Of Silages. J. applied bacteriol. 75: 512-518.

[106] Broberg A, Jacobsson J, Strom K, Schnurer J (2007) Metabolite Profiles Of Lactic Acid Bacteria In Grass Silage. Applied environm. microbiol. 73: 5547-5552.

[107] McDonald P, Henderson AR, Heron SJE (1991) The Biochemistry Of Silage, 2nd ed. Chalcombe Publications, Aberystwyth, UK. pp. 184–236.

[108] Weinberg ZG, Muck RE, Weimer PJ, Chen Y, Gamburg M (2004) Lactic Acid Bacteria Used In Inoculants For Silage As Probiotics For Ruminants. Appl. biochem. biotechnol. 118: 1-9.

[109] Gillor O, Etzion A, Riley MA (2008) The Dual Role Of Bacteriocins As Anti- And Probiotics. Appl. microbiol. biotechnol. 81: 591-606.

[110] Lavermicocca P, Valerio F, Evidente A, Lazzaroni S, Corsetti A, Gobbetti M (2000) Purification And Characterization Of Novel Antifungal Compounds From The Sourdough Lactobacillus Plantarum Strain 21B. Appl. environ. microbiol. 66: 4084–4090.

[111] Falagas ME, Betsi GI, Athanasiou S (2007) Probiotics For The Treatment Of Women With Bacterial Vaginosis. Clin. microbiol. infect. 13: 657-664.

[112] Rajaram G, Manivasagan P, Thilagavathi B, Saravanakumar A (2010) Purification And Characterization Of A Bacteriocin Produced By *Lactobacillus Lactis* Isolated From Marine Environment. Adv. j. food sci. technol. 2: 138-144.

[113] Surwase SS, Adsul GG, Jadhav DS (2011) Anti-microbial Activity Associated With Bacteriocin From *Lactobacillus Acidophilus*. J. res. antimicrob. 1: 5-8.

[114] Dobson A, Cotter PD, Ross RP, Hill C (2012) Bacteriocin Production: A Probiotic Trait? Appl. Environ. Microbiol. 78: 1-6.

[115] Kos B, Beganovic J, Jurasic L, Svadumovic M, Pavunc AL, Uroic K, Suskovic J (2003) Coculture-Inducible Bacteriocin Biosynthesis Of Different Probiotic Strains By Dairy Starter Culture *Lactococcus Lactis*. Mljekarstvo 61: 273-282.

[116] Savadogo A, Ouattara CAT, Bassole IHN, Traore AS (2006) Bacteriocins And Lactic Acid Bacteria – A Minireview. African J. Biotechnol. 5: 678-683.

[117] Gordon DM, Oliver E, Littlefield-Wyer J (2007) The Diversity Of Bacteriocins In Gram-Negative Bacteria. In: Riley MA, Chavan M, editors. Bacteriocins: Ecology and evolution. Springer, Berlin, Germany. pp. 5-18.

[118] Heng NCK, Wescombe PA, Burton JP, Jack RW, Tagg JR (2007) The diversity of bacteriocins in Gram-positive bacteria. In: Riley MA, Chavan M, editors. Bacteriocins: Ecology and evolution. Springer, Berlin, Germany. pp. 45-92.

[119] Zendo T, Sonomoto K (2011) Classification And Diversity Of Bacteriocin. In: Sonomoto K, Yokota A, editors. Lactic Acid Bacteria and Bifidobacteria: Current Progress in Advanced Research. Caister Academic Press, Portland, USA.

[120] Dobson AE, Sanozky-Dawes RB, Klaenhammer TR (2007) Identification Of An Operon Inducing Peptide Involved In The Production Of Lactacin B By *Lactobacillus Acidophilus*. J. appl. microbiol. 103: 1766-1778.

[121] Cintas LM, Casaus MP, Herranz C, Nes IF, Hernandez PE (2001) Review: Bacteriocins Of Lactic Acid Bacteria. Food sci. tech. int. 7: 281-305.

[122] Juodeikiene G, Bartkiene E, Viskelis P, Urbonaviciene D, Eidukonyte D, Bobinas C (2012) Fermentation processes using lactic acid bacteria producing bacteriocins for preservation and improving functional properties of food products. In: Petre M, editor. Advances in applied biotechnology. Intech, Croatia. pp. 63-100.

[123] Galvez A, Abriouel H, Lopez RL, Omar NB (2007) Bacteriocin-Based Strategies For Food Biopreservation. Int. j. food microbiol. 120: 51-70.

[124] Beasley SS, Saris PEJ (2004) Nisin-producing *Lactococcus Lactic* Strains Isolated From Human Milk. Appl. environ. microbiol. 70: 5051-5053.

[125] Deegan LH, Cotter PD, Hill C, Ross P (2006) Bacteriocins: Biological Tools For Bio-Preservation And Shelf-Life Extension. Int. dairy j. 16: 1058-1071.

[126] Parada JL, Caron CR, Medeiros ABP, Soccol R (2007) Bacteriocins From Lactic Acid Bacteria: Purification, Properties And Use As Biopreservatives. Brazilian arch. biol. technol. 50: 521-542.

[127] Bonos EM, Christaki EV, Florou-Paneri PC (2010) Performance And Carcass Characteristics Of Japanese Quail As Affected By Sex Or Mannan Oligosaccharides And Calcium Propionate. South African j. animal sci. 40: 173-184.

[128] O'Connor EB, Barrett E, Fitzgerald G, Hill C, Stanton C, Ross RP (2005) Production Of Vitamins, Exopolysaccharides And Bacteriocins By Probiotic Bacteria. In: Tamime AY, editor. Probiotic Dairy Products. Blackwell Publishing, Oxford, UK.

[129] LeBlanc JG, Laino JE, Juarez del Valle M, Vannini V, van Sinderen D, Taranto MP, Font de Valdez G, Savoy de Giori G, Sesma (2011) B-Group Vitamin Production By Lactic Acid Bacteria – Current Knowledge And Potential Applications. J. applied microbiol. 111: 1297-1309.

[130] LeBlanc JG, de Giori GS, Smid EJ, Hugenholtz J, Sesma F (2007) Folate Production By Lactic Acid Bacteria And Other Food-Grade Microorganisms. Commun. curr. res. educ. top. trends appl. microbiol. 1: 329-339.

[131] Eitenmiller RR, Landen WO (1999) Folate. In: Eitenmiller RR, Landen WO, editors. Vitamin Analysis For The Health And Food Sciences. CRC Press LLC, Boca Raton. p. 411-465.

[132] Wouters JTM, Ayad EHE, Hugenholtz J, Smit G (2002) Microbes From Raw Milk For Fermented Dairy Products. Int. dairy j. 12: 91-109.

[133] Rossi M, Amaretti A, Raimondi S (2011) Folate Production By Probiotic Bacteria. Nutrients 3: 118-134.

[134] Quesada-Chanto A, Afschar AS, Wagner F (1994) Microbial Production Of Propionic Acid And Vitamin B_{12} Using Molasses Or Sugar. App. microbiol. biotechnol. 41: 378-383.

[135] Taranto MP, Vera JL, Hugenholtz J, De Valdez GF, Sesma F (2003) *Lactobacillus Reuteri* CRL1098 Produces Cobalamin. J. bacteriol. 185: 5653-5647.

[136] Molina V, Medici M, Taranto MP, Font de Valdez G (2008) Effects Of Maternal Vitamin B12 Deficiency From End Of Gestation To Weaning On The Growth And Haematological And Immunological Parameters In Mouse Dams And Offspring. Arch. anim. nutr 62: 162–168.

[137] Olson RE (1984) The Function And Metabolism Of Vitamin K. Ann. rev. nutr. 4: 281-337.

[138] Morishita T, Tamura N, Makino T, Kudo S (1999) Production Of Menaquinones By Lactic Acid Bacteria. J. dairy sci. 82: 1897-1903.

[139] Wilson JA (1983) Disorders Of Vitamins: Deficiency, Excess And Errors Of Metabolism. In: Petersdorf RG, Harrison TR. Harrison's Principles of Internal Medicine. McGraw-Hill Book Co. New York, USA. pp. 461–470.

[140] Capozzi V, Menga V, Digesu AM, De Vita P, van Sinderen D, Cattivelli L, Fares C, Spano G (2011) Biotechnological Production Of Vitamin B2-Enriched Bread And Pasta. J. agric. food chem. 59: 8013–8020.

[141] Naidu AS, Bidlack WR, Clemens RA (1999) Probiotic Spectra of Lactic Acid Bacteria (LAB). Crit. rev. food sci. nutr. 38: 13-126.

[142] Tamang JP (2011) Prospects Of Asian Fermented Foods In Global Markets. 11th ASEAN Food Conference, Bangkok, Thailand.

[143] Matthews A, Grimaldi A, Walker M, Bartowsky E, Grbin P, Jiranek V (2004) Lactic Acid Bacteria As A Potential Source Of Enzymes For Use In Vinification. Appl. environ. mcrobiol. 70: 5715-5731.

[144] Mtshali PS (2007) Screening And Characterisation Of Wine-related Enzymes Produced By Wine-associated Lactic Acid Bacteria. MSc. Thesis, Stellenbosch University.

[145] Mogensen G (1993). Starter Cultures. In: Smith J, editor. Technology of reduced-additive foods London: Blackie Academic & Professional, UK. pp. 1–25.

[146] Guldfeldt LU, Sorensen KI, Stroman P, Behrndt H, Williams D, Johansen E (2001). Effect Of Starter Cultures With A Genetically Modified Peptidolytic Or Lytic System On Cheddar Cheese Ripening. Int. dairy j. 11: 373–382.

[147] Gonzalez L, Sacristan N, Arenas R, Fresno JM, Tornadijo ME (2010) Enzymatic Activity Of Lactic Acid Bacteria (With Antimicrobial Properties) Isolated From A Traditional Spanish Cheese. Food microbiol. 27: 592–597.

[148] Ganzle MG, Schwab C (2009) Ecology Of Exopolysaccharide Formation By Lactic Acid Bacteria: Sucrose Utilisation, Stress Tolerance, And Biofilm Formation. In: Ullrich M, editor. Bacterial Polysaccharides: Current Innovations And Future Trends: Caister Academic Press, Bremen, Germany.

[149] Ruas-Madiedo P, Hugenholtz J, Zoon P (2002) An Overview Of The Functionality Of Exopolysaccharides Produced By Lactic Acid Bacteria. Int. dairy j. 12: 163-171.

[150] Ruas-Madiedo P, Salazar N, de los Reyes-Gavilan (2009) Biosynthesis and Chemical Composition Of Exopolysaccharides Produced By Lactic Acid Bacteria. In: Ullrich M, editor. Bacterial Polysaccharides: Current Innovations And Future Trends: Caister Academic Press, Bremen, Germany.

[151] De Vuyst L, de Vin F, Vaningelgem F, Degeest B (2001) Recent Developments In The Biosynthesis And Applications Of Heteropolysaccharides From Lactic Acid Bacteria. Int. dairy j. 11: 687-707.

[152] Monsan P, Bozonnet S, Albenne C, Jouela G, Willemot RM, Remaud-Simeon M (2001) Homopolysaccharides From Lactic Acid Bacteria. Int. dairy j. 11: 675-685.

[153] Duboc P, Mollet B (2001) Application Of Exopolysaccharides In The Dairy Industry. Int. dairy j. 11: 759-768.

[154] Waldherr F, Vogel RF (2009) Commercial Exploitation Of Homo-exopolysaccharides In Non-dairy Food Systems. In: Ullrich M, editor. Bacterial Polysaccharides: Current Innovations And Future Trends: Caister Academic Press, Bremen, Germany.

[155] Zhang Y, Li S, Zhang C, Luo Y, Zhang H, Yang Z (2011) Growth And Exopolysaccharide Production By *Lactobacillus Fermentum* F6 In Skim Milk. African j. biotechnol. 10: 2080-2091.

[156] De Vuyst, L, Degeest B (1999) Heteropolysaccharides From Lactic Acid Bacteria. FEMS microbiol. rev. 23: 153-177.

[157] Freitas F, Alves VD, Reis AM (2011) Advances In Bacterial Exopolysaccharides: From Production To Biotechnological Applications. Trends biotechnol. 29: 388-398.

[158] Gibson R, Roberfroid MB (1995) Dietary Modulation Of The Human Colonic Microbiota: Introducing The Concept Of Prebiotics. J. nutr. 125: 1401-1412.

[159] Ruijssenaars HJ, Stingele F, Hartmans S (2000) Biodegradability Of Food-associated Extracellular Polysaccharides. Curr. microbiol. 40: 194-199.

[160] German B, Schiffrin E, Reniero R, Mollet B, Pfeifer A, Neeser JR (1999) The development of functional foods: Lessons from the gut. Trends biotechnol. 17: 492–499.

[161] Welman AD, Maddox IS (2003) Exopolysaccharides From Lactic Acid Bacteria: Perspectives And Challenges. Trends biotechnol. 21: 269-274.

[162] Bouzar F, Cerning J, Desmazeaud M (1996) Exopolysaccharide Production In Milk By *Lactobacillus Delbrueckii Ssp. Bulgaricus* CNRZ 1187 And By Two Colonial Variants. J. dairy sci. 79: 205-211.

[163] Sutherland IW (1999) Polysaccharases For Microbial Exopolysaccharides. Carbohydr. polym. 38: 319-328.

[164] Patel S, Majumder A, Goyal A (2012) Potentials Of Exopolysaccharides From Lactic Acid Bacteria. Indian j. microbiol. 52: 3-12.

[165] Patra F, Tomar SK, Arora S (2009) Technological And Functional Applications Of Low-Calorie Sweeteners From Lactic Acid Bacteria. J. food sci. 74: 16-21.

[166] Monedero V, Perez-Martínez G, Yebra MJ (2010) Perspectives Of Engineering Lactic Acid Bacteria For Biotechnological Polyol Production. Appl. microbiol. biotechnol. 86: 1003–1015.

[167] Soetaert W (1990) Production Of Mannitol With *Leuconostoc Mesenteroides*. Med. fac landbouwwet rijksuniv gent. 55: 1549–52.

[168] Kim P (2004) Current Studies On Biological Tagatose Production Using L-Arabinose Isomerase: A Review And Future Perspective. Appl. microbiol. biotechnol. 65: 243–249.

Chapter 12

Dynamic Stresses of Lactic Acid Bacteria Associated to Fermentation Processes

Diana I. Serrazanetti, Davide Gottardi, Chiara Montanari and Andrea Gianotti

Additional information is available at the end of the chapter

1. Introduction

Despite their negligible mass the microbial agents, starters and non starters, play a profound role in the characterization of the fermented foods in terms of chemical and sensorial properties. In fact, fermented foods may be defined as foods processed through the activity of microorganisms. Fermentation processes take a special place in the evolution of human cuisine, by altering the taste experience of food products, as well as extending the storage period. In particular, foods fermented with lactic acid bacteria (LAB) have constituted an important part of human diet and of fermentation processes (involving various foods, including milk, meat, vegetables and fruits) [1] since ancient times. They have played an essential role in the preservation of agricultural resources and in the improvement of nutritional and organoleptic properties of human foods and animal feed. Moreover, these organisms nowadays are increasingly used as health promoting probiotics, enzyme and metabolite factories and vaccine delivery vehicles [2].

It is interesting to outline how the changes of food characteristics during the fermentation process can be described as dynamic fluctuations of the food environment itself and, at the same time, stress source for the microorganisms involved [3, 4], such as LAB. In fact, whenever autochthonous bacteria are adapted and competitive in their respective environment, the environment can be described as stressful for LAB [5, 4]. The fermentation parameters, including temperature, water activity (Aw), oxygen, pH, as well as the concentration of starter cultures, affect the regulatory mechanism and the response mechanisms of LAB, as well as their effects on the final products properties [4].

When LAB are added to food formulations, several factors that may influence the ability of those microorganisms to survive, growth and become active in the new matrix have to be considered [6]. These factors include: 1) the physiological state of the LAB used as starters (whether the cells are from the logarithmic or the stationary growth phase); 2) the physical conditions of product ripening and storage (eg. temperature); 3) the chemical composition of the matrix (eg. acidity, available carbohydrates content, nitrogen source, mineral content, water activity and oxygen concentration); 4) possible interactions of the starter cultures with probiotics and other microorganisms naturally occurring or added to the system [6].

In figure 1 the main factors affecting the viability and the responses of LAB from production to storage are described [7].

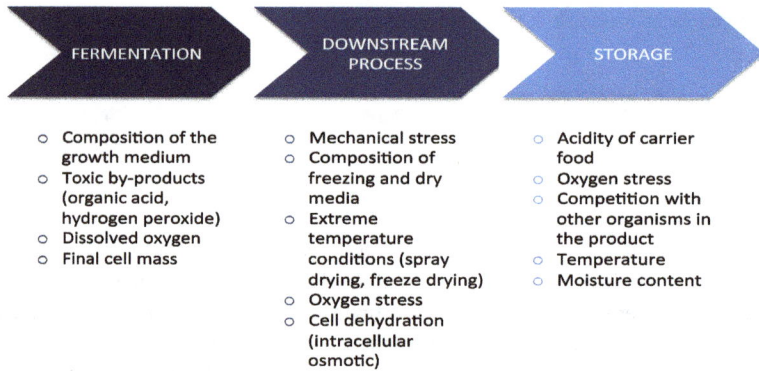

FERMENTATION

- Composition of the growth medium
- Toxic by-products (organic acid, hydrogen peroxide)
- Dissolved oxygen
- Final cell mass

DOWNSTREAM PROCESS

- Mechanical stress
- Composition of freezing and dry media
- Extreme temperature conditions (spray drying, freeze drying)
- Oxygen stress
- Cell dehydration (intracellular osmotic)

STORAGE

- Acidity of carrier food
- Oxygen stress
- Competition with other organisms in the product
- Temperature
- Moisture content

Figure 1. Factors affecting the viability and the responses of LAB to the various fermented foods production steps.

To better elucidate what happens to LAB during fermentation processes, we decided to use a model (defined "virtual food") that mimics various steps occurring during processing and that can affect LAB performances or viability.

2. Lactic acid bacteria and stress: Basic concepts

"Stress results from interactions between subjects and their environment that are perceived as straining or exceeding their adaptive capacities and threatening their well-being. The element of perception indicates that human stress responses reflect differences in personality, as well as differences in physical strength or general health" [8].

Stress has driven evolutionary changes (the development and natural selection of species over time). Thus, the species that adapted best to the causes of stress (stressors) have survived and evolved into the plant and animal kingdoms we now observe. The same evolutionary process regarded microorganisms. In fact, bacteria, irrespective of natural habitat, are exposed to constant fluctuations in their growth conditions. Consequently they have developed sophisticated responses, modulated by the re-modelling of protein complexes and by phosphorylation dependent signal transduction systems, to adapt and to survive to a variety of insults. To ensure survival to environmental adversities, bacteria may adapt to changes in their immediate vicinity by responding to the imposed stress. These responses are different and vast and depend on the microorganism nature and on the environmental stress and are accomplished by changes in the patterns of gene expression for those genes whose products are required to combat the deleterious [3]. In particular, cellular metabolic pathways are closely related to stress responses and the flux of particular metabolites to understand the hypothetically shifts and implications in the food systems has been studied in LAB [9-13, 4, 14, 15].

LAB are a functionally related group of organisms known primarily for their bioprocessing roles in food and beverages [16]. LAB play a crucial role in the development of the organoleptic and hygienic quality of fermented products. These microorganisms are used as starter cultures in many fermented products (i.e. beer, milk, dough, sausages and wine). Therefore, the reliability of starter cultures in terms of quality and functional properties (important for the development of aroma and texture), but also in terms of growth performance and robustness, has become essential for successful fermentations [17]. There have been some reports describing the physiological stress responses in LAB, particularly *Lactobacillus* species, which have a broad biodiversity [17-21, 13, 22, 4, 14, 15].

LAB evolved specific mechanisms to respond and to survive to environmental stresses and changes (stress-sensing system and defences). In fact, microorganisms could have specific regulators tailored to each of their regulated genes and adapt their expression according to environment. Stress defences are good examples of such integrated regulation systems. Bacterial stress responses rely on the coordinated expression of genes that alter different cellular processes (cell division, DNA metabolism, housekeeping, membrane composition, transport, etc.) and act in concert to improve the bacterial stress tolerance. The integration of these stress responses is accomplished by networks of regulators that allow the cells to react to various and complex environmental shifts. LAB respond to stress in a very specific way dependent on the species, on the strains and on the type of stress. The best-studied stresses are acid, heat, oxidative and cold stresses, although for the latter most of the studies focused on a specific family of proteins instead of analyzing the whole response [4].

Despite the extensive use of LAB, there is a paucity of information concerning the stress-induced mechanisms studied *in vivo* for improving the survival of these organisms during real food processing. A better knowledge of the adaptive responses of LAB is important because the fermentation processes often expose these microorganisms to adverse environmental conditions. LAB should resist to adverse conditions encountered in industrial processes, for example during starter handling and storage (freeze drying, freezing or spray-drying) and during the fermentation environment dynamic changes. These phenomena reinforce the need for robust LAB since they may have to survive and grow in different unfavorable conditions expressing specific functions (for example during stationary phase or storage) [17].

3. Principal responses to the most common stresses

Heat shock response: The effect of heat shock and the induction of a stress response in *Lactobacillus* spp. have been studied for *Lactobacillus delbrueckii* subsp. *bulgaricus* [23] and *Lactobacillus paracasei* [24, 25], *Lactobacillus acidophilus*, *Lactobacillus casei* and *Lactobacillus helveticus* [26], *Lactobacillus collinoides* [27] , *Lactobacillus sakei* [28], *Lactobacillus johnsonii* [29], *Lactobacillus rhamnosus* [30], *Lactobacillus plantarum* [31-33] and *Lactobacillus salivarius* [34]. The heat resistance of LAB is a complex process involving proteins with different roles in cell physiology, including chaperone activity, ribosome stability, stringent response mediation, temperature sensing and control of ribosomal functions [31]. The time taken to initiate the stress response is different for different treatments and different strains. The major problem encountered by cells at high temperature is the denaturation of proteins and their subsequent aggregation. In addition Earnshaw et al. [35], , Texeira et al. [36] and Hansen et al. [37] described also as response to heat stress the destabilization of macromolecules as ribosomes and RNA as well as alterations of membrane fluidity.

Heat stress response is characterized by the transient induction of general and specific proteins and by physiological changes. In every strain tested the involvement of Heat Shock Proteins (HSPs such as DnaK, GroEL and GroES during the heat stress was clear) [23-38]. The role of these stress proteins is complex; in fact, the bind substrate proteins in a transient non-covalent manner prevent premature folding and promote the attainment to the correct state *in vivo*. The resistance to heat stress is higher when the cells were previously exposed and adapted to this type of stress in the stationary phase, otherwise, when pre-adapted in exponential phase, the cells are more sensitive. In particular, the storage stability of the culture that was heat shocked after stationary phase was superior to that of culture heat shocked after log phase [34, 23, 30].

Cold shock response: It is very important to improve knowledge about LAB behavior in cold environment. In fact, during industrial processes, like in cheese ripening and refrigerated storage of fermented products, these microorganisms are subjected to different temperatures far below the optimal growth temperature. When LAB living cells are exposed to these cold environments, important physiological changes occur, such as decrease in membrane fluidity and stabilization of secondary structures of RNA and DNA, resulting in a reduced efficiency of translation, transcription and DNA replication. The response of microorganisms to these effects is termed cold-shock response during which a number of Cold Induced Proteins (CIPs) are synthesized. The roles of these proteins are at the levels of membrane fluidity, DNA supercoiling and transcription and translation. Few papers have described cold shock proteins and mechanisms in LAB, in particular they have focused on *Lactococcus lactis* and *L. plantarum* [39-42]. Kim et al. [39, 40] tested different LAB to evaluate cold shock effects on cryotolerance. Improved understanding of cold-shock-induced cryotolerance may contribute to the development of environmental conditions that allow improved viability/activity of frozen or freeze-dried commercial LAB starter cultures. The results showed that, as with heat stress, there is also an improvement of the viability of the tested strains as concerning the cryotolerance after a cold shock. The process of freezing appeared to have different effects on different LAB as well as different effects on strains within the same genus. Moreover, the freezing response of the strains depends on the time of the cold shock process and the induction of cryotolerance appears to be dependent on the growth phase in which the cold shock took place [43-47].

Another interesting study regarding LAB response to sub-lethal cold stress was developed by Montanari et al.[14]. These Authors separated and quantified the cell cyclopropane fatty acids lactobacillic (C19cyc11) and dehydrosterculic (C19cyc9) to study the adaptive response to sub-lethal acid and cold stresses in *L. helveticus* and *Lactobacillus sanfranciscensis*. These microorganisms showed different fatty acids composition and environmental adaptation to short term cold and acidic stresses. In *L. helveticus* C19cyc11 dramatically increased after 2 h at 10°C and with the pH decrease, particularly in micro-aerobic conditions, in the presence of tween 80, and in anaerobic conditions. The increase of lactobacillic acid in *L. helveticus* is necessary to maintain the cell membrane in a suitable state of fluidity. Moreover, cyclopropane fatty acids confer resistance to ozonolysis, singlet oxygen and mild oxidative treatments [48, 49], suggesting a cross protection and response of LAB cell membrane to physicochemical stresses. A combined analysis of the genome-wide transcriptome and metabolism was performed with a dairy *Lactococcus lactis* subsp. *lactis* under dynamic conditions similar to the conditions encountered during the cheese-making process. Specific responses to acid and cold stresses were identified, but also the induction of unexpected pathways was determined. In particular, the induction of purine biosynthesis and prophage [50].

Oxidative stress response: LAB are facultative anaerobic microorganisms that have in common the reduction of part of pyruvate produced to lactate production in order to regenerate NAD+ from NADH formed during glycolysis. They do not require oxygen for growth and, in fact, a negative effect of oxygen on the development of these bacteria has often been observed. It was generally believed that these bacteria could under no condition use oxygen as the terminal electron acceptor [17]. However, many LAB have NADH oxidase and some can even express a functionally active respiratory chain in the presence of heme [51-57]. Respiration-competent LAB differ from the features of *Escherichia coli* and *Bacillus subtilis*, since they carry limited equipment for respiration. All respiring LAB carry genes encoding electron donor (NADH dehydrogenase) and a single electron acceptor (cytochrome bd oxidase) [58]. Addition of heme to the system activates respiration chain NADH oxidase activity, but none of the tested LAB synthesize heme [01].

When for some reasons the generation of free radicals is higher than the rate of their detoxification the cells are exposed to a constraint called "oxidative stress" [59]. For the food-associated LAB a still fragmented picture of the resistance mechanisms present emerges. Representatives of the different mechanisms have been described in different LAB [60-64]. Apart from the toxic effects of oxygen, aeration can induce important changes in the sugar metabolism of LAB. In fact, the presence of oxygen is a factor that greatly affects the outcome of a fermentation process. In general, LAB tolerate oxygen but grow better under nearly anaerobic conditions. However, in the presence of heme and oxygen LAB start respiration metabolism, by which the cell metabolism is reprogrammed so that pH, oxygen status, growth capacity and survival are markedly altered [56]. In the presence of oxygen and during the fermentation metabolism, H_2O_2 is formed. Numerous species of LAB contain peroxidase and/or catalase to prevent and eliminate these deleterious effects [17]. Concerning the prevention of reactive oxygen species (ROS) formation, the scope of the reactions is the eliminations of free oxygen. In a study on *L. helveticus* the fatty acids composition in the cell membrane changed in response to oxidative stress. In fact, the activity of oxygen consuming desaturase system increased to reduce the free radical damage to the cell [19]. Generally, the response to oxidative stress of LAB is similar, but also depends on the species, on the strains and, with regard to catalase action, on the bacterial density [4]. In *L. lactis* several genes have been identified and the respective encoded proteins have been shown to contribute to oxidative stress resistance. Moreover, the induction of these genes is growth phase-dependent (exponential or stationary) and their products confer multi-stress resistance [52]. General stress resistance mechanisms may also confer resistance to oxidative stress. In fact, in a model system several acid resistant mutants of *L. lactis* that appeared also more resistant to oxidative stress were isolated [64].

Acid stress response: Understanding the acid resistance mechanism used by LAB to survive to by-products of their own metabolism (i.e. homofermentative *L. lactis* converts 90% of metabolized sugar to lactic acid) and the response available in low-pH foods is of great importance. In LAB one of the most effective mechanisms for resistance in acid stress environment is the glutamate decarboxylase (GAD). In fact, few years ago, it was proposed that amino acid decarboxylase functions to control the pH of the bacterial environment by consuming hydrogen ions as part of carboxylation reaction [65]. LAB are also capable of inducing an Acid Tolerance Response (ATR) in response to mild acid treatments. The system induced includes pH homeostatis, protection and repair mechanisms. Genes and proteins, involved in pH homeostasis and cell protection or repair, play a role in acid adaptation, but this role can also extend to more general acid tolerance mechanisms. A more specific study was developed on the effects of lactic acid stress on *L. plantarum* by transcription profiling [66]. The difference, in terms of stress response, into the dissociated or undissociated forms of lactic acid has been highlighted. The toxicity of organic acids depends on their degree of dissociation and thus on the pH. For LAB end product inhibition by lactic acid could result in a disturbance of the regeneration of cofactor NAD+, especially under anaerobic conditions, in which the cell does not have the possibility of NAD+ regeneration by NADH oxidase. The response at membrane fatty acids level to acid stress was studied in *L. helveticus* and *L. sanfranciscensis* [14]. The relevant proportion of dodecanoic acid in the latter species under acid stress suggests that carbon chain shortening is the principal strategy of *L. sanfranciscensis* to modulate fluidity or chemico-physical properties of the membranes in the presence of acid stress. Moreover, a specific shift in leucine catabolic pathway at pH 3.6 was identified in *L. sanfranciscensis* [15]. In fact, the acid stress induced a metabolic shift toward overproduction of 3-methylbutanoic and 2-methylbutanoic acids, accompanied by sugar reduced consumption and primary carbohydrate metabolite production. The metabolites coming from branched chain amino acids (BCAAs) catabolism increased up to seven times under acid stress. While the overproduction of 3-methylbutanoic acid under acid stress can be attributed to the need to maintain redox balance, the rationale for the production of 2-methylbutanoic acid from leucine can be found in a newly proposed biosynthetic pathway leading to 2-methylbutanoic acid and 3 mol of ATP per mol of leucine. Leucine catabolism to 3-methylbutanoic and 2-methylbutanoic acids suggests that the switch from sugar to amino acid catabolism supports growth of *L. sanfranciscensis* in restricted environments such as sourdough, characterized by acid stress and recurrent carbon starvation.

Osmotic stress response: In the various applications in food and feed industry LAB can be exposed to osmotic stress when important amounts of salts or sugars are added to the product [17]. In fact, in most of the food habitats where lactobacilli live, they are confronted with salt [67] and sugar stress [68]. Study on the differences between salt and sugar osmotic stress revealed that the hyperosmotic conditions imposed by sugar stress are much less detrimental and only transient (transient osmotic stress), because the cells are able to balance the extra and the intracellular concentrations of lactose and sucrose [17]. Bacteria need to adapt to this change in their environment in order to survive [69], and they can do it by accumulating (by uptake or synthesis) compatible solutes, generally of organic origin, under hyperosmotic conditions [17]. The compatible solutes are defined as osmoprotectants. The main strategy to adapt to high osmolarity of non-halophilic bacteria is associated with the enhancement of the osmotolerance [68]. Moreover, the osmoprotectants can also stabilize enzymes and provide protection not only against osmotic stress but also against other type of stresses (high temperature, freezing and drying). The intracellular accumulation of compatible solutes prevents the loss of water caused by high external osmolarity and allows the maintenance of turgor [68]. The accumulation of carnitin, betain and proline was determined in LAB grown in MRS and complex diluted MRS medium (DMRS medium) [70]. Moreover, a specific response mechanism to osmotic stress was identified in a sourdough model system [13]. In particular, the growth of *L. sanfranciscensis* under osmotic stress resulted in a relevant accumulation of 3-methylbutanoic acid. Its synthesis is associated with the BCCAs., is NAD+ dependent and produces NADH during the reaction [71]. The accumulation of 3-methylbutanoic acid as predominant metabolite has been also observed in model systems simulating sourdough as a consequence of osmotic, acid or oxidative stress [12, 15].

High pressure stress response: High-pressure processing (HPP) or high pressure homogenization (HPH) are non-thermal processes capable of inactivating and eliminating pathogenic and food spoilage microorganisms in specific foods [11, 72], and it represents an exceptional stimulus for most mesophilic bacteria. Several proteins are induced after high pressure treatment and some of these have also been involved in the response to other various stresses [8]. The responses to HHP stress have been studied in particular on *L. sakei* and *L. sanfranciscensis* [73, 18]. These Authors suggested the presence of *de novo* protein synthesis as a consequence of HHP stress [73]. As concerning HPH several interesting studies on the responses on *Lactobacillus* spp., at the level of proteolytic and metabolic activities point of view have been conducted [11, 21, 22, 74]. HPH treatment positively affects the proteolytic activity of some of *Lactobacillus* strains, but the activation and the quantitative and qualitative changes of the metabolic activity appear to be the most promising results. The pre-treatment at different pressure was able to induce relevant changes in term of fermentation dynamics and metabolism with respect to the untreated cells [11]. The same approach was applied on *L. acidophilus* and *L. paracasei* to improve the technological performances of probiotic strains [21, 22, 74]. The sub-lethal treatment with HPH enhanced the capacity of some *in vitro* probiotic features (i.e. hydrophobicity and tolerance to simulated gastric acidity) in a strain dependant way. *L. paracasei* A13 enhanced cellular hydrophobicity and auto-aggregation capacity after HPH treatment at 50 MPa. On the contrary, the HPH treatment decreased these features in the other strains considered. Highest values of hydrophobicity were found for *L. acidophilus* DRU and its bile-resistant derivative *L. acidophilus* DRU+, while lower values were obtained for *L. paracasei* strain [74]. Moreover, the stress responses enable survival under more severe conditions, enhancing resistance to subsequent processing conditions [75]. HPH treatment at 50 MPa can favour the maintenance of cell viability during a refrigerated storage in buttermilk, a suitable medium to maintain the cell viability during refrigeration [76]. The increased viability can be attributed to the increased precocious availability of low molecular weight peptides and free fatty acids such as oleic acid [21, 22].

Competition and communication: Food fermentations are typically carried out by mixed cultures consisting of multiple strains or species [77]. Mixed-culture food fermentations are of primary economic importance. The performance of these cultures, consisting of LAB, yeasts, and/or filamentous fungi, is not the simple result of "adding up" the individual single-strain functionalities, but is largely determined by interactions at the level of substrates, exchange of metabolites and growth factors or inhibiting compounds [77].

General microbial interference is an effective non-specific control mechanism common to all populations and environments including foods. It represents the inhibition of the growth of certain microorganisms by other members of the habitat.

The mechanisms involved are common to all genera and include [78]:

1. Nutrient competition,
2. Generation of unfavorable environment,
3. Competition for attachment/adhesion sites.

Most substrates for food fermentations have a highly heterogeneous physicochemical composition, which offers the possibility for the simultaneous occupation of multiple niches by "specialized" strains, for instance, through the utilization of different carbon sources. In these substrates, coexisting strains often interact through trophic or nutritional relations via multiple mechanisms [77].

Carbon sources are often present at high concentrations in food substrates, and therefore competition concerns the rapid uptake of nutrients and conversion into biomass. In dairy fermentations nitrogen is limiting, and initially organisms compete for the free amino acids and small peptides available. While in the later stages of fermentation, they compete for the peptides released by the actions of proteolytic enzymes [77].

In a cell-density-dependent quorum-sensing system, bacteria produce extracellular signaling molecules such as peptides or post-translationally modified peptides that act as inducers for gene expression when concentrations of these molecules exceed a certain threshold value [79]. These changes might eventually lead to competitive advantages for the population, more effective adaptation and responses to changing environmental conditions, or the co-ordination of interactions between bacteria and their abiotic and biotic environments [7]. In fact, microorganisms produce diffusible chemicals for the purpose of communication and it has been reported that the stress caused by the exposure of microbial cells to their own cell free conditioned media, containing metabolites and bioactive compounds including "quorum sensing" molecules, including 2(5H)-furanones, promotes cell differentiation, autolysis and overproduction of specific metabolites [12, 80, 9, 10]. In this way the microbial cultures used in food fermentations can also contribute (by "secondary" reactions and relations) to the formation of flavor and texture [81].

4. General steps regarding a virtual fermented food process

In the figure 2, the steps that mainly interest food fermentation are reported. A model virtual fermented food was identified to resume the common denominator of the fermented foods dynamics, particularly focused on the reciprocal influences between environmental fluctuation and LAB fermentation.

Whatever kind of food we want to produce, fermented or not, the first step of the process is the formulation: in this phase the main raw materials (meat, milk, fruit and vegetables or their derivatives) are mixed with other ingredients, that have different roles: salts or sugars to improve taste, spices to give specific sensorial quality and as antimicrobials, additives or other substances able to affect physical and structural properties, preservatives to improve microbial stability and shelf life. The addition of those ingredients can be perceived as stress. In fermented products, proper microorganisms, mainly yeasts and LAB, are also added as starter cultures, in order to start and lead the fermentation and to obtain a stable and standard final product. As a consequence, the microorganisms, naturally occurring or added as starter cultures, have to cope with a completely different system: in particular, naturally occurring microflora have to face the changes induced by the ingredients, while the starter cultures, deriving from growth media or added as lyophilized cultures, have to adapt to a real food system, where different sources of stresses are often present.

In particular, the first sub-lethal stress, which LAB face, regards the difference between the growth medium composition and the real food. Generally, LAB lyophilized cultures can be added to the ingredients after a reactivation and subsequently added to the product. This procedure identify the presence of a stress for the LAB cells. Starter cultures are added to the raw materials in large numbers and incubated under optimal conditions, but the adaptation to substrate or raw material is always necessary [82]. It is very important to consider the physiological state of the LAB before the inoculum. This state strongly depends on the time of harvesting of the culture (whether during the logarithmic or stationary phase of growth), on the conditions leading to transition to the stationary phase, on the treatment of the culture during and after harvesting and on the chemical composition of the environment. Therefore it is important during formulation and technological processes to consider also these factors, mainly for those products where microorganisms are added as starter cultures.

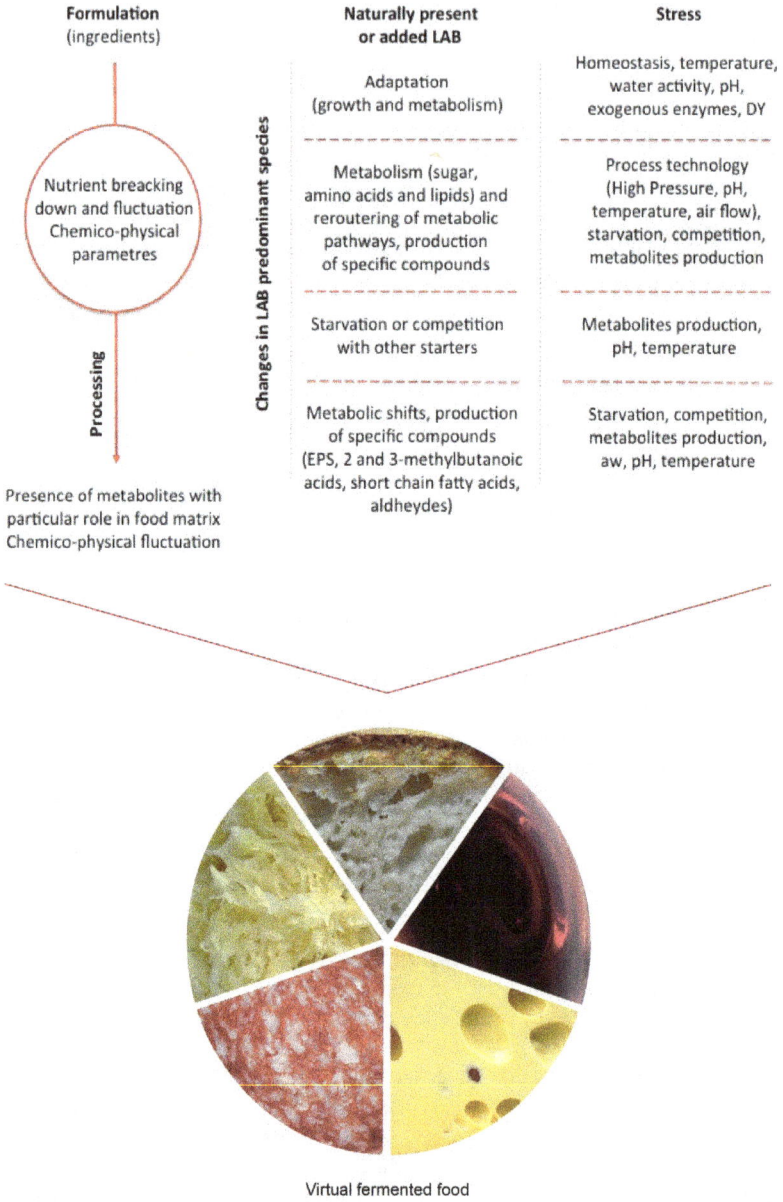

Figure 2. Fermented food model: reciprocal influences between environmental fluctuation and lactic acid bacteria fermentation.

The interaction between the starters and the ingredients and between the starters and the naturally present microbial population can trigger few important mechanisms that will influence the quality and the characteristics of the fermented product. Analogously, many food processes and formulations have been tested for safety by challenge test inoculating pathogen bacterial cells at different growth phases, and the results proved that cells grown to the stationary phase or adapted to various stresses have greater resistance than exponential cells [83].

Other ingredients usually added to obtain safe and stable products are food preservatives, including:

a. Antioxidants,
b. Anti-browning agents
c. Antimicrobials.

These latter are arbitrarily classified into two groups: traditional or "regulatory approved" and naturally occurring [84]. The former includes acidifiers such as acetic acid, lactic acid and citric acid and antimicrobials such as benzoic acid and benzoates, propionate, nitrites and nitrates, sorbic acid and sorbates and sulfites. The latter includes compounds from microbial, plant and animal sources that are, for the most part, only proposed for use in foods as antimicrobials (e.g. lactoferrin, lysozyme, nisin). Throughout the ages, food antimicrobials have been used primarily to prolong shelf-life and preserve quality of foods through inhibition of spoilage microorganisms, while only few are used exclusively to control the growth of specific foodborne pathogens (e.g. nitrite, used for hundreds of years to inhibit growth and toxin production of *Clostridium botulinum* in cured meats). In food formulation antimicrobials are part of a multiple intervention system that involves the chemical along with environmental (extrinsic) and food related (intrinsic) stresses and processing steps. Some of these substances (for example lactic acid and citric acid) provoke a direct acidification of a food or food ingredient, and therefore challenge the microflora inducing and increase of acid resistance of the microflora itself. In fermented food the situation can be somewhat different, because the pH is gradually lowered by LAB creating a pH gradient, more likely than a sharp alteration in the pH due to direct acidification.

A good model describing the shock related to the inoculum of LAB in the raw complex material has been described during the production of fermented sausages [85]. The relatively high pH of raw meat rapidly decreases during the initial fermentation phase because organic acids, mainly lactate, are formed by LAB and the water activity is reduced during ripening, because of the addition of salt as well as drying. Furthermore, adjuvants, such as potassium or sodium nitrite and/or nitrate, are mostly added to optimize the fermentation process.

Generally strains used as starter cultures must tolerate these kinds of stresses and exhibit a high ecologic performance in the stressful food environment. Genes related to stress response are induced when *L. sakei* is inoculated in the raw meat system [86]. In fact, ctsR, a gene that coded for a class III heat shock proteins repressor associated with the environmental stress response of Gram positive bacteria, increased its expression when *L.*

sakei starts to adapt to the raw environment. This mechanism demonstrated that the sudden changes in the environment conditions are perceived as stress by *Lactobacillus* species. In particular, in the case of *L. sakei*, added to raw meat and spices, the principal stress response regarded high osmolarity and temperature shifts. Moreover, the presence of curing salt is regarded as one of the major hurdles in the initial phase of sausages fermentation. Because nitrite was found to be the effective for growth inhibition of pathogens, nitrite was also hypothesized as a stressor for *L. sakei* [85] and the exposure of this strain to stresses can induce changes in metabolic activities in a food environment [4]. The metabolic changes in *L. sakei* resulted in enhanced exploitation of available nutrients or increased activity of glycolytic enzymes, leading to the accelerated production of lactic acid by stress-treated *L. sakei* cells [85]. However, the exposition of *L. sakei* to low temperature and high osmolarity gives rise to the repression of phosphofructokinase and consequently to a decreased flux through the glycolytic pathway [87].

Moreover, it is important to consider that some ingredients can be also antimicrobials because of their own characteristics: in fact, if the recipe includes herbs and spices (aromatic plants, pepper), garlic and onions, an effect on microorganisms can be exerted by specific compounds characterizing these products, such as essential oils, terpenes and sulfur compounds [88].

Another essential aspect affecting the performances and metabolism of LAB are the intrinsic characteristics of raw materials that sometimes act in a synergic way with other ingredients. Considering for example fermented vegetables, the microflora of the starting fresh vegetables is typically dominated by Gram negative aerobic bacteria and yeasts, while LAB make up a minor portion of the initial population [89] and therefore they would not be able to start and lead a fermentation process. However, if anaerobic conditions are settled and salts are added, LAB can have a competitive advantage and induce spontaneous lactic acid fermentation. The growth of specific LAB is dependent on the chemical (substrate, salt concentration, pH) and physical (vegetable type, temperature) environments. As the environments change during fermentation, so can the dominant organisms, often leading to a specific and reproducible succession of bacteria.

In sauerkraut [89, 90] the presence of 1.8-2.2% of NaCl and a temperature of 18°C inhibits many strains of LAB, with the exception of *Leuconostoc mesenteroides* that initiates the fermentation; however this species is sensitive to acid conditions, so after a few days, when the concentration of lactic acid increases, *L. mesenteroides* is replaced by more acid resistant LAB such as *Lactobacillus brevis* and *L. plantarum*, able to further lower the pH up to 3-3.5, stabilizing the final product.

Considering olives fermentation is possible to outline the characteristics of the product affecting LAB: while the brine provides a good environment for LAB growth, with glucose, fructose and mannitol as the main source of fermentable sugars, the presence of high levels phenols (such as oleuropein) exert an antimicrobial activity, inhibiting some strains and selecting the types of organisms that predominate during the fermentation [91-93]. These LAB have to be resistant not only to phenols, but also to lye treatments and water washes, that can be performed during the processing and increase the initial pH, reducing also the nutrients content on the olive surface. The species able to face these kind of stresses usually belong to the genera *Pediococcus*, *Leuconostoc* and *Lactococcus*; after the first stage of fermentation, when the pH reaches 6, *L. plantarum* rapidly grows and dominate the fermentation, that goes on until the fermentable sugars are depleted. The viability and vigor of *L. plantarum* can be encouraged also by yeasts that are still present in this stage of fermentation and can produce vitamins [94].

Moreover, the presence of some gases can modify the growth performances of LAB. That is also influenced by the mixing step of the ingredients in some food processes (e.g. dough mixing). In fact, in bread making process, the continuous agitation of the dough can increase the microbes exposure to oxygen, and this can be a source of oxidative stress, mainly for LAB that are usually anaerobic or facultative anaerobic. Also in these cases the bacteria can react in different ways, activating metabolic and transcriptional responses in order to detoxify ROS, as previously described.

For the fermented vegetables ,above reported, the rapid consumption of oxygen due to the presence of yeasts and aerobic bacteria in the first stage of fermentation has a positive effect on LAB. In fact, they are exposed only for a short time to oxidative stress and, due to their competitive advantage, they rapidly and intensively grow in the food system.

After formulation, the technological processes involving LAB include a fermentation process.

It is reported that various beneficial phenotypic traits of LAB in food fermentations such as rapid acidification, selective proteolysis, tolerance of osmotic and stresses, resistance to ROS, and ability to thrive in nutrient poor conditions and at low temperatures are influenced by stress responses in various species of LAB [95, 96]. The knowledge of these mechanisms, and mainly of the stress responses activated by the fermentation process parameters can be useful in order to develop strains with optimal fermentation characteristics [83].

The first metabolic reaction regards the oxidation of carbohydrates (this reaction depends on the hetero-fermentative or homo-fermentative species involved) that give rise to acids, alcohols and CO_2. These metabolites are directly involved in flavor, aroma and texture of the product and in a second time can influence the production and the availability of other metabolites such as vitamins and antioxidant compounds [78]. Moreover, the LAB interactions with the ingredients increase also the digestibility and decrease the glycemic index, enhancing the healthy features of the fermented foods [97].

At the same time with carbohydrates oxidation, other metabolic mechanisms interest LAB cells such as proteolysis and lipolysis. The first reaction produces polypeptides with interesting characteristics as antimicrobial compounds, salt substitutes (the oligopeptides are able to increase the palatability of the system), and amino acids deriving aromatic compounds. On the other hand lipolysis produces medium chain fatty acids, with important antimicrobial properties. All these reactions (carbohydrates oxidation, lipolysis and proteolysis) generate precursors for other mechanisms in the cells and in the food matrix that give rise to the dynamic environment characteristics of fermented foods. It is important to outline that the compounds produced by the cells, metabolizing the substrate, can modify the system, producing also compounds that can stimulate the growth of symbiotic species or inhibit the growth of antagonistic microorganisms.

The conversion of carbohydrates to metabolites as acetic acid, lactic acid or CO_2 implies the acidification of the system. The contemporary pH decrease and the presence of sugar (osmotic stress) stimulate the exopolysaccharides (EPSs) production. In fact, in sourdough EPSs can be involved in acid tolerance of sourdough LAB [98]. EPSs are long-chain polysaccharides consisting of branched, repeating units of sugars or sugar derivatives. These sugar units are mainly glucose, galactose and rhamnose, in different ratios [99]. The presence of EPSs in the system can create a novel stress to the cells. The inclusion of cells within biofilm can increase their resistance to unfavorable environmental factors such as extreme temperature, low pH and osmolarity, the changes in the texture can induce in LAB also specific stress responses.

For example in yogurt production, the acidification by LAB implies proteins coagulation and thereby changes in the viscosity of the milk. In *L. bulgaricus*, during the acid adaptation present in the fermentation milk to obtain yogurt, some cellular changes were observed: the chaperones GroES, GroEL, HrcA, GrpE, DnaK, DnaJ, ClpE, ClpP and ClpL were induced and ClpC was repressed [100]. Some genes involved in the biosynthesis of fatty acids were induced (*fabH, accC, fabI*), while the genes involved in the mevalonate pathway of isoprenoid synthesis (*mvaC, mvaS*) were repressed [101, 102]. The changes in Aw value are depending not only on EPSs production by LAB after the exposition to acidic and osmotic stress, but also on the ingredients composition and on the step of fermentation.

Considering cheese, the Aw decreases during manufacture and ripening as a result of dehydration, salting, and production of water-soluble solutes from glycolysis, proteolysis, and lipolysis; the cheese Aw values range from 0.70 for extra hard cheeses to 0.99 for fresh, soft cheeses, such as cottage cheese, while semi-hard cheeses have Aw values of around 0.90. The cheese pH also decreases during manufacture and ripening [103]. The effects of different Aw and pH on *L. lactis* simulating cheese ripening have been analyzed [103]. The results evidenced that at low Aw, particularly at low pH, the growth and lactose utilization rates decreased and lactose fermentation to L-(1)-lactate switched to a pathway involving nontraditional saccharide products rather than the traditional lactococcal heterofermentative products.

In *L. plantarum* WCFS1 the addition of 300 mM and 800 mM of NaCl induced mild osmotic stress and osmotic stress respectively. In the presence of 800 mM of NaCl several genes showed an increased expression with respect to the control culture. In particular, those genes were associated with various stress responses in prokariotes, i.e. genes encoding Clp protease, an excinuclease, catalase (peroxide stress) and Dpr-like protein (peroxide stress). These differences in the gene expression were also identified in the presence of acid stress. These results suggest that lactic acid stress in *L. plantarum* WCFS1 also induces a more general stress response (as above described for different *Lactobacillus* species). An overlap between the stimulus for lactic acid and those for peroxide and UV radiation has also been reported for *L. lactis* [104, 66]. The response of *L. sanfranciscensis* to osmotic stress (saccarose 40%) gives rise to the overproduction of 3-methylbutanoic acid and gamma-decalactones when *L. sanfranciscensis* was co-inoculated with yeasts, simulating a sourdough environment. The production of lactones can be indicated as unfavourable environment for microbial growth and metabolism. In fact, these compounds have both particular aromatic and antimicrobial features [13].

The ability of the target strains to dominate the fermentation is related not only to the ingredients (as above described), but also to the fermentation conditions, mainly temperature and atmosphere. If the fermentation is not performed at the optimal growth temperature for the microorganisms, they could be unable to compete with naturally occurring microflora, and consequently the whole process could be compromised. On the contrary, some microbial species have developed specific thermal resistance mechanisms, and they can easily adapt to these unfavorable conditions without implications for the fermentation processes. Moreover, the adaptation to thermal stresses often leads to tolerance to other stresses, in a mechanism usually define "cross protection", as reported for *L. lactis* [105]. The ability of commercial *L. lactis* ssp. *lactis* and *L. lactis* ssp. *cremoris* to withstand freezing at –60°C for 24 h was significantly improved by a prior 25 min heat shock at ~40°C or by a 2 h cold shock at 10°C, opening interesting perspectives for the production on resistant starter cultures, both frozen or lyophilized [105].

Other Authors with regard to different stresses reported the "cross protection" mechanism: for example the mechanisms of multiple adaptations to hops of two different strains of *L. brevis* have been characterized [106]. Hop resistance of lactobacilli requires multiple resistance mechanisms. This is consistent with the stress conditions acting on bacteria in beer, which mainly consist of acid stress and the antimicrobial effect of the hop compounds, in addition to ethanol stress and starvation. The effect of interaction of acid stress and presence/absence of oxygen in the system on *L. helveticus* and *L. sanfranciscensis*, in particular on their cell membrane composition, has been reported [14]. Upon acid stress the level of cyclopropane fatty acids increased at the expense of the level of long-chain unsaturated fatty acids. *L. helveticus* and *L. sanfranciscensis*, exposed to acid sub lethal stress demonstrated the same increase in cyclopropane fatty acids. In particular, *L. helveticus* presented higher concentration of C19cyc11 at pH 4 and pH 3, while *L. sanfranciscensis* presented more C19cyc9 at pH 3 in microaerophilic condition without tween 80, at pH 3.6 in anaerobiosis with tween 80, and at pH 4 in anaerobiosis without tween 80. These results demonstrated the same behavior in front of multiple stresses by LAB membrane [106, 14]

Consider the atmosphere, i.e. the presence or not of oxygen, as another important variable during fermentation, it is known that oxygen can inhibit the growth of LAB, especially in the first stages. However, the food system is usually a consortium of different microorganisms: for example in bakery products and in fermented sausages the fermentation is carried out both by yeasts and LAB; the formers can therefore consume the amount of oxygen present in the mix, allowing the growth of LAB. The same thing happens for fermented vegetables, where naturally occurring Gram negative bacteria and yeast rapidly remove the oxygen, promoting the rapid predominance of Lactobacilli.

Some secondary metabolites such as bacteriocins can play a role in LAB performances and metabolism, affecting also the total population and ecology of fermented foods [107, 108]. Bacteriocins are antimicrobial peptides or proteins produced by bacteria that can be active on different microorganisms, depending on their structure. LAB belonging to the genera *Lactococcus, Pediococcus, Lactobacillus, Leuconostoc, Carnobacterium, Propionibacterium* are known to produce bacteriocins with both narrow and broad inhibitory spectra [109]. The use of functional LAB starter cultures (eg. bacteriocinogenic starter cultures), well adapted to the environment and the process conditions applied, may contribute to the development of better controllable and more efficient production processes [110]. An example can be nisin, a peptide produced by *L. lactis* ssp. *lactis*, that has a narrow spectrum affecting primarily only Gram-positive bacteria and their spores, including lactic acid bacteria, *Bacillus, Clostridium, Listeria,* and *Streptococcus*. However some LAB such as *Streptococcus thermophilus* and *L. plantarum* are able to produce the enzyme nisinase, which neutralizes the antimicrobial activity of the peptide [111]. Therefore these LAB could be suitable for a co-fermentation with *L. lactis*.

Another interesting case of bacteriocin production, as a consequence of oxidative stress and carbon dioxide exposure, has been reported [110]: oxidative stress and carbon dioxide are involved in the production of a specific bacteriocin, amylovorin L, by *Lactobacillus amylovorus*, able to inhibit other LAB species. During traditional sourdough fermentation, a decrease in redox potential of the rather firm mixture occurs. The oxygen initially present is consumed by *Candida* spp. or converted into hydrogen peroxide or water, thereby creating microaerophilic or anaerobic environment in which the growth of the desired LAB is favored. While in a large-scale sourdough type II fermentation currently the use of dough mixture with high dough yield is exploited. This sourdough has to be stirred to liberate part of the carbon dioxide produced to prevent running over. During mixing, oxygen is incorporated into the dough. Also, the development of yeast and hence the production of carbon dioxide is favored in continuously stirred sough mixtures with high water content. Elevation of the airflow rates leading to oxidative stress conditions resulted in an enhanced specific amylovorin L production. Growth in the presence of carbon dioxide also increased the specific bacteriocin production. Mild aeration or a controlled supply of oxygen as well as growth in an environment containing high amounts of carbon dioxide might thus contribute to the competitiveness of *L. amylovorus* DCE471 in a sourdough ecosystem [110]. The production of plantaricin A by *L. plantarum* was also demonstrated in relation to a quorum sensing mechanism [79].

Another example of the influence of the process on LAB metabolism has been widely described [112]. These Authors monitored the evolution of the gene expression of *L. plantarum* IMDO 130201 during a sourdough process. In particular, the genes and the metabolites related to acidic stress were analyzed. It is interesting to highlight that during the pH decrease (production of lactic acid by *L. plantarum*) the genes coding for plantaricin production had higher levels of expression at low pH values, indicating that the bacteriocin production was activated under acid stress conditions by *L. plantarum* IMDO 130201 strain. The presence of the pheromone plantaricin A (PlnA) in a system inoculated with *L. plantarum* DC400 was also reported [79]. Biosynthesis of PlnA was variously stimulated depending on the microbial partner. In fact, *L. sanfranciscensis* DPPMA174 induced the highest synthesis of PlnA, which, in turn, determined lethal conditions for it. The proteome of *L. sanfranciscensis* DPPMA174 responded to the presence of PlnA. The up-regulation of 31 proteins related to stress response, amino acid metabolism, energy metabolism, membrane transport, nucleotide metabolism, regulation of transcription and cell redox homeostasis was found. At the same time, other proteins such as cell division protein (FtsZ), glutathione reductase (LRH_11212) and response regulator (rrp11) were down-regulated. These results demonstrated a hypothetically and interesting waterfall of events all related with stresses response and with the typical fermentation products dynamics (Figure 3). At the same time, the low pH values implied a poor expression of the genes involved in carbohydrate degradation in *L. plantarum* IMDO 130201. The bacterium was directed toward survival at low pH by amino acid conversions rather than by relying on growth [112]. The same behavior was identified in *L. sanfranciscensis* LSCE1 response to pH 3.6 [15]. Under the adopted experimental conditions, which did not produce any decrease in viability of *L. sanfranciscensis* LSCE1, the acid stress, within 2 h, was accompanied by a reduction of the carbohydrate metabolism, as shown by the decrease of ethanol, acetate, and lactate. This mechanism suggests the existence of a switch from sugar to amino acid catabolism that supports survival and growth also in specific and restricted environments, such as sourdoughs, characterized by acid stress and recurrent carbon starvation. Under the acid conditions (pH 3.6) and in the presence of specific nutrients 3-methylbutanoic acid was the predominant metabolite among those detected by solid phase micro-extraction gas chromatographic analysis and mass spectrometry (GC-MS-SPME), released after 2 h of acid stress exposure [15]. The acid stress implied less carbohydrate utilization and ethanol, lactate, and acetate production, but high amino acids catabolism that confers a different and characteristic metabolites pattern. Stress resistance assume great importance as one of the adaptation factors to gastrointestinal tract of probiotic strains as reported in a detailed review [113].

5. Stress resistance of probiotic LAB

There are two main categories of factors that contribute to the optimal functioning of probiotic lactobacilli: factors that allow optimal adaptation to the new niches that they temporarily encounter in the host (adaptation factors) and factors that directly contribute to the health-promoting effects (probiotic factors) [113].

Adaptation factors include stress resistance, active metabolism adapted to the host environment, and adherence to the intestinal mucosa and mucus.

In fact, probiotic lactobacilli encounter various environmental conditions upon ingestion by the host and during transit in the gastro intestinal tract (GIT). They need to survive to: 1) the harsh conditions of the stomach secretion generating a fasting pH of 1.5, increasing to pH 3 to 5 during food intake; 2) the bile excreted by liver in small intestine represents another challenge for bacteria entering the GIT. Bile salts also seem to induce an intracellular acidification so that many resistance mechanisms are common for bile and acid stress. Indeed, the protonated form of

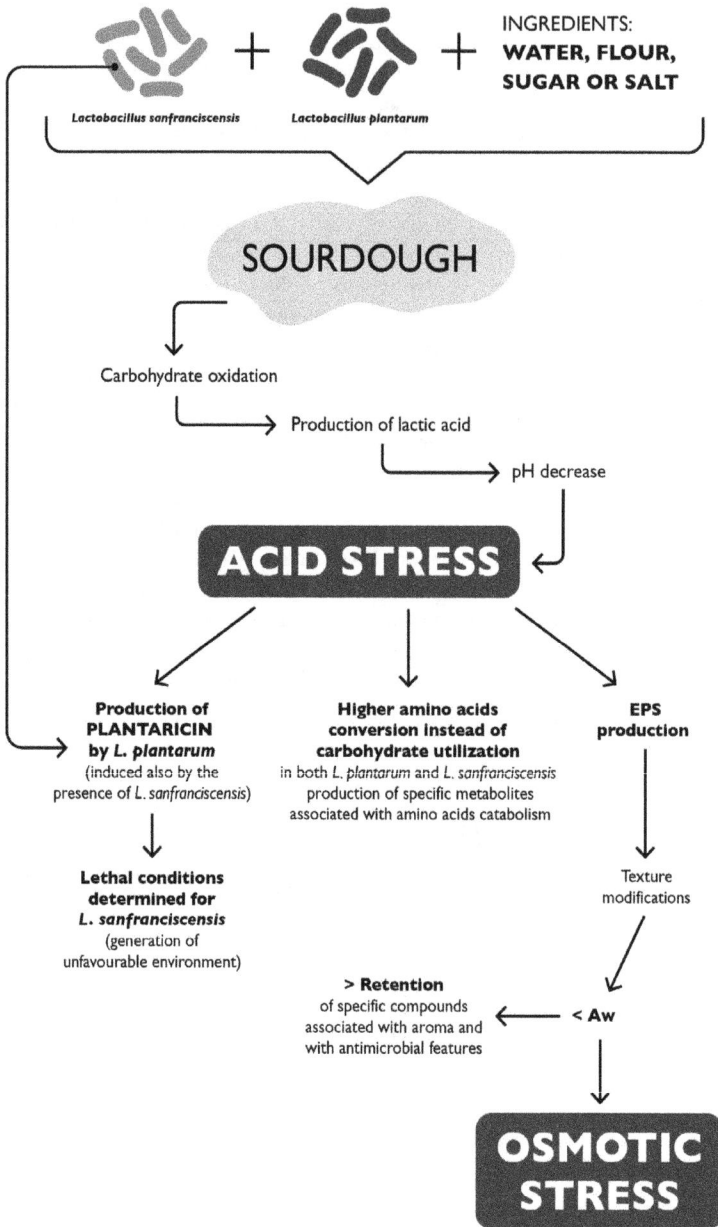

Figure 3. Sourdough fermentation dynamics. Case of possible parallel phenomena interesting acid and osmotic stress.

bile salts is thought to exhibit toxicity through intracellular acidification in a manner similar to those of organic acids like the lactic acid produced by the lactobacilli themselves. For a detailed overview of acid, bile, and other stress resistance mechanisms of lactobacilli, the reader is referred to more extensive review [113]. 3) In analogy to the stresses encountered by intestinal pathogens, they also encounter oxidative and osmotic stress in GI tract. 4) Interactions with other microbes and 5) Interactions with cells of the host immune system and the various antimicrobial products that they produce can also impose a serious threat for the probiotic microbes. Analogously to what described in food LAB, the phenomenon of cross-adaptation is often observed, i.e., that adaptation to one stress condition also protects against another stress factor, implying some common mechanisms. In this respect, also for probiotic LAB non-actively-growing stationary-phase cells are generally more resistant to various stressors than early-log-phase cells.

5.1. Maintaining integrity of the cell envelope

The different macromolecules constituting the cell membranes and cell walls of lactobacilli have been shown to contribute to maintaining cell integrity during stress to various degrees. For example, low pH caused a shift in the fatty acid composition of the cell membrane of an oral strain of *L. casei*. Similarly, bile salts have been shown to induce changes in the lipid cell membrane of *Lactobacillus reuteri* CRL1098.

The role of EPS in acid and bile resistance is less clear. However, EPS production has not been studied in detail after exposure to bile. In fact, to our knowledge, phenotypic analyses of dedicated *Lactobacillus* mutants affected in EPS biosynthesis genes have not yet been performed. Homopolysaccharides (HoPSs) from *L. reuteri* have been reported to have a more established role in stress resistance by the maintenance of the cell membrane in the physiological liquid crystalline phase under adverse conditions.

5.2. Repair and protection of DNA and proteins

A number of proteins that play a role in the protection or repair of macromolecules such as DNA and proteins also seem to be essential for acid and bile resistance. Intracellular acidification can result in a loss of purines and pyrimidines from DNA. Bile acids have also been shown to induce DNA damage and the activation of enzymes involved in DNA repair. Perhaps even more vital in the general stress response are chaperones that intervene in numerous stresses for important tasks such as protein folding, renaturation, protection of denatured proteins, and removal of damaged proteins.

5.3. Two-component and other regulatory systems

Mechanisms to specifically sense the presence of certain stress factors and regulate gene expression in response to these stimuli are also crucial for bacterial survival under adverse conditions. Although these mechanisms are not well characterized for lactobacilli, they often involve two-component regulatory systems (2CRSs). 2CRSs allow bacteria to sense and respond to changes in their environment after receiving an environmental signal through transmembrane sensing domains of the histidine protein kinase (HPK).

6. Methodological approaches to study the effects of stress on LAB

The study of stress responses by LAB is getting closer and closer to the different "omic" fields: genomic, proteomic and metabolomic. Other traditional approaches regarding the membrane cells composition and modifications, both from a structural (cellular fatty acids composition by gas-chromatographic method) and morphological (membrane and wall modification by electronic microscopy) point of view are still used.

Genes implicated in LAB stress responses are numerous and the levels of characterization of their actual role and regulation differ widely between species. The studies concerning stress responses in LAB sometimes benefit from the knowledge already acquired in other bacteria. For example, parts of the studies on heat response have been focused on specific genes because of their major role demonstrated in other microorganisms [17]. The cheapest and easiest way to study a stress response in LAB is to follow some specific genes related to stresses such as heat shock, salts and acids [114, 115]. This type of study is useful especially if the entire genome sequence of some LAB is still unknown. However, nowadays the study of whole trascriptome (the total set of RNAs) is one of the most exhaustive ways to study modifications of gene expression as a result of a stress condition. The transcriptome of a cell contains information about the biological state of the cell and the genes that play a role under specific circumstances. The principal technique used to study the trascriptome is microarray [116].

DNA microarray technology has been used in numerous experiments to analyze gene expression: one example is the evaluation of the general stress response of *B. subtilis* [117] or the investigation of the transcription profiles of *L. plantarum* grown in steady-state cultures that varied in lactate/lactic acid concentration, pH, osmolarity [66, 104]. This approach is useful also to study the behaviour of bacteria in a real food system. Hüfner et al. [5] studied the global transcriptional response of *L. reuteri* to sourdough environment, showing a significant changes of mRNA levels for 101 genes involved in diverse cellular processes, from carbohydrate and energy metabolism, to cell envelope biosynthesis, exopolysaccharide production, stress responses, signal transduction and cobalamin biosynthesis.

The gene expression dynamics of *L. casei* during fermentation in soymilk when grown up to lag phase, late logarithmic phase, or stationary phase were also studied. Comparisons of different transcripts close to each other revealed 162 and 63 significantly induced genes, in the late logarithmic phase and stationary phase, whose expression was at least threefold up-regulated and down-regulated, respectively. Approximately 38.4% of the up-regulated genes were associated with amino acid transport and metabolism, followed by genes/gene clusters involved in carbohydrate transport and metabolism, lipid transport and metabolism, and inorganic ion transport and metabolism [118].

The study of trascriptome is a good approach that gives a good overview of the changes that can occur inside a stressed bacterium. A limitation of this technique is that it is expensive and requires that the genome sequences of the organisms under study should be available for designing the oligonucleotides for the microarray [119].

A different but, at the same time, related point of view regards the study of proteins and proteome. The most common method to obtain this information is to extract total proteins and separate them by a sodium dodecyl sulphate polyacrylamide gel electrophoresis (SDS-PAGE) followed by a western blotting (in the first case) or a two dimesional electrophoresis (2D-E) analysis (in the second case). Also in this case if the study is focused on a single protein, it is necessary to know before the characteristic of the target protein to optimize the analytical conditions. 2D-Electrophoresis can provide more than 10000 detectable protein spots in a single gel run. Thus, proteins with post-translational modifications (PTMs), such as processing, phosphorylation and glycosylation, can be easily detected as separate spots. A spot separated by 2D-E theoretically consists of an almost homogeneous protein, and thus can be identified following digestion with a sequence-specific protease by peptide mass fingerprinting (PMF) approaches, typically using matrix-assisted laser desorption ionization (MALDI)- time-of-flight (TOF) mass spectrometers. The same level of automation is also available for proteomic approaches involving tandem mass spectrometry (MS-MS) analysis, extremely useful when studying organisms with incomplete or partial genomic information [120].

This kind of approach was used to investigate the cell surface proteins of a typical strain of *L. casei* in response to acidic growth conditions [121]. They demonstrated that growth of *L. casei* under acidic conditions caused molecular changes at the cell surface in order to accomplish an adaptive strategy, resulting in slower growth at low pH. Moreover, the proteomic approach was useful to study the heat shock response respectively on *L. helveticus* PR4 and *L. plantarum* [26, 31]. The cold adaptation of *Lactococcus piscium* strain CNCM I-4031 was studied with the same approach [122]. This analysis could be also performed to compare the effects that new technologies produce on bacteria comparing with the normal stress conditions. In fact, the HHP stress response of *L. sanfranciscensis* was compared with cold, heat, salt, acid and starvation stresses responses [18].

Due to increasingly available bacterial genomes in databases, proteomic tools have recently been used to screen proteins expressed by microorganisms in food, in order to better understand their metabolism *in situ*. While up to now the main objective has been the systematic identification of proteins, the next step will be to bridge the gap between identification and quantification of these proteins [123]. Proteomics has also been used to analyse the proteins released during the ripening of Emmentaler cheese. In an innovative study, proteomics was used to prepare a reference map of the different groups of proteins found in cheese [124]. These authors were able to categorize these proteins into five classes: those involved in proteolysis, glycolysis, stress response, nucleotide repair and oxidation-reduction. In addition, information was obtained regarding the peptidases released into the cheese during ripening process. This study enabled the Authors to differentiate between the various casein degradation mechanisms present, and to suggest that the streptococci within the cheese matrix are involved in peptide degradation and together with the indigenous lactobacilli contribute to the ripening process. Using proteomics these Authors were able to get a greater understanding of the microbial succession involved in the ripening of Emmentaler cheese, which information could not have been obtained using other protein separation techniques. This example illustrates the power of proteomics as a tool for analyzing the composition of a complex mixture of proteins and peptides [119].

The global identification of stress-induced proteins in a given organism has technical limitations. Membrane proteins, for example, are rarely detected by this method. Secondly, it may be that changes in membrane proteins composition result from long-term adaptation processes, while short-term responses may primarily be accounted for the activation (and/or stabilization) of proteins already present. The latter hypothesis is valid especially in the case of transport systems, although for some of the systems studied a transcriptional induction has also been observed [17]. The use of this technique is not as widespread as that of DNA microarrays due to the challenges associated with the purification and separation of complex mixtures of proteins found in cell extracts. At the same time the study of the only transcriptome should take into consideration that a lot of post-transcriptional processes may act on RNA (ex. RNA interference, polyadenilation ecc) [125].

As reported above, the stress responses of LAB are studied also through the analysis of membrane composition, structure and integrity. Not unexpectedly, in fact, the cell membrane plays an important role in stress resistance. First of all, the membrane itself can change in adaptation to environmental conditions and these changes contribute to the protection of the bacteria [17]. The adaptive response to sub-lethal acid and cold stresses in *L. helveticus* and *L. sanfranciscensis* has been analyzed (as described above) [14]. The extraction and identification by GC-MS of lipid fatty acids and free fatty acids could give an overview of the membrane fluidity state. In the same article they developed a gas chromatographic method to separate and quantify the cell cyclopropane fatty acids lactobacillic (C19cyc11) and dehydrosterculic (C19cyc9) demonstrating different responses of the strains tested in terms of cyclopropane fatty acids production, probably due to the different original optimal environment. The comparison between the wild type and the acid-resistant mutant *L. casei* LBZ-2 evidenced in the latter higher membrane fluidity, higher proportions of unsaturated fatty acids, and higher medium chain length. In addition, cell integrity analysis showed that the mutant maintains a more intact cellular structure and lower membrane permeability after environmental acidification [126].

The last but not least approach used to study the stress response of LAB is the metabolic one. The study of the metabolites released, as a consequence of the stress exposure, can contribute to the understanding of the mechanisms that regulate the microbial interactions and the metabolic alterations induced by stress conditions. Moreover, these approaches can be exploited to identify which technological conditions induce microorganisms to produced desirable metabolites [4, 15].

With this perspective the use of GC-MS-SPME as a potent and easy tool to study the generation of volatile metabolite compounds such as flavoring molecules or aroma precursors was widely adopted [9,11-13, 15] and contributed to rationalize the process and optimize the products. In particular, the effects of HPH on different species of *Lactobacillus* involved in dairy product fermentation and ripening, monitoring the changes in volatile compounds as indicators of metabolic profiles has been studied [11].

Analysing the oxidative and heat stresses in L. *helveticus* two new 2[5H]-furanones released by this strain both as a possible signalling molecules and as possible important flavouring compounds has been identified by GC-MS-SPME [9]. On the contrary the study of non-volatile metabolites can be performed by normal chromatographic technique (HPLC), especially for amino acids and sugars [15], or by Fast Protein Liquid Chromatography (FPLC) separation for peptides, followed by a mass spectrometry identification [127]. An NMR approach to evaluate the effects on the growth of L. *plantarum* raising the medium molarity by high concentrations of KCl or NaCl and iso-osmotic concentrations of non-ionic compounds was performed [128].

Since all the techniques described above, if used alone, do not allow a total comprehension of stress responses, a lot of studies are trying to combine two or more approaches together. Combined transcriptomic and proteomic analyses were used to evaluate the glucose-limited chemo-stat in *Enterococcus faecalis* V583 [129] or to study the effect of bile salts in the growth of L. *casei* [130]. A combined physiological and proteomic approach, instead, was followed to unravel lactic-acid-induced alterations in L. *casei* [131].

Therefore it is possible to understand, from the references above, that techniques used to study the stress responses of LAB are taking more and more "omic" approach. This comports an accumulation of a huge number of data that it is not easy to manage and to compare. For this reason the use of new programs of data analysis is required. One of these approaches could be the use of heat maps, a technique born as a tool to understand microarray results [66]. Nowadays it could be useful also to manage the data from other fields: in fact, a heat maps was used to show the correlation between metabolites produced, the relative gene expression of specific genes and stress conditions [15]. The same useful tool, combined with other statistical analyses, has been also applied [132].

7. Conclusion

It is known that LAB can adapt to stress with different mechanisms widely studied in model and real systems. An overview of those responses has been described and reported in this chapter.

Stress not only induces changes enabling better survival, but also different performances in a system. In fermented food, the knowledge of the mechanisms that regulate LAB metabolic changes and their effects gain importance especially when those responses can be exploited in order to improve the food properties [4]. In particular, fermented foods are dynamic systems subjected to continuous evolution of their physico-chemical characteristics. The complex fluctuation of the food environment itself, during processing, is stress source for every microorganism involved and the changes that affect the fermented food habitats, can be perceived by LAB as stress.

In this chapter examples of the dynamic fluctuation effect on LAB metabolism have been described in order to outline that every reaction can cause a waterfall of metabolic events influencing the sensorial quality, the shelf-life and the bioactive compounds production of fermented foods.

The subjects of those events are LAB, indicating the importance of metabolism of these microorganisms in food. The cell physiology is crucial to ensure that cells are well suited to survival during downstream processes and that they exhibit high performances.

The production and exploitation of naturally adapted strains can be interesting for companies because of the absence of ethical and legal concerns. The adapted strains are not considered genetically modified microorganisms (GMOs) and therefore they can be applied in food processing without legal restrictions and, more important, without affecting the consumer perception, currently (in Europe) not ready to introduce in his diet foods produced with GMOs.

Individual stresses used in food processing and preservation may render probiotic LAB more resistant to further and different stresses, including those encountered in the human body, e.g. those encountered during gastro-intestinal passage (pH of the stomach, exposure to bile salts in small intestine etc.). A positive correlation has been recently observed between EPS production and resistance to bile salt and low pH stress in *Bifidobacterium* species isolated from breast milk and infant faeces [128].

This knowledge can open interesting perspectives to improve at the same time the performances of LAB, the quality of fermented food and the health-promoting properties of the LAB used.

Moreover, it will be interesting to identify the gastrointestinal tract also as a complex and dynamic system in which LAB need to adapt to adverse conditions, responding with metabolic shifts provided with interesting technological an healthy features.

The "omics" technologies could be particularly useful for identifying the mechanism leading to LAB stress responses. These approaches could also help to identify the mechanisms for cell fitness and stress adaptation that will be needed to develop more generic and science based technologies [7].

Author details

Diana I. Serrazanetti *, Davide Gottardi, Chiara Montanari and Andrea Gianotti
Department of Food Science, Alma Mater Studiorum, University of Bologna, Bologna, Italy

Inter-Departmental Center of Industrial Agri-Food Research (CIRI Agroalimentare), Cesena, Italy

Acknowledgement

We thank Prof.ssa Maria Elisabetta Guerzoni for her enormous scientific support and Luca Vagnini for his graphic abilities (http://www.lucavagnini.com).

* Corresponding Author

8. References

[1] Poolman B, Ruhdal J, Gruss A (2011) LAB Physiology and energy metabolism. In: Ledeboer A, Hugenholtz J, Kok J, Konings W, Wouters J. Thirty years of research on lactic acid bacteria.Rotterdam: 24 Media Labs. pp. 77-101.

[2] Ledeboer A, Hugenholtz J, Kok J, Konings W, Wouters J (2011) In: Ledeboer A, Hugenholtz J, Kok J, Konings W, Wouters J. Thirty years of research on lactic acid bacteria.Rotterdam: 24 Media Labs. pp. v-vi.

[3] Marles-Wright J, Lewis R (2007) Stress response of bacteria. Curr. Opin. Struct. Biol. 17:755–760.

[4] Serrazanetti DI, Guerzoni ME, Corsetti A, Vogel RF (2009) Metabolic impact and potential exploitation of the stress reactions in lactobacilli. Food Microbiol. 26:700–711.

[5] Hüfner E, Britton RA, Roos S, Jonsson H, Hertel C (2008) Global transcriptional response of *Lactobacillus reuteri* to the sourdough environment. Syst. Appl. Microbiol. 31:323–338.

[6] Heller KJ (2001) Probiotic bacteria in fermented foods: product characteristics and starter organisms. Am. J. Clin. Nutr. 73:374S–379S.

[7] Lacroix C, Yildirim S (2007) Fermentation technologies for the production of probiotics with high viability and functionality. Curr. Opin. Biotechnol. 18:176–183.

[8] Serrazanetti DI (2009) Effects of acidic and osmotic stresses on flavor compounds and gene expression in *Lactobacillus sanfranciscensis*. Ph.D. thesis. University of Teramo, Teramo, Italy.

[9] Ndagijimana M, Vallicelli M, Cocconcelli PS, Cappa F, Patrignani F, Lanciotti R, Guerzoni ME (2006) Two 2[5H]-furanones as possible signaling molecules in *Lactobacillus helveticus*. Appl. Environ. Microbiol. 72:6053–6061.

[10] Vannini L, Ndagijimana M, Saracino P, Vernocchi P, Corsetti A, Vallicelli M, Cappa F, Cocconcelli PS, Guerzoni ME (2007) New signaling molecules in some Gram-positive and Gram-negative bacteria. Int. J. Food Microbiol. 120:25–33.

[11] Lanciotti R, Patrignani F, Iucci L, Saracino P, Guerzoni ME (2007) Potential of high-pressure homogenization in the control and enhancement of proteolytic and fermentative activities of some *Lactobacillus* species. Food Chem. 102:542–550.

[12] Guerzoni ME, Vernocchi P, Ndagijimana M, Gianotti A, Lanciotti R (2007) Generation of aroma compounds in sourdough: effects of stress exposure and lactobacilli–yeasts interactions. Food Microbiol. 24:139–148.

[13] Vernocchi P, Ndagijimana M, Serrazanetti DI, Gianotti A, Vallicelli M, Guerzoni ME (2008) Influence of starch addition and dough microstructure on fermentation aroma production by yeasts and lactobacilli. Food Chem. 108:1217–1225.

[14] Montanari C, Sado-Kamdem SL, Serrazanetti DI, Etoa FX, Guerzoni ME (2010) Synthesis of cyclopropane fatty acids in *Lactobacillus helveticus* and *Lactobacillus sanfranciscensis* and their cellular fatty acids changes following short term acid and cold stresses. Food Microbiol. 27:493–502.

[15] Serrazanetti DI, Ndagijimana M, Sado SL, Corsetti A, Vogel RF, Ehrmann M, Guerzoni ME (2011) Acid stress-mediated metabolic shift in *Lactobacillus sanfranciscensis* LSCE1. Appl .Environ. Microbiol. 77:2656-2666.

[16] Klaenhammer TR, Barrangou R, Buck BL, Azcarate-Peril MA, Altermann E (2005) Genomic features of lactic acid bacteria effecting bioprocessing and health. FEMS Microbiol. Rev. 29:393–409.

[17] van de Guchte M, Serror P, Chervaux C, Smokvina T, Ehrlich SD, Maguin E (2002) Stress responses in lactic acid bacteria. Antonie Leeuwenhoek. 82:187–216.

[18] Hörmann S, Scheyhing C, Behr J, Pavlovic M, Ehrmann M, Vogel RF (2006) Comparative proteome approach to characterize the high-pressure stress response of *Lactobacillus sanfranciscensis* DSM 20451^T. Proteomics. 6:1878–1885.

[19] Guerzoni ME, Lanciotti R, Cocconcelli PS (2001) Alteration in cellular fatty acid composition as a response to salt, acid, oxidative and thermal stresses in *Lactobacillus helveticus*. Microbiol. 147:2255–2264.

[20] Pavlovic M, Hörmann S, Vogel RF, Ehrmann MA (2008) Characterisation of a piezotolerant mutant of *Lactobacillus sanfranciscensis*. Z. Naturforsch. 63:791–797.

[21] Burns P, Patrignani F, Serrazanetti DI, Vinderola GC, Reinheimer JA, Lanciotti R, Guerzoni ME. (2008) Probiotic crescenza cheese containing *Lactobacillus casei* and *Lactobacillus acidophilus* manufactured with High Pressure-Homogenized Milk. J Dairy Sci. 91:500–512.

[22] Patrignani F, Burns P, Serrazanetti DI, Vinderola G, Reinheimer J, Lanciotti R, Guerzoni ME (2009) Suitability of high pressure-homogenized milk for the production of probiotic fermented milk containing *Lactobacillus paracasei* and Lactobacillus acidophilus. J. Dairy Res. 5:1-9.

[23] Gouesbert G, Jan G, Boyaval P (2002) Two-dimensional electrophoresis study of *Lactobacillus delbrueckii* subsp. *bulgaricus* thermotolerance. Appl. Environ. Microbiol. 68:1055–1063.

[24] Desmond C, Fitzgerald GF, Stanton C, Ross RP (2004) Improved stress tolerance of GroESL-overproducing *Lactococcus lactis* and probiotic *Lactobacillus paracasei* NFBC 338. Appl. Environ. Microbiol. 70:5929–5936.

[25] Corcoran BM, Ross RP, Fitzgerald GF, Dockery P, Stanton C (2006) Enhanced survival of GroESL-overproducing *Lactobacillus paracasei* NFBC 338 under stressful conditions induced by drying. Appl. Environ. Microbiol. 72:5104–5107.

[26] Di Cagno R, De Angelis M, Limitone A, Fox PF, Gobbetti M (2006) Response of *Lactobacillus helveticus* PR4 to heat stress during propagation in cheese whey with a gradient of decreasing temperatures. Appl. Environ. Microbiol. 72:4503–4514.

[27] Laplace JM, Sauvageot N, Harke A, Auffray Y (1999) Characterization of *Lactobacillus collinoides* response to heat, acid and ethanol treatments. Appl. Microbiol. Biotechnol. 51:659–663.

[28] Schmidt G, Hertel C, Hammes WP (1999) Molecular characterisation of the dnaK operon of *Lactobacillus sakei* LTH681. Syst. Appl. Microbiol. 22:321–328.

[29] Zink R, Walker C, Schmidt G, Elli M, Pridmore D, Reniero R (2000) Impact of multiple stress factors on the survival of dairy lactobacilli. Sci. Aliment. 20:119–126.

[30] Prasad J, McJarrow P, Gopal P (2003) Heat and osmotic stress responses of probiotic *Lactobacillus rhamnosus* HN001 (DR20) in relation to viability after drying. Appl. Environ. Microbiol. 69:917–925.

[31] De Angelis M, Gobbetti M (2004) Environmental stress responses in *Lactobacillus*: a review. Proteom. 4:106–122.

[32] Bucio A, Hartemink R, Schrama JW, Verreth J, Rombouts FM (2005) Survival of Lactobacillus plantarum 44a after spraying and drying in feed and during exposure to gastrointestinal tract fluids in vitro. J. Gen. Appl. Microbiol. 51:221–227.

[33] Castaldo C, Siciliano RA, Muscariello L, Marasco R, Sacco M (2006) CcpA affects expression of the groESL and dnaK operons in Lactobacillus plantarum. Microb. Cell Fact. 5:35.

[34] Gardiner GE, O'Sullivan E, Kelly J, Auty MAE, Fitzgerald GF, Collins JK, Ross RP, Stanton C (2000) Comparative survival rates of human-derived probiotic Lactobacillus paracasei and L. salivarius strains during heat treatment and spray drying. Appl. Environ. Microbiol. 66:2605–2612.

[35] Earnshaw RG, Appleyard J, Hurst RM (1995) Understanding physical inactivation processes: combined preservation opportunities using heat, ultrasound and pressure. Int. J. Food Microbiol. 28:197–219.

[36] Teixeira P, Castro H, Mohacsi-Farkas C, Kirby RJ (1997) Identification of sites of injury in Lactobacillus bulgaricus during heat stress. Appl. Microbiol. 83:219–226.

[37] Hansen PJ, Drost M, Rivera RM, Paula-Lopes FF, AI-Katananit YM, Krininger CE, Chase CC (2001) Adverse impact of heat stress on embryo production: causes and strategies for mitigation. Theriogenol. 55:91–103.

[38] Kilstrup M, Jacobsen S, Hammer K, Vogensen FK (1997) Induction of heat shock proteins DnaK, GroEL, and GroES by salt stress in Lactococcus lactis. Appl. Environ. Microbiol. 63:1826–1837.

[39] KimWS, Dunn NW (1997) Identification of a cold shock gene in lactic acid bacteria and the effect of cold shock on cryotolerance. Curr. Microbiol. 35:59–63.

[40] KimWS, Khunajakr N, Dunn NW (1998) Effect of cold shock on protein synthesis and on cryotolerance of cells frozen for long periods in Lactococcus lactis. Cryobiol. 37:86–91.

[41] Wouters JA, Jeynov B, Rombouts FM, de Vos WM, Kuipers OP, Abee T (1999) Analysis of the role of 7 kDa cold-shock proteins of Lactococcus lactis MG1363 in cryoprotection. Microbiol. 145:3185–3194.

[42] Derzelle S, Hallet B, Francis KP, Ferain T, Delcour J, Hols P (2000) Changes in cspL, cspP, and cspC mRNA abundance as a function of cold shock and growth phase in Lactobacillus plantarum. J. Bacteriol. 182:5105–5113.

[43] Fonseca F, Béal C, Corrieu G (2001) Operating Conditions That Affect the Resistance of Lactic Acid Bacteria to Freezing and Frozen Storage. Cryobiol. 43:189–198.

[44] Zhang G, Fan M, Li Y, Wang P, Lv Q (2012) Effect of growth phase, protective agents, rehydration media and stress pretreatments on viability of Oenococcus oeni subjected to freeze-drying Afr. J Microbiol. Res. 6:1478-1484.

[45] Bâati L, Fabre-Gea C, Auriol D, Blanc PJ (2000) Study of the cryotolerance of Lactobacillus acidophilus: effect of culture and freezing conditions on the viability and cellular protein levels. Int. J. Food. Microb. 59:241–247.

[46] Panoff JM, Thammavongs B, Guéguen M. (2000) Cryotolerance and cold stress in lactic acid bacteria. Sci. Aliment. 20:105-110.

[47] Panoff JM, Thammavongs B, Laplace JM, Hartke A, Boutibonnes P, Auffray Y (1995) Cryotolerance and Cold Adaptation in Lactococcus lactis subsp. lactis IL1403. Cryobiol. 32:516–520.

[48] Grogan DW, Cronan JE (1986) Characterization of Escherichia coli mutants completely detective in synthesis of cyclopropane fatty acids. J. Bacteriol. 166:872.

[49] Grogan D, Cronan J (1997) Cyclopropane ring formation in membrane lipids of bacteria. Microbiol. Mol. Biol. Rev. 61:429-441.

[50] Raynaud S, Perrin R, Cocaign-Bousquet M, Loubiere P (2005) Metabolic and Transcriptomic Adaptation of *Lactococcus lactis* subsp. *lactis* biovar *diacetylactis* in Response to Autoacidification and Temperature Downshift in Skim Milk. Appl. Environ. Microbiol. 71:8016-8023.

[51] Wolf G, Arendt EK, Pfähler U, Hammes WP (1990) Heme-dependent and hemeindependent nitrite reduction by lactic acid bacteria results in different N-containing products. Int. J. Food Microbiol. 10:323–329.

[52] Duwat P, Cesselin B, Sourice S, Gruss A (2000) *Lactococcus lactis*, a bacteriamodel for stress responses and survival. Int. J. Food Microbiol. 55:83–86.

[53] Rochat T, Gratadoux JJ, Gruss A, Corthier G, Maguin E, Langella P, van de Guchte M (2006) Production of a heterologous nonheme catalase by *Lactobacillus casei*: an efficient tool for removal of H_2O_2 and protection of *Lactobacillus bulgaricus* from oxidative stress in milk. Appl. Environ. Microbiol. 72:5143–5149.

[54] Yamamoto Y, Poyart C, Trieu-Cuot P, Lamberet G, Gruss A, Gaudu P (2005) Respiration metabolism of Group B Streptococcus is activated by environmental haem and quinone and contributes to virulence. Mol. Microbiol. 56:525–534.

[55] Vido K, Diemer H, Van Dorsselaer A, Leize E, Juillard V, Gruss A, Gaudu P (2005) Roles of thioredoxin reductase during the aerobic life of *Lactococcus lactis*. J. Bacteriol. 187:601–610.

[56] Rezaïki L, Cesselin B, Yamamoto Y, Vido K, van West E, Gaudu P, Gruss A (2004) Respiration metabolism reduces oxidative and acid stress to improve long-term survival of *Lactococcus lactis*. Mol. Microbiol. 53:1331–1342.

[57] Lechardeur D, Cesselin B, Fernandez A, Lamberet G, Garrigues C, Pedersen M, Gaudu P, Gruss A (2011) Using heme as an energy boost for lactic acid bacteria. Curr. Opin. Biotechnol. 22:143-149.

[58] Brooijmans RJW, de Vos WM, Hugenholtz J (2009) *Lactobacillus plantarum* WCFS1 Electron Transport Chains. Appl. Environ. Microbiol. 75:3580-3585.

[59] Lushchak VI (2001) Oxidative stress and mechanisms of protection against it in bacteria. Biochem. Moscow. 66:476–489.

[60] Hertel C, Schmidt G, Fischer M, Oellers K, Hammes WP (1998) Oxygendependent regulation of the expression of the catalase gene katA of *Lactobacillus sakei* LTH677. Appl. Environ. Microbiol. 64:1359–1365.

[61] Miyoshi A, Rochat T, Gratadoux JJ, Le Loir Y, Costa Oliveira S, Langella P, Azavedo V (2003) Oxidative stress in *Lactococcus lactis*. Genet. Mol. Res. 2:348–359.

[62] Bruno-Bárcena JM, Andrus JM, Libby SL, Klaenhammer TR, Hassan HM (2004) Expression of a heterologous manganese superoxide dismutase gene in intestinal lactobacilli provides protection against hydrogen peroxide toxicity. Appl. Environ. Microbiol. 70:4702–4710.

[63] Jänsch A, Korakli M, Vogel RF, Ganzle MG (2007) Glutathione reductase from *Lactobacillus sanfranciscensis* DSM20451^T: contribution to oxygen tolerance and thiol exchange reactions in wheat sourdoughs. Appl. Environ. Microbiol. 73:4469–4476.

[64] Rallu F, Gruss A, Ehrlich SD, Maguin E (2000) Acid- and multistress-resistant mutants of *Lactococcus lactis*: identification of intracellular stress signals. Mol. Microbiol. 35:517–528.

[65] Cotter PD, Hill C (2003) Surviving the acid test: responses of Gram-positive bacteria to low pH. Microbiol. Mol. Biol. Rev. 67:429–453.

[66] Pieterse B, Leer RJ, Schuren FHJ, van der Werf MJ (2005) Unravelling the multiple effects of lactic acid stress on *Lactobacillus plantarum* by transcription profiling. Microbiol. 151:3881–3894.

[67] Piuri M, Sanchez-Rivas C, Ruzal SM (2005) Cell wall modifications during osmotic stress in *Lactobacillus casei*. J. Appl. Microbiol. 98:84–95.

[68] Piuri M, Sanchez-Rivas C, Ruzal SM (2003) Adaptation to high salt in *Lactobacillus*: role of peptides and proteolytic enzymes. J. Appl. Microbiol. 95:372–379.

[69] Jordan S, Hutchings MI, Mascher T (2008) Cell envelope stress response in Gram-positive bacteria. FEMS Microbiol. Rev. 32:107–146.

[70] Kets EPW, Teunissen PJM, Bont JAM (1996) Effect of Compatible Solutes on Survival of Lactic Acid Bacteria Subjected to Drying. Appl. Environ. Microbiol. 62:259-261.

[71] Ward DE, van der Wejden CC, van der Merwe MJ, Westerhoff HV, Claiborne A, Snoep JL (2000) Branched-chain α-keto acid catabolism via the gene products of the bkd operon in *Enterococcus faecalis*: A new, secreted metabolite serving as a temporary redox sink. J Bacteriol. 182:3239–3246.

[72] Considine KM, Kelly AL, Fitzgerald GF, Hill C, Sleator RD (2008) High-pressure processing – effects on microbial food safety and food quality. FEMS Microbiol. Lett. 281:1–9.

[73] Jofré A, Champomier-Vergès M, Anglade P, Baraige F, Martín B, Garriga M, Zagorec M, Aymerich T (2007) Protein synthesis in lactic acid and pathogenic bacteria during recovery from a high pressure treatment. Res. Microbiol. 158:512–520.

[74] Tabanelli G (2011) Use of sub-lethal high pressure homogenization (HPH) treatments to enhance functional properties of Lactic acid bacteria probiotic strains. Ph.D. thesis. University of Bologna, Bologna, Italy.

[75] Chung HJ, Bang W, Drake MA (2006) Stess response of *Escherichia coli*. Compr Rev Food Sci F. 5:52-64

[76] Burns P, Vinderola G, Molinari F, Reinheimer J (2008) Suitability of whey and buttermilk for the growth and frozen storage of probiotic lactobacilli. Int. J. Dairy Technol. 61:156-164.

[77] Sieuwerts S, de Bok FAM, Hugenholtz J, van Hylckama Vlieg JET (2008) Unraveling Microbial Interactions in Food Fermentations: from Classical to Genomics Approaches. Appl. Environ. Microbiol. 74:4997-5007

[78] Caplice E, Fitzgerald GF (1999) Food fermentations: role of microorganisms in food production and preservation. Int. J. Microbiol. 50:131-149

[79] Di Cagno R, De Angelis M, Calasso M, Vincentini O, Vernocchi P, Ndagijimana M, De Vincenzi M, Dessì MR, Guerzoni ME, Gobbetti M (2010) Quorum sensing in sourdough *Lactobacillus plantarum* DC400: induction of plantaricin A (PlnA) under co-cultivation with other lactic acid bacteria and effect of PlnA on bacterial and Caco-2 cells. Proteomics. 10:2175-2190.

[80] Lorenz MC, Cutler NS, Heitman J (2000) Characterization of Alcohol-induced Filamentous Growth in *Saccharomyces cerevisiae*. Mol. Biol. Cell 11:183–199.

[81] Hansen PJ, Drost M, Rivera RM, Paula-Lopes FF, AI-Katananit YM, Krininger CE, Chase CC (2001) Adverse impact of heat stress on embryo production: causes and strategies for mitigation. Theriogenology. 55: 91–103.

[82] Giraffa G (2004) Studying the dynamics of microbial populations during food fermentation. FEMS Microbiol. Rev. 28:251-260.

[83] Johnson EA (2002) Microbial Adaptation and Survival in Foods. In: Microbial Stress Adaptation and Food Safety. Ed. A.E. Yousef and V.K Juneja. CRC Press

[84] Davidson PM (2001) Chemical preservatives and natural antimicrobial compounds, p. 593–627. In: Food Microbiology: Fundamentals and Frontiers, 2nd ed. M.P. Doyle.

[85] Hüfner E, Hertel C (2008) Improvement of raw sausage fermentation by stress-conditioning of the starter organism *Lactobacillus sakei*. Curr. Microbiol. 57:490-496.

[86] Hüfner E, Markieton T, Chaillou S, Crutz-Le Coq AM, Zagorec M, Hertel C (2007) Identification of *Lactobacillus sakei* genes induced during meat fermentation and their role in survival and growth. Appl. Environ. Microbiol. 73:2522-2531.

[87] Marceau A, Zagorec M, Chaillou S, Mera T, Champomier-Verges MC (2004) Evidence for involvement of at least six proteins in adaptation of *Lactobacillus sakei* to cold temperatures and addition of NaCl. Appl. Environ. Microbiol. 70:7260-7268.

[88] Kamdem S, Patrignani F, Guerzoni ME (2007) Shelf-life and safety characteristics of Italian Toscana traditional fresh sausage (Salsiccia) combining two commercial ready-to-use additives and spices. Food Control. 18: 421 - 429.

[89] Schneider M (1988) Zur mikrobiologie von sauerkraut bei der vergärung in verkaufsfertigen kleinbehältern. Dissertation Hohenheim University, Stuttgart.

[90] Pederson CS, Albury MN (1969) The sauerkraut fermentation. New York State Agricultural Experiment Station Technical Bulletin 824, Geneva, New York.

[91] Ruiz-Barba JL, Garrido-Fernàndez A, Jiménez-Diaz R (1991) Bactericidal action of oleuropein extracted from green olives against *Lactobacillus plantarum*. Let. Appl. Microbiol. 12:65-68.

[92] Ciafardini G, Marsilio V, Lanza B, Possi N (1994) Hydrolysis of oleuropein by *Lactobacillus plantarum* strains associated with olive fermentation. Appl. Environ. Microbiol. 60:4142-4147

[93] Lavermicocca P, Valerio F, Lonigro SL, De Angelis M, Morelli L, Callegari ML, Rizzello CG, Visconti A (2005) Study of Adhesion and Survival of *Lactobacilli* and *Bifidobacteria* on Table Olives with the Aim of Formulating a New Probiotic Food.Appl. Environ. Microbiol. 71:4233–4240.

[94] Ruiz-Barba JL, Jiménez-Diaz R (1995) Availability of essential B-group vitamins to *Lactobacillus plantarum* in green olive fermentation brines. Appl. Environ. Microbiol. 61:1294-1297.

[95] O'Sullivan E, Condon S (1997) Intracellular pH is a major factor in the induction of tolerance to acid and other stresses in *Lactococcus lactis*. Appl. Environ. Microbiol. 63:4210–4215.

[96] Sanders JW, Venema G, Kok J (1999) Environmental stress responses in *Lactococcus lactis*. FEMS Microbiol. Rev. 23:483–501.

[97] Guerzoni ME, Gianotti A, Serrazanetti DI (2011) Fermentation as a tool to improve healthy properties of bread". In V. R. Preedy, R. R. Watson, & V. B. Patel, (Eds.), Flour and breads and their fortification in health and disease prevention (pp.385-393). London, Burlington, San Diego: Academic Press, Elsevier.

[98] Gänzle MG, Schwab C (2009). Ecology of exopolysaccharide formation by lactic acid bacteria: sucrose utilisation, stress tolerance, and biofilm formation. In: Ulrich, M. (Ed.), Bacterial Polysaccharides – Current Innovation and Trends. Horizon Press.

[99] Welman AD, Maddox IS (2003) Exopolysaccharides from lactic acid bacteria: perspectives and challenges. Trends Biotechnol. 21:269-274

[100] Lim EM, Ehrlich SD, Maguin E (2000). Identification of stress-inducible proteins in *Lactobacillus delbrueckii* subsp. *bulgaricus*. Electrophoresis. 21:2557–2561.

[101] Streit F, Corrieu G, Beal C (2007) Acidification improves cryotolerance of *Lactobacillus delbrueckii* subsp. *bulgaricus* CFL1. J. Biotechnol. 128:659–667.

[102] Streit F, Delettre J, Corrieu G, Beal C (2008) Acid adaptation of *Lactobacillus delbrueckii* subsp. *bulgaricus* induces physiological responses at membrane and cytosolic levels that improves cryotolerance. J. Appl. Microbiol. 105:1071–1080.

[103] Liu SQ, Asmundson RV, Gopal PK, Holland R, Crow VL (1998) Influence of Reduced Water Activity on Lactose Metabolism by *Lactococcus lactis* subsp. *cremoris* at Different pH Values. Appl. Environ. Microbiol. 64:2111-2116

[104] Hartke A, Bouche S, Giard JC, Benachour A, Boutibonnes P, Auffray Y (1996) The lactic acid stress response of *Lactococcus lactis* subsp. *lactis*. Curr. Microbiol. 33:194–199.

[105] Broadbent JR, Lin C (1999). Effect of heat shock or cold shock treatment on the resistance of *Lactococcus lactis* to freezing and lyophilization. Cryobiology. 39: 88–102.

[106] Behr J, Gänzle MG, Vogel RF (2006) Characterization of a highly hop-resistant *Lactobacillus brevis* strain lacking hop transport. Appl. Environ. Microbiol. 72:6483-92.

[107] Atrih A, Rekhif N, Michel M, Lefebvre G (1993) Detection and characterization of a bacteriocin produced by *Lactobacillus plantarum* C19. Can. J. Microbiol. 39:1173-1179.

[108] Jiménez-Diaz R, Rios-Sànchez RM, Desmazeaud M, Ruiz-Barba JL, Piard JC (1993) Plantaricin S and T, two new bacteriocins produced by *Lactobacillus plantarum* LPCO10 isolated from a green olive fermentation. Appl. Environ. Microbiol. 59:1416-1424.

[109] Klaenhammer TR (1988) Bacteriocins of lactic acid bacteria. Biochim. 70:337-349.

[110] Neysens P, De Vuyst L (2005) Carbon dioxide stimulates the production of amylovorin L by *Lactobacillus amylovorus* DCE 471, while enhanced aeration causes biphasic kinetics of growth and bacteriocin production. Int. J. Food Microbiol. 105:191–202.

[111] Davidson PM, Harrison MA (2002) Microbial Adaptation to Stresses by Food Preservatives, in Microbial Stress Adaptation and Food Safety. Ed. A.E. Yousef and V.K Juneja. CRC Press

[112] Vrancken G, De Vuyst L, Rimaux T, Allemeersch J, Weckx S (2011) Adaptation of *Lactobacillus plantarum* IMDO 130201, a Wheat Sourdough Isolate, to Growth in Wheat Sourdough Simulation Medium at Different pH Values through Differential Gene Expression. Appl. Environ. Microbiol. 77:3406–3412

[113] Lebeer S, Vanderleyden J, De Keersmaecker SCJ (2008) Genes and Molecules of Lactobacilli Supporting Probiotic Action. Microbiol. Mol. Biol. R. 72:728–764

[114] Jobin MP, Garmyn D, Divies C, Guzzo J (1999) Expression of the *Oenococcus oeni* trxA gene is induced by hydrogen peroxide and heat shock. Microbiol. 145:1245-1251.

[115] Capozzi V, Arena MP, Crisetti E, Spano G, Fiocco D (2011) The hsp 16 Gene of the Probiotic *Lactobacillus acidophilus* Is Differently Regulated by Salt, High Temperature and Acidic Stresses, as Revealed by Reverse Transcription Quantitative PCR (qRT-PCR) Analysis. Int. J. Mol. Sci. 12:5390-5405.

[116] Pieterse B (2006) Transcriptome analysis of the lactic acid and NaCl-stress response of *Lactobacillus plantarum* Ph.D. thesis. Wageningen University, Wageningen, The Netherlands

[117] Petersohn A, Brigulla M, Haas S, Hoheisel JD, Volker U, Hecker M (2001) Global Analysis of the General Stress Response of *Bacillus subtilis*. J. Bacteriol. 183:5617-5631.

[118] Wang JC, Zhang WY, Zhong Z, Wei AB, Bao QH, Zhang Y, Sun TS, Postnikoff A, Meng H, Zhang HP (2012) Transcriptome analysis of probiotic *Lactobacillus casei* during fermentation in soymilk. J. Ind. Microbiol. Biotech. 39:191-206.

[119] Anandan S (2006) Genomic and Proteomic Approaches for Studying Bacterial Stress Responses. In: Advances in Microbial Food Safety. V. Juneja, J.P. Cherry and M.H Tunick eds. ACS Books.

[120] Renzone G, D'Aambrosio C, Arena S, Rullo R, Ledda L, Ferrara L, Scaloni A (2005) Differential proteomic analysis in the study of prokaryotes stress resistance. Ann. Ist. Super. Sanità. 41:459-468.

[121] Nezhad MH, Knight M, Britz ML (2012) Evidence of changes in cell surface proteins during growth of *Lactobacillus casei* under acidic conditions. Food. Sci. Biotech. 21:253-260.

[122] Garnier M, Matamoros S, Chevret D, Pilet MF, Leroi F, Tresse O (2010) Adaptation to Cold and Proteomic Responses of the Psychrotrophic Biopreservative *Lactococcus piscium* Strain CNCM I-4031. Appl. Environ. Microbiol. 76:8011–8018.

[123] Jardin J, Mollé D, Piot M, Lortal S, Gagnaire V (2012) Quantitative proteomic analysis of bacterial enzymes released in cheese during ripening. Int. J. Food Micro. 155:19–28.

[124] Gagnaire V, Piot M, Camier B, Vissers JPC, Gwenael J, Leonil J (2004) Survey of bacterial proteins released in cheese: a proteomic approach. Int. J. Food Microbiol. 94:185-201.

[125] Güell M, Yus E, Lluch-Senar M, Serrano L (2011) Bacterial transcriptomics: what is beyond the RNA horiz-ome? Nat. Rev. Microbiol. 9:658-669.

[126] Wu C, Zhang J, Wang M, Du G, Chen J (2012) *Lactobacillus casei* combats acid stress by maintaining cell membrane functionality. J. Ind. Microbiol. Biotechnol. Accepted

[127] Coda R, Rizzello CG, Pinto D, Gobbetti M (2012) Selected Lactic Acid Bacteria Synthesize Antioxidant Peptides during Sourdough Fermentation of Cereal Flours, Appl. Environ. Microbiol. 78:1087-1096.

[128] Glaasker E, Tjan FSB, Ter Steeg PF, Konings WN, Poolman B (1998) Physiological Response of *Lactobacillus plantarum* to Salt and Nonelectrolyte Stress. J. Bacteriol. 180:4718-4723.

[129] Mehmeti I, Faergestad EM, Bekker M, Snipen L, Nes IF, Holo H (2012) Growth Rate-Dependent Control in *Enterococcus faecalis*: Effects on the Transcriptome and Proteome, and Strong Regulation of Lactate Dehydrogenase. Appl. Environ. Microbiol. 78:170-176.

[130] Alcántara C, Zúñiga M (2012) Proteomic and transcriptomic analysis of the response to bile stress of *Lactobacillus casei* BL23. Microbiol. Accepted

[131] Wu C, Zhang J, Chen W, Wang M, Du G, Chen J (2011) A combined physiological and proteomic approach to reveal lactic-acid-induced alterations in *Lactobacillus casei* and its mutant with enhanced lactic acid tolerance. Appl. Microbiol. Biotechnol. 93:707-722.

[132] Jozefczuk S, Klie S, Catchpole G, Szymanski J, Cuadros-Inostroza A, Steinhauser D, Selbig J, Willmitzer L (2010) Metabolomic and transcriptomic stress response of *Escherichia coli*. Mol. Sys. Biol. 6:364.

Chapter 13

Lactic Acid Bacteria in Hydrogen-Producing Consortia: On Purpose or by Coincidence?

Anna Sikora, Mieczysław Błaszczyk,
Marcin Jurkowski and Urszula Zielenkiewicz

Additional information is available at the end of the chapter

1. Introduction

Hydrogen is both a valuable energy carrier and a feedstock for various branches of the chemical industry. It is thought to be one of the most important energy carriers of the future, an alternative to conventional fossil fuels. Water vapor and heat energy are the sole products of hydrogen burning. Therefore, the use of hydrogen to generate energy does not contribute to ozone depletion, the greenhouse effect, climate changes or acid rains. Hydrogen is a highly efficient energy source; its specific energy equals 33 Wh/g, which is the highest among all fuels. For comparison, the specific energy of methane is 14.2 Wh/g and coal, 9.1 Wh/g. Hydrogen can be used as a fuel in hydrogen fuel cells or burn directly in internal combustion engines. In the chemical industry, hydrogen is used for syntheses of ammonia, alcohols, aldehydes, hydrogen chloride and for the hydrogenation of edible oils, heavy oils or ammonia, for removal of oxygen traces in prevention against metal oxidation and corrosion processes (Nath & Das, 2003; Logan, 2004; Antoni et al., 2007; Piela & Zelenay, 2004).

Conventional methods of hydrogen production, such as gasification of coal, steam reforming of natural gas and petroleum, and electrolysis of water, are based on fossil fuels. Therefore, these methods are regarded as energy expensive and cause environmental pollution (Nath & Das, 2003; Logan, 2004; Nath & Das, 2004).

Considering the limited reserves of fossil fuels, environmental pollution and global warming, there is great interest in biological methods of producing fuels, such as bio-hydrogen, biogas (methane), ethanol or diesel. Among the known biological processes leading to hydrogen production are dark fermentation, photofermentation, direct and indirect biophotolysis, as well as anaerobic respiration of sulphate-reducing bacteria under conditions of sulphate depletion. Taking under account potential applications, microbial hydrogen production has been focused on: (i) photolysis of water using algae and *Cyanobacteria*, (ii) photofermentation of organic compounds by photosynthetic bacteria, and (iii) dark fermentation of organic compounds using anaerobic bacteria.

Members of the *Clostridiales* and *Enterobacteriaceae* are well-recognized hydrogen-producers during the process of dark fermentation. For future applications, dark fermentation seems to be the most promising concept. However, low hydrogen yields and generation of large quantities of non-gaseous organic products remain key problems of dark fermentation. The theoretical maximum hydrogen yield during dark fermentation is 4 moles of H_2/mole of glucose (~33% substrate conversion), but the actual yield is only 2 moles of H_2/mole of glucose (~17% conversion). Currently, many investigations are focused on improving the hydrogen yield during fermentation as an alternative method of hydrogen production and combining dark fermentation with other processes, like methanogenesis, photofermentation or microbial electrolysis of cells, to achieve more effective substrate utilization (Li & Fang, 2007; Das & Veziroglu, 2008; Hallenbeck & Ghosh, 2009; Lee et al., 2010; Hallenbeck, 2011). Biohydrogen fermentations may be carried out in different batch types, continuous or semi-continuous bioreactors, where mixed microbial consortia develop. In the most effective systems, consortia are selected for growth and dominance under non-sterile conditions and usually show high stability and resistance to transient unfavorable changes in the bioreactor environment. Depending on the bioreactor type and growth conditions, consortia form various structures which ensure retention and accumulation of the active biomass. These include microbial-based biofilms and macroscopic aggregates of microbial cells, such as flocs and granules (Campos et al., 2009; Hallenbeck & Ghosh, 2009). A good understanding of the structure of hydrogen-producing microbial communities, symbiotic relationships within the consortia as well as factors favoring hydrogen production is vital for optimizing the process.

Interestingly, lactic acid bacteria (LAB) are often detected in mesophilic hydrogen-producing consortia as bacteria that accompany hydrogen producers. In this chapter, we discuss the issue of whether LAB are bad or good (positive or negative) components of hydrogen-producing consortia. We present different opinions about the potential significance and the role of LAB in hydrogen-producing communities.

2. Hydrogen-producing bacteria

Fermentation is an anaerobic type of metabolic process of low energy gain in which organic compounds are degraded in the absence of external electron acceptors and a mixture of oxidized and reduced products are formed. Products, namely organic compounds and gasses (hydrogen and carbon dioxide), determine the type of fermentation. Main hydrogen yielding fermentations are butyric acid fermentation (saccharolytic clostridial-type fermentation) and mixed-acid fermentation (enterobacterial-type fermentation). The first step of both fermentations is the Embden-Meyerhof pathway or glycolysis in which glucose is converted into pyruvate and NADH is formed.

In the clostridial-type fermentation pyruvate is oxidized to acetyl-CoA by pyruvate:ferredoxin oxidoreductase (PFOR) in the presence of ferredoxin (Fd) (See Equation 1).

$$\left(\text{PFOR}\right)$$
$$\text{Pyruvate} + \text{CoA} + \text{Fd} \rightarrow \text{acetyl}-\text{CoA} + \text{FdH} + \text{CO}_2 \tag{1}$$

Reduced ferredoxin is also formed in the reaction with NADH catalyzed by NADH:ferredoxin oxidoreductase (NFOR) (See Equation 2).

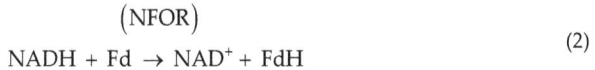

$$\left(\text{NFOR}\right)$$
$$\text{NADH} + \text{Fd} \rightarrow \text{NAD}^+ + \text{FdH} \tag{2}$$

Hydrogen is released by hydrogenases that catalyze proton reduction using electrons from ferredoxin. The activity of PFOR and NFOR enzymes is thermodynamically regulated by the hydrogen concentration. Partial hydrogen pressure >60 Pa inhibits the NFOR activity and favors formation of non-gaseous end-products from acetyl-CoA including acetate, butyrate, ethanol, butanol and lactate. PFOR is active at hydrogen concentrations up to 3×10^4 Pa (Angenent et al., 2004; Girbal et al., 1995; Hallenbeck, 2005; Kraemer & Bagley, 2007; Lee et al., 2011).

The theoretical maximum hydrogen yield during clostridial-type fermentation is 4 moles of hydrogen per mole of glucose, when all of the substrate is converted to acetic acid (See Equation 3).

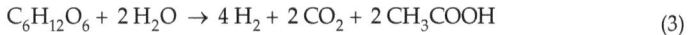

$$C_6H_{12}O_6 + 2\,H_2O \rightarrow 4\,H_2 + 2\,CO_2 + 2\,CH_3COOH \tag{3}$$

This gives the maximal possible level of hydrogen yield during dark fermentation. When the glucose is converted to butyrate the hydrogen yield drops to 2 moles (See Equation 4).

$$C_6H_{12}O_6 + 2\,H_2O \rightarrow 2\,H_2 + 2\,CO_2 + CH_3CH_2CH_2COOH \tag{4}$$

Formation of other non-gaseous end products of fermentation causes further decrease in hydrogen yields. The scheme of the clostridial-type fermentation is presented in Figure 1 (Papoutsakis, 1984; Saint-Amans et al., 2001).

The described type of fermentation is the most characteristic for spore-forming representatives of the *Clostridium* as well as *Bacillus* genera and others, such as the rumen bacteria e.g. *Ruminococcus albus*. Among the fermentative anaerobes, clostridia have been well known and extensively studied for their capability to produce hydrogen from various carbohydrates (Kalia & Purohit, 2008; Lee et al., 2011). The hydrogen yields of pure *Clostridium* cultures, including *C. acetobutylicum*, *C. bifermentans*, *C. butyricum*, *C. kluyveri*, *C. lentocellum*, *C. paraputrificum*, *C. pasteurianum*, *C. saccharoperbutylacetonicum*, *C. thermosuccinogenes*, and *C. thermolacticum* were examined. The optimum hydrogen yields observed for these bacteria varied between 1.1 moles of H_2/mole of hexose and 2.6 moles of H_2/mole of -hexose, dependent on the organism per se as well as environmental conditions (for review see Lee et al., 2011).

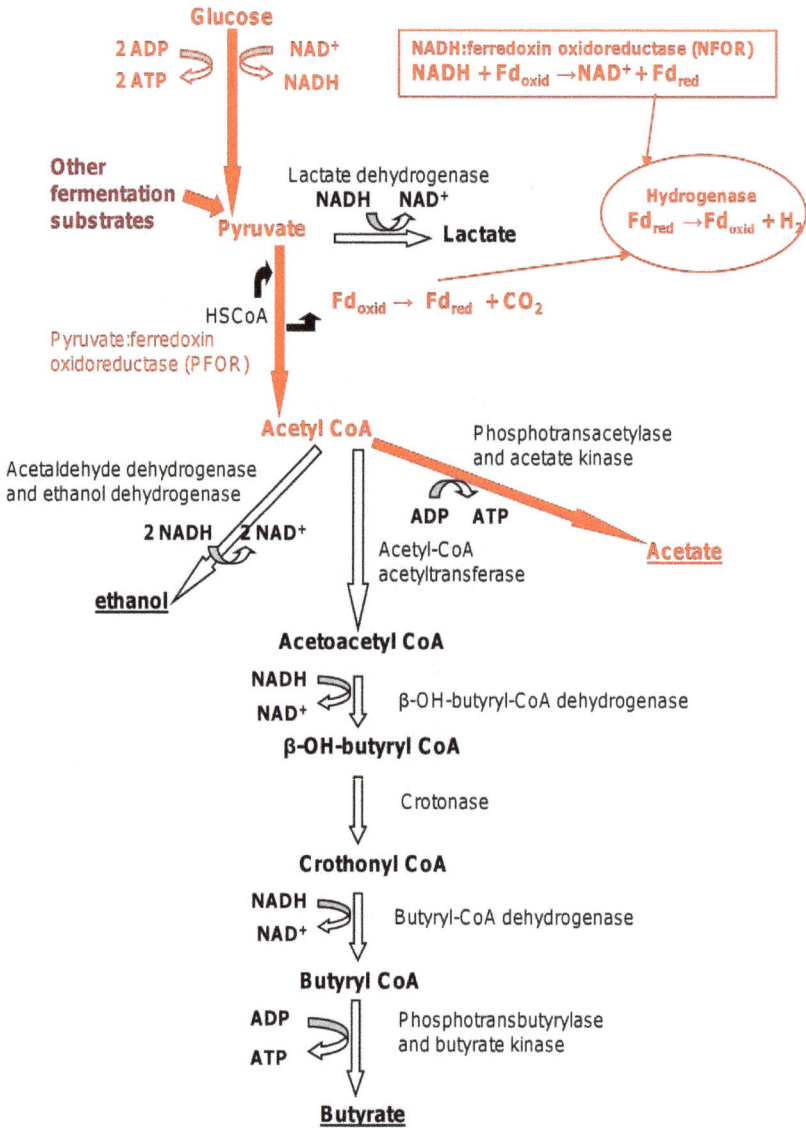

Figure 1. The scheme of clostridial-type fermentation. The pathway leading to the theoretical maximum hydrogen yield of 4 moles of hydrogen per 1 mole of glucose, when all of the substrate is converted to acetic acid is labeled in red.

In the mixed acid-fermentation (also known as formic acid fermentation) pyruvate formate-lyase (PFL) converts pyruvate to acetyl-CoA and formic acid (See Equation 5).

$$PFL$$
$$\text{Pyruvate} + \text{CoA} \rightarrow \text{acetyl} - \text{CoA} + \text{formic acid} \tag{5}$$

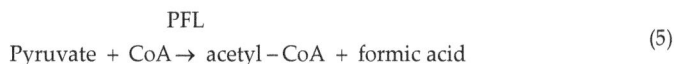

The formic acid can be degraded into hydrogen and carbon dioxide by formate hydrogen-lyase (FHL) (See Equation 6).

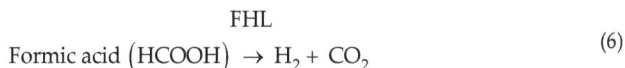

$$FHL$$
$$\text{Formic acid} \left(HCOOH\right) \rightarrow H_2 + CO_2 \tag{6}$$

There are two types of mixed-acid fermentations. In the first type ethanol and a complex mixture of acids, particularly acetic, lactic, succinic and formic acids are produced. This pattern is seen in *Escherichia*, *Salmonella*, *Proteus* and other genera. The second type is characteristic for *Enterobacter*, *Serratia*, *Erwinia* and some species of *Bacillus*. In this type of fermentation, acetoin, 2,3-butanediol, ethanol and lower amount of acids are formed.

The theoretical hydrogen yields during mixed acid fermentation are lower than those described for the clostridial-type fermentation. Hydrogen yields of *Escherichia* spp., as obtained for the pure culture of *E. coli* NCIMB 11943, are in a range of 0.2–1.8 moles of H₂/mole of hexose, when glucose or starch hydrolysate are substrates, whereas hydrogen yields determined for pure *Enterobacter* spp. cultures are much higher, ranging from 1.1 moles of H₂/mole of hexose to ca. 3.0 moles of H₂/mole of hexose (Lee et al., 2011). It is known that in the *Enterobacter*-type fermentation hydrogen is also generated through oxidation of NADH by NFOR in reactions similar to those described for the clostridial-type fermentation (Nakashimada et al., 2002; Sawers, 2005; Maeda et al., 2007).

The pathway of the mixed-acid fermentation is presented in Figure 2.

3. Lactic acid bacteria in hydrogen-producing consortia

3.1. Lactic acid bacteria – General information

Lactic acid bacteria are Gram-positive bacteria, producing lactic acid as the main product of carbohydrate fermentation. Two types of lactic acid fermentation are distinguished: homolactic and heterolactic fermentation. In homolactic acid fermentation, two molecules of pyruvate that are formed during glycolysis are converted to lactate. In heterolactic acid fermentation, one molecule of pyruvate is converted to lactate; the other is converted to ethanol and carbon dioxide.

At present, nearly 400 LAB species have been recognized. They include bacteria belonging to the order *Lactobacillales* classified into seven families: *Lactobacillaceae* (genera: *Lactobacillus* and *Pediococcus*); *Aerococcaceae* (genus *Aerococcus*); *Carnobacteriaceae* (genera: *Alloiococcus*, *Carnobacterium*, *Dolosigranulum*, *Granulicatella* and *Lactosphaera*); *Enterococcaceae* (genera: *Enterococcus*, *Tetragenococcus* and *Vagococcus*); *Leuconostocaceae* (genera: *Leuconostoc*, *Oenococcus* and *Weisella*); *Streptococcaceae* (genera: *Streptococcus*, *Lactococcus* and *Melissococcus*); *Microbacteriaceae* (genus *Microbacterium*). Extremely varied among lactic acid bacteria is genus *Lactobacillus* which comprises over 145 species. Genera *Bifidobacterium* and *Propionibacterium* (class: *Actinobacteria*) as well as spore forming rods belonging to the order *Bacillales*, family *Sporolactobacillaceae*, genus *Sporolactobacillus* constitute further groups of LAB. With the exception of bacteria belonging to the genera *Lactobacillus, Carnabacterium, Weissella* and *Sporolactobacillus* which are rods, other species of lactic acid bacteria are cocci (de Vos et al., 2009).

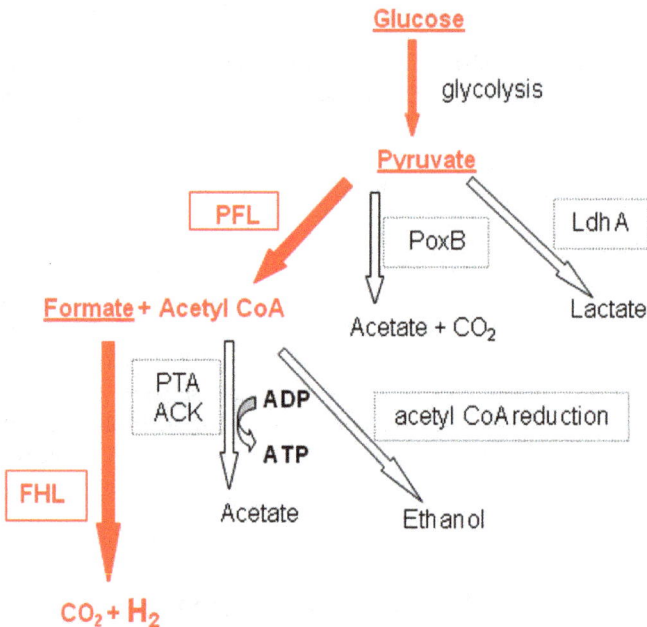

LdhA – lactate dehydrogenase, PoxB – pyruvate oxydase, PTA – phosphotransacetylase, ACK – acetate kinase, PFL – pyruvate formate lyase, FHL – formate hydrogen lyase.

Figure 2. The scheme of mixed-acid fermentation (*Escherichia coli*-type). The pathway leading to hydrogen production is shown in red.

LAB are microorganisms ubiquitous in the environment. Due to their high nutritional requirements, they are usually found in environments rich in carbohydrates, amino acids and nucleotides. On the other hand, they show considerable adaptation to the harsh conditions, which allows them to inhabit a range of various niches (Korhonen, 2010).

The digestive tracts of man and animals are among the environments where LAB occur. They have been reported in saliva, the small intestine and colon (Korhonen, 2010). The development of the gastrointestinal microflora in infants is influenced by contact with diverse microflora of the mother and of the closest surrounding. The main species found in both infants and adults are *Lactobacillus ruminis, L. salivarius, L. gasseri, L. reuteri* as well as *Bifidobacterium longum* and *B. breve* (Salminen et al., 2005; Ishibashi et al., 1997). The diversity of lactic acid bacteria colonizing the human digestive system is high; however, the species composition is constantly changing as most of the species colonize the gastrointestinal tract for only a short period (Korhonen, 2010). Microorganisms in the adult intestine outnumber by 10-fold cells constituting the human body. The microbial composition for each individual is unique, depending on age, diet, diseases and environmental factors (Qin J. et al., 2010). LAB have been widely used as probiotic bacteria in the human gastrointestinal tract, contributing to pathogen inhibition and immunomodulation (Zhang et al., 2011).

The natural occurrence of lactic acid bacteria on plants (fruits, vegetables and grains) as well as in milk permitted their use in biotechnology (Makarova et al., 2006). *Lactobacillus, Pediococcus, Leuconostoc* and *Oenococcus* which reside on grapes, enable fruit fermentation and wine production (de Nadra, 2007). Also, LAB can occur naturally or be intentionally added as starter cultures during plant, meat and dairy fermentation (Korhonen, 2010).

In marine environments LAB play a role in the breakdown of organic matter. In the last decade LAB belonging to the following genera: *Amphibacillus, Alkalibacterium, Marinilactibacillus, Paraliobacillus, Halolactibacillus* were isolated from the samples taken from the sea and oceanic as well as from animals that inhabit these ecosystems. These bacteria were named "marine LAB" (Ishikawa et al., 2005).

3.2. Lactic acid bacteria – Influence on hydrogen producers

Interestingly, lactic acid bacteria are often detected in mesophilic hydrogen-producing consortia as bacteria that accompany hydrogen producers. The technique most commonly used for analyzing the diversity of hydrogen-producing microbial communities is polymerase chain reaction-denaturing gradient gel electrophoresis (PCR-DGGE), followed by either direct sequencing or cloning and sequencing of DGGE bands. One of the disadvantages of this method is underestimation of the true bacterial diversity due to the fact that only the most prominent DGGE bands are analyzed. Various studies have shown that DGGE bands representing LAB are one of the most dominant bands (Fang et al., 2002; Kim et al., 2006; Li et al., 2006; Wu et al., 2006; Hung et al., 2007; Ren et al., 2007; Jo et al., 2007; Lo et al., 2008; Sreela-or et al., 2011). Another method of analyzing the biodiversity of hydrogen-producing consortia is cloning and sequencing of the 16S rDNA gene amplified on the total DNA isolated from the culture probes. Also with this method, sequences related to lactic acid bacteria have been detected (Yang et al., 2007). An alternative method used by our group for the first time to perform metagenomic analysis of hydrogen-producing microbial communities is 454-pyrosequencing. Our results showed that *Clostridiaceae, Enterobacteriaceae* and heterolactic fermentation bacteria, mainly *Leuconostocaeae,* were the most dominant bacteria in hydrogen-producing consortia under optimal condition for gas production (Chojnacka et al., 2011).

The aim of the chapter is a provocative discussion on the true role of LAB in hydrogen-producing bioreactors and their influence on hydrogen producers. Table 1 presents a set of selected studies which examine the possible influence of lactic acid bacteria on hydrogen production during dark fermentation.

Subject of examination	Results and suggested influence of LAB on hydrogen producers	References
A.	Negative role of LAB	
Investigation of the effects of LAB on hydrogen fermentation of bean curd manufacturing waste in a series of co-cultures of *Clostridium butyricum* and two strains of *C. acetobutylicum* with *Lactobacillus paracasei* and *Enterococcus durans*.	Inhibition of hydrogen producers by LAB due to (i) substrate competition (replacement of hydrogen fermentation by lactic acid fermentation); (ii) excretion of bacteriocins.	Noike et al., 2002
Fermentative hydrogen production from molasses in continuous stirred-tank reactors and DG-DGGE (double gradient denaturating gradient gel electrophoresis) analysis of bacterial community structure.	*C. pasteurianum, Lactococcus* sp., *Desulfovibrio ferrireducens, Actinomyces* sp., *Klebsiella oxytoca, Acidovorax* sp., uncultured *Actinobacterium* and *Bacteroidetes* were detected in the bioreactor where the main non-gaseous end products were ethanol, butyric acid and acetic acid. Negative role of *Lactococcus* species: inhibition of hydrogen production by substrate competition (competitive ethanol production).	Ren et al., 2007
DGGE examination of microbial community during unstable hydrogen production from food waste of kimchi in a continuous culture.	Conversion of hydrogen fermentation to lactic acid fermentation due to shifts in the microbial community structure from *Clostridium* spp. to *Lactobacillus* spp. Negative role of LAB: substrate competition.	Jo et al., 2007
Investigation of hydrogen production from food waste in batch fermentation by anaerobic mixed cultures and DGGE analysis of microbial community.	*Clostridium* species (*C. butyricum, C. acetobutylicum, C. beijerinckii, Clostridium* sp.) were the dominant hydrogen producers. Negative role of LAB representatives (*Lactobacillus* sp., *Enterococcus* sp.): inhibition of hydrogen production by substrate competition (competitive ethanol and lactic acid production).	Sreela-or et al., 2011

Subject of examination	Results and suggested influence of LAB on hydrogen producers	References
B	Role of LAB in hydrogen-producing consortia not discussed	
Fermentative hydrogen production from sucrose-containing wastewater in a well-mixed reactor and DGGE analysis of bacterial community structure of the granular sludge.	*Clostridium* species (*C. pasteurianum, C. tyrobutyricum, C. acidisoli*) and *Sporolactobacillus racemicus* were detected in the bioreactor. A high-rate fermentative hydrogen production was observed. The role of LAB (*Sporolactobacillus racemicus*) in the microbial community is not discussed.	Fang et al., 2002
Fermentative hydrogen production from sucrose in a continuously stirred anaerobic bioreactor seeded with silicone-immobilized sludge and DGGE analysis of bacterial community structure of the granular sludge.	DGGE analysis revealed the presence of representatives of the following genera and species: *Clostridium* (*C. intestinale* and *C. pasteurianum*), *Escherichia coli*, *Streptococcus* sp., *Klebsiella pneumoniae*. A high-rate fermentative hydrogen production was observed. The role of LAB (*Streptococcus* sp.) in the microbial community is not discussed.	Wu et al., 2006
Fermentative hydrogen production from sucrose in a continuous stirred tank reactor and DGGE analysis of bacterial community structure of the granular sludge.	*Clostridium cellulosi, Clostridium* sp., *Klebsiella ornithinolytica, Prevotella* sp. and *Leuconostoc pseudomesenteroides* were detected in the bioreactor. A high-rate fermentative hydrogen production was observed. The role of LAB (*L. pseudomesenteroides*) in the microbial hydrogen-producing community is not discussed.	Li et al., 2006
Fermentative hydrogen production from sucrose or xylose in a continuous dark fermentation bioreactor and DGGE analysis of the bacterial community structure.	*Clostridium* species (*C. butyricum, C. pasteurianum* on sucrose and *C. celerecrescens* on xylose), *Klebsiella pneumoniae, K. oxytoca, Streptococcus* sp., *Escherichia* sp., *Pseudomonas* sp. *Dialister* sp., *Bacillus* sp., *Bifidobacterium* sp. were detected in the bioreactor. The role of LAB (*Streptococcus* sp. and *Bifidobacterium* sp.) in the microbial community is not discussed.	Lo et al., 2008

Subject of examination	Results and suggested influence of LAB on hydrogen producers	References
C.	Positive role of LAB	
Fermentative hydrogen production from glucose in anaerobic agitated granular sludge bed bioreactors and DGGE and FISH analyses of the granular sludge.	The DGGE analysis showed that the bacterial community was mainly composed of *Clostridium* sp., *Klebsiella oxytoca* and *Streptococcus* sp. A high-rate fermentative hydrogen production was observed. The FISH images suggested that *Streptococcus* cells acted as seeds for granule formation.	Hung et al., 2007
Fermentative hydrogen production from cheese whey wastewater by mixed continuous cultures and molecular analysis of the consortium by cloning and sequencing of the 16S rDNA gene amplified on the total DNA isolated from the culture probe.	The most prevalent bacteria, representing approximately 50% of the total sequences analyzed, were representatives of the genus *Lactobacillus*. Remaining sequences belonged to the genera *Olsenella*, *Clostridium* and *Prevotella*. Decrease in hydrogen production was accompanied by the reductions in the number of detected bacteria from the genus *Lactobacillus*. Authors declare isolation of *Lactobacillus* bacteria capable of hydrogen production in the process of lactose fermentation.	Yang et al., 2007
Fermentative hydrogen production from molasses in packed bed bioreactors and metagenomic analysis of bacterial biofilms and granules by 454-pyrosequencing.	Metagenomic analysis of microbial consortia by 454-pyrosequencing of amplified 16S rDNA fragments revealed that the most dominant bacteria were the representatives of the *Firmicutes* (*Clostridiaceae* and *Leuconostocaeae*) and *Gammaproteobacteria* (*Enterobacteriaceae*). Bacteria of heterolactic fermentation were one of the predominant microbes in hydrogen-producing consortia. The speculation that LAB may favor hydrogen production is discussed. For details see Tables 2-4, Figures 3-5 and description in the text.	Chojnacka et al., 2011

Table 1. A set of selected studies demonstrating the contribution of LAB in hydrogen-producing cultures and presenting their possible influence on hydrogen production.

Some studies argue that development of LAB in bioreactors may inhibit hydrogen production (Table 1, part A). Cessation of hydrogen generation by LAB was suggested to be due to (i) substrate competition and/or (ii) excretion of bacteriocins inhibiting growth of other bacteria. These observations derive from examinations of both batch (Sreela-or et al., 2011) and continuous (Ren et al., 2007; Jo et al., 2007) mixed cultures as well as co-cultures where one component was a representative of clostridia and the second one of lactic acid bacteria (Noike et al., 2002). Heat treatment was proposed as a method of eliminating lactic acid bacteria (Noike et al., 2002; Baghchehsaraee et al., 2008).

Substrate competition includes changes in the type of fermentation occurring in the bioreactors during long-term continuous processes and replacement of hydrogen fermentation by lactic acid or ethanol fermentation (Noike et al., 2002; Jo et al., 2007; Ren et al., 2007; Sreela-or et al., 2011). In all of the studies decrease in hydrogen production was observed with simultaneous increase of lactic acid and ethanol concentrations in the effluents or fluid phase of the culture.

The hypothesis that bacteriocins may act as inhibitors of hydrogen production was postulated by Noike and co-workers (2002), who showed in a series of co-cultures experiments that cessation of hydrogen production by *C. acetobutylicum* and *C. butyricum* was caused by both the presence of *Enterococcus durans* and *Lactobacillus paracasei* as well as supernatants from their culture media. Moreover, treatment of the supernatants with trypsin recovered normal hydrogen production by selected clostridial strains.

Studies listed in part B of Table 1 determined the presence of lactic acid bacteria in hydrogen-producing consortia; yet, their role in these microbial communities is not discussed. It is noteworthy that (i) those papers discuss efficient systems of biohydrogen production and (ii) studies were performed under optimal conditions for hydrogen production (Fang et al., 2002; Kim et al., 2006; Li et al., 2006; Wu et al., 2006; Hung et al., 2007).

Part C of Table 1 presents the only so far available studies arguing that LAB could play a positive role in hydrogen-producing microbial communities and stimulate hydrogen production.

Hung and colleagues (2007) studied the efficiency of fermentative hydrogen production from glucose in anaerobic agitated granular sludge bed bioreactors under different substrate concentration and hydraulic retention times (HRT). PCR-DGGE and FISH methods were used to analyze the biohydrogen-producing microbial community of the granular sludge. The bacterial community was composed of *Clostridium* sp. (possibly *C. pasteurianum*), *Klebsiella oxytoca* and *Streptococcus* sp. The percentage of *Streptococcus* sp. contributing to the microbial community was dependent on the HRT. The shorter HRT, meaning the faster the flow of the medium and increased dilution rate, the higher the contribution of *Streptococcus* sp. in the bacterial consortium was observed. Formation of granular sludge enables biomass retention. FISH analysis revealed that *Streptococcus* cells are located inside granules surrounded by *Clostridium* cells. Authors postulate that *Streptococcus* cells may act as the seed for sludge granule formation.

According to Yang et al. (2007) some LAB are able to produce hydrogen. They declare isolation of strains from the genus *Lactobacillus* capable of hydrogen production during lactose fermentation.

3.3. Fermentative hydrogen production and microbial analysis of bacterial biofilms and granular sludge formed in packed bed bioreactors

We developed an effective system of bacterial hydrogen production based on long-term continuous cultures (from an inoculum of a lake bottom sediment) grown on sugar beet molasses in packed bed reactors filled with granitic stones (Chojnacka et al., 2011). In separate cultures, two consortia of anaerobic fermentative bacteria producing hydrogen-rich gas developed on the stones as biofilms. Furthermore, in one of the cultures a granular sludge was also observed (Figures 3 and 4). Cultures were named, respectively, (i) the culture with stone biofilm only and (ii) granular sludge culture. Both cultures were regularly renewed by removal of an excess of biomass.

Analysis of the surface topography of biofilms from both cultures revealed their porous, irregular structure with many cavities and channels. Bacteria appeared to be suspended in and surrounded by a matrix substance. The granules were white and light cream in color, with a diameter between 0.2 – 2 mm, and of hard structure, resistant to squashing or crumbling. Moreover, the granules were clustered in structures resembling bunches of grapes with a noticeable net of channels. Similar to the bacterial biofilm, the granules consisted of bacterial cells surrounded by a matrix.

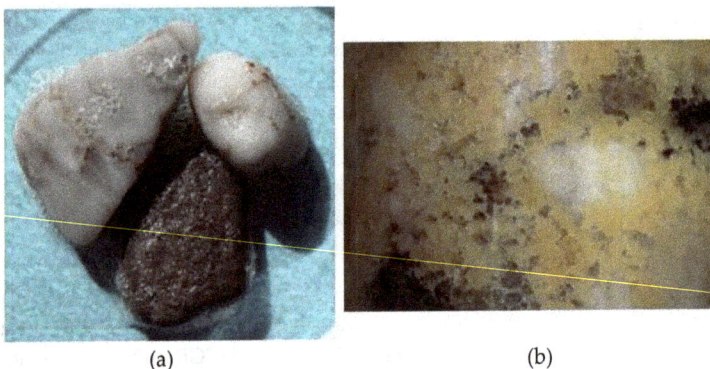

(a)   (b)

Figure 3. Images of the two structures formed by selected consortia of fermentative bacteria grown in a bioreactor on M9 medium containing molasses: (a) stones covered with bacterial biofilm (b) the granular sludge.

Metagenomic analysis of microbial communities by 454-pyrosequencing of amplified 16S rDNA fragments revealed that the overall biodiversity of hydrogen-producing cultures was quite small. Stone biofilm from the culture without the granular sludge was dominated by *Clostridiaceae* and heterolactic fermentation bacteria, mainly *Leuconostocaeae*. Representatives of *Leuconostocaeae* and *Enterobacteriaceae* were dominant in both the granules and the stone biofilm formed in the granular sludge culture. The granular sludge contained bacteria of heterolactic fermentation, dominated by *Leuconostoc* species as well as unclassified *Streptococcaceae* and unclassified *Enterobacteriaceae*. Surprisingly, sequences representing the *Clostridiaceae* were in a relative minority (Table 2).

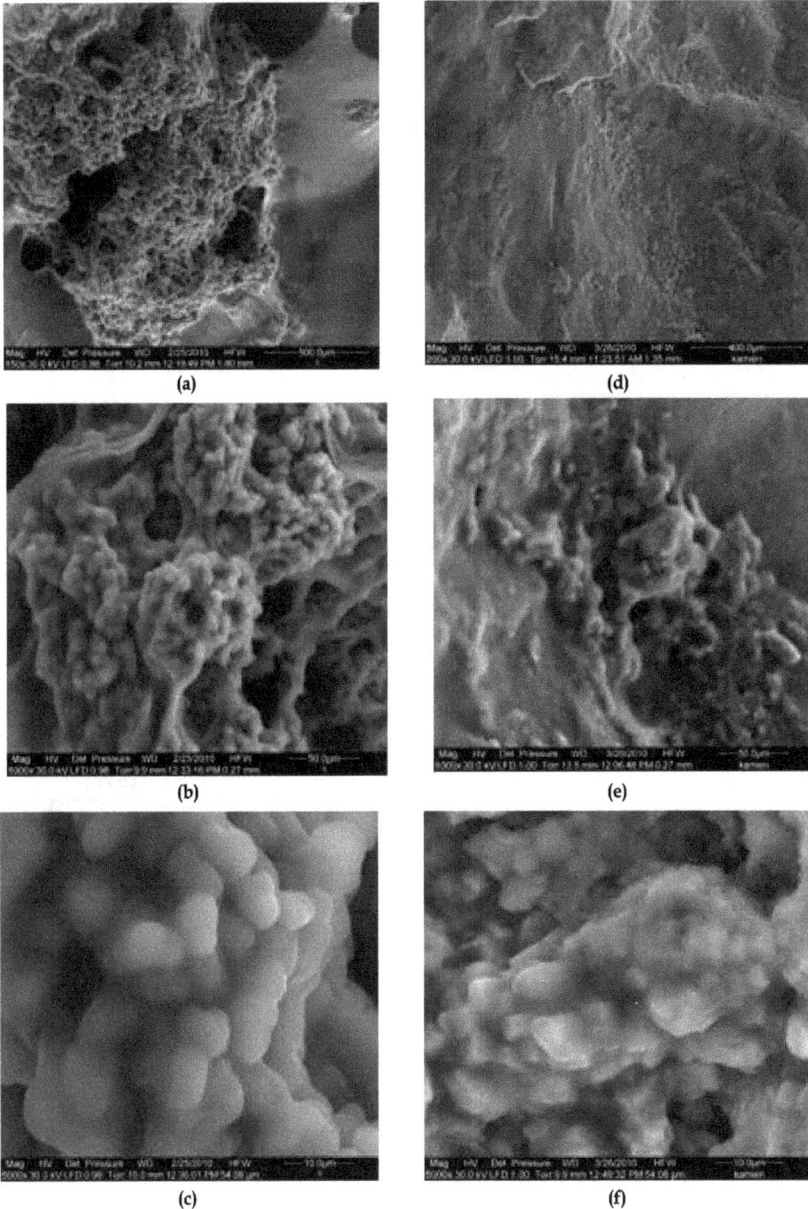

Figure 4. Scanning electron micrographs of structures formed by selected consortia of fermentative bacteria grown on M9 medium containing molasses: (a – c) granules; (d – f) bacterial biofilm formed on the granitic stones filling the bioreactor in the granular sludge culture.

taxon	B	G	Bg
Bacteria	4596	8578	26066
Firmicutes	4410	6473	10216
Bacilli	1564	5425	7586
Bacillales	119	17	96
Bacillaceae	1	0	11
Sporolactobacillaceae	94	6	44
Sporolactobacillus	94	6	44
Lactobacillales	1343	5302	7272
Enterococcaceae	5	16	417
Enterococcus	2	2	123
Lactobacillaceae	96	3	790
Lactobacillus	92	3	725
Leuconostocaceae	826	4661	4589
Leuconostoc	826	4634	4586
Streptococcaceae	4	37	95
Lactococcus	4	17	74
Clostridia	2627	579	2240
Clostridiales	2572	443	2023
Clostridiaceae	2131	134	1100
Clostridium	1182	66	593
Proteobacteria	168	1974	15755
Gammaproteobacteria	168	1970	15744
Enterobacteriales	156	1769	15528
Enterobacteriaceae	156	1769	15528
Enterobacter	61	319	2130
Raoultella	6	50	23
Pseudomonadales	6	8	137
Moraxellaceae	5	0	128
Acinetobacter	2	0	68
Pseudomonadaceae	1	8	9
Pseudomonas	1	7	9

B – stone biofilm from the culture without granular sludge; Bg – stone biofilm from the granular sludge culture; G – granules from the granular sludge culture.

Table 2. Number of reads assigned to respective taxonomic branches of 16S rRNA gene fragments amplified from the total DNA pool from bacterial communities formed in bioreactors.

Results of the metagenomic analysis by 454-pyrosequencing were confirmed by FISH (Fluorescence-In-Situ-Hybridization) analysis (Fig. 5) as well as by isolatating of lactic acid bacteria from the culture (Table 3).

Both, the stone biofilm and granules are composed of bacteria of many different shapes. As judged from fluorescence *in situ* hybridization, the relative abundance of selected bacterial groups varied during the rounds of bioreactor cycles. At the very beginning of biofilm development clostridial and lactobacilli cells were detected only sporadically among gammaproteobacteria (Fig. 5A a-c). In the growing biofilm systematic increase of Firmicutes (especially lactobacilli) cells was observed (Fig. 5Bd-e).

A

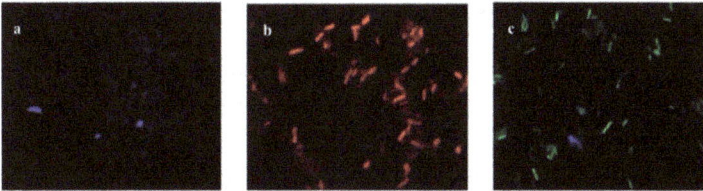

B

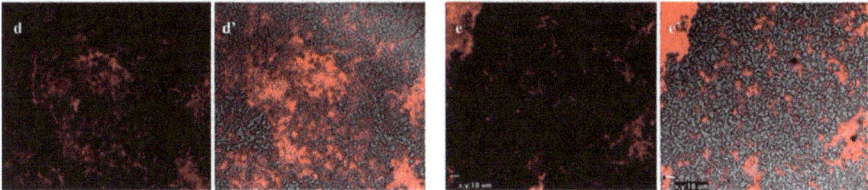

A: young biofilm; a-clostridia/dyLight405, b-lactobacilli/TAMRA; c-gammaproteobacteria/CY3,
B: mature biofilm; d-firmicutes/CY5; e-lactobacilli/TAMRA; d', e'-in combination with phase contrast.

Figure 5. FISH image of the hydrogen-producing biofilm from the granular sludge culture described previously (Chojnacka et al., 2011) analyzed by confocal laser fluorescence microscopy. The sample was stained with fluorescently labeled specific probes.

A cultivable approach with the use of media promoting the growth of lactic acid bacteria (MRS, M17) revealed that the bioreactor was inhabited by a vast number of these bacteria. Similarly to the metagenomic data, the majority of growing colonies represented *Leuconostoc* or *Lactobacillus* genera. All in all, six different species listed in Tab. 3 were isolated. It was determined that heterofermentative species (*Leuconostoc*, *L. brevis*, *L. rhamnosus*) slightly outnumbered homofermentatives.

Isolate	Isolation ratio	Homofermenters:heterofermenters ratio
Lactobacillus plantarum	46,7%	
Enterococcus casseliflavus	0,5%	
Leuconostoc mesenteroides	46,7%	
Leuconostoc mesenteroides ssp. mesenteroides	0,5%	0,89
Lactobacillus brevis	5,1%	
Lactobacillus rhamnosus	0,5%	

Table 3. The species of LAB isolated from the hydrogen-producing bioreactor.

Samples were collected from both, stone (biofilm) and liquid phase of hydrogen-producing culture, and plated on selective media for lactic acid bacteria. Plates were incubated under anaerobic conditions. Obtained colonies were tested for Gram positivity and lack of catalase enzyme. For strains which gave positive results, the V3 fragment of the 16S rRNA gene was amplified. Subsequently, fragments were analyzed using MSSCP technique. Strains with unique or representative gel patterns were chosen for further studies based on amplification and sequencing of 16S rRNA gene. Resulting sequences were identified by comparison to known sequences using the NCBI database. Names of homofermentative species are written in bold.

Formation of granular sludge rich in heterolactic bacteria significantly enhanced hydrogen production. Table 4 presents a list of parameters describing and comparing the two bacterial cultures that were the subject of the study of Chojnacka et al. (2011), under optimal conditions for hydrogen production. Significantly higher total gas production was observed for the culture containing granular sludge than for the biofilm-only culture (9.5 vs. 6.6 cm^3/min/working volume of the bioreactor). Furthermore, the percentage contribution of hydrogen was almost 49 and 36 %, whereas of carbon dioxide 47 and 60%, in the former and latter cultures, respectively. The granular sludge culture produced hydrogen at the rate of 6649 cm^3/day/working volume of the bioreactor, whereas the biofilm-only culture at the rate of 3393 cm^3/day/working volume of the bioreactor. Fermentation gas produced by both cultures contained 0.0004% methane, meaning that it was practically methane-free. Consequently, under optimal conditions, the culture containing granular sludge rich in heterolactic bacteria was two-fold more effective in producing hydrogen than that containing biofilm only: 5.43 moles of H_2 vs. 2.8 moles of H_2/mole of sucrose from molasses, respectively.

It is known that butyrate is the predominant metabolite during butyric acid fermentation at pH 5.0 – 5.5 (Li and Fang, 2007). The analysis of the non-gaseous fermentation products in both cultures in the study of Chojnacka et al. (2011) revealed that butyric acid was the main metabolite with partial contribution of ethanol. Concentration of butyric acid was almost 1.8-fold higher in the culture containing granular sludge than in the biofilm-only culture. No net production of lactic and propionic acids was observed in the granular sludge culture, whereas these were the second and third most abundant fermentation products in the cultures containing only biofilm. The formic and acetic acids present in the medium were utilized by both cultures. It is noteworthy that in the granular sludge culture rich in heterolactic bacteria showing very good performance in hydrogen production and a high content of butyric acid, the number of *Clostridiales* sequences was significantly lower than in the biofilm-only culture.

Based on our results presented in the study of Chojnacka et al. (2011) we speculate that LAB may possibly play a significant but not fully understood and perhaps underestimated role in the hydrogen producing communities. This hypothesis is based on two observations: (i) the higher the number of LAB in the hydrogen-producing community, the more efficiently hydrogen is produced; (ii) complete consumption of lactic acid, significantly increased concentration of butyric acid as well as larger hydrogen yield in the culture containing granular sludge than in that with just the biofilm.

Parameter	Culture without granular sludge	Culture containing granular sludge
Total gas production (cm³/min/working volume of the bioreactor)	6.6	9.5
Composition of fermentation gas (%):		
Hydrogen	35.7 %	48.6 %
Carbon dioxide	60%	47.1 %
Water vapor	~4.3%	~4.3 %
Methane	0.0004 %	0.0004%
Others (NH3, H2S, formic, acetic, propionic and butyric acids)	~1%	~1%
Hydrogen production (cm³/day/working volume of the bioreactor)	3393	6649
Yield of hydrogen (moles H₂/mole of sucrose)	2.8	5.43
Net production of the non-gaseous end products (mg/L):		
Lactic acid	2419 ± 42.6	0
Formic acid	0	0
Acetic acid	0	0
Propionic acid	248 ± 0.7	0
Butyric acid	4331 ± 60.0	7641 ± 33.1
Ethanol (%)	0.06 ± 0.002	0.1 ± 0.004

Table 4. Parameters describing two cultures of hydrogen-producing bacteria under optimal conditions for hydrogen production based on the study of Chojnacka et al. (2011).

3.4. Enhancement of hydrogen production by lactic acid

Based on the study of Chojnacka et al. (2011), for the culture containing granular sludge rich in heterolactic bacteria no net production of lactic acid was observed, indicating complete consumption of this metabolite, whereas its concentration in the biofilm-only culture was quite high (Table 4). Noticeable is the fact that molasses - a fermentative substrate in this study, also contains acetic and lactic acids at concentrations of about 800 mg/L each. Furthermore, a significantly higher concentration of butyric acid was detected in the culture containing granular sludge than in biofilm-only culture.

There are studies arguing that lactic acid and acetic acid mixed with the substrate stimulate biohydrogen production. Baghchehsaraee et al. (2009) showed that the addition of lactic acid to a mixed culture grown on starch-containing medium increased both hydrogen production and butyric acid formation. Furthermore, complete consumption of lactic acid produced by the culture was observed. When lactic acid was the only carbon source, the level of hydrogen production was very low (0.5% substrate conversion efficiency). Therefore, authors claimed that the addition of lactic acid to the medium probably alters the metabolic pathways in bacterial cells.

In the study of Kim et al. (2012), the effects of different lactate concentrations on hydrogen production from glucose in batch and continuous cultures were examined. Lactic acid was determined to be a factor increasing the efficiency of hydrogen production in a proper range of concentrations. The key issue was to establish the optimal lactic acid concentration. FISH analyses revealed that *Clostridium* sp. was the dominant hydrogen producer in the examined system.

Matsumoto and Nishimura (2007) examined fermentative hydrogen production from sweet potato sho-chu post-distillation slurry that contained large amounts of organic acids. Hydrogen production was accompanied by a decrease in the concentrations of acetic ad lactic acids and co-production of butyric acids. The authors isolated a clostridial strain, *Clostridium diolis* JPCC H-3, capable of effective hydrogen production from the slurry solution and a mixture of acetic and lactic acids in an artificial medium.

The ability to produce hydrogen from lactic and acetic acids seems to be widely conserved in the genus *Clostridium* and other hydrogen-producing bacteria capable of butyric acid fermentation of carbohydrates. It was shown that the *Clostridium acetobutylicum* strain P262 and *Butyribacterium methylotrophicum* utilized lactate and acetate and converted them to butyrate, carbon dioxide and hydrogen in the absence of carbohydrates in the medium. Cell extracts from bacteria grown on acetate and lactate showed a higher activity of NAD-independent lactate dehydrogenase than these from bacteria grown on carbohydrate-rich medium (Diez-Gonzales et al., 1995; Shen et al., 1996). The authors presented potential biochemical pathways leading to butyrate and hydrogen production from lactate and acetate. Conversion of lactate and acetate to butyrate and symbiotic interactions between LAB and clostridial species in animal intestinal tracts are intensively studied and discussed in section 4. Therefore, also the biochemical routes leading to butyrate and hydrogen production from lactate and acetate are presented in the same section.

In the study of Matsumoto and Nishimura (2007) the process of hydrogen production by *C. diolis* from both the slurry solution and a mixture of acetic and lactic acids in an artificial medium occurred to be pH-dependent and was observed in a range of pH (~5.8 – 7.4). Juang et al. (2011) also observed utilization of lactate and acetate for biohydrogen and butyrate production during their studies on hydrogen and methane production from organic residues of ethanol fermentation from tapioca starch by mixed bacteria culture. Lactate and acetate came from maltose fermentation, the main carbohydrate of ethanol fermentation residues. The optimal hydrogen production was observed at pH 5.5 – 6.0. Jo et al. (2008) showed that conversion of lactate and acetate to butyrate and hydrogen by *Clostridium tyrobutyricum* was inhibited due to pH decrease from 5.5 to 4.6. The pH values were dependent on HRT and organic loading rate. At high organic loading rate accumulation of lactate, pH decrease and a lower efficiency of hydrogen production were observed.

Matsumoto and Nishimura (2007), Jo et al. (2008) and Juang et al. (2011) point to pH values as a critical factor for hydrogen production from lactate and acetate. Various optimal pH for hydrogen production are observed. The differences may depend on the microbial system applied for hydrogen production and the initial substrate. It is speculated that unfavorable changes in pH could be the main reason of inhibiting hydrogen production that could be incorrectly attributed to the presence of lactic acid bacteria in hydrogen-producing consortia. In the study of Chojnacka et al. (2011), the optimal pH was around 5.0. Any change in pH, a decrease below 4.5 or increase above 5.5, caused a significant decline in fermentative gas production. Changes in pH may either be the reason or the results of disturbing the "homeostasis" of hydrogen-producing microbial communities in bioreactors.

4. Interactions between LAB and clostridial species in the animal intestinal tract

Microflora of the mammalian intestine is composed of a diverse population of both aerobic and anaerobic bacteria. Symbiotic relationships occur between different intestinal species or groups of species, among which are interactions between LAB and clostridial species. Numerous observations arising from different models describe lactate conversion to butyrate by intestinal bacteria and enhancement of butyrate production by LAB (Hashizume et al., 2003; Duncan et al. 2004; Bourriaud, et al., 2005; Meimandipour et al., 2009; Abbas, 2010; Munoz-Tamayo, et al., 2011).

The microbial community of the human colon contains many bacteria that produce lactic acid including lactobacilli, bifidobacteria, enterococci and streptococci. However, lactate is normally detected only at very low concentration (<5 mM) in feces of healthy individuals due its rapid conversion to short chain fatty acids (SCFAs; acetate, propionate and butyrate) by acid-utilizing bacteria. Therefore, lactate is thought to be a precursor of the formation of various SCFAs (Hashizume et al., 2003; Duncan et al., 2004; Bourriaud et al., 2005; Munoz-Tamayo et al., 2011). Bourriaud and colleagues (2005) performed convincing experiments exploring the lactate metabolism and short fatty acids production. They incubated three human microfloras with media containing ^{13}C-labelled lactate and detected the labeled products of fermentation by ^{13}C NMR spectrometry. Results revealed that butyrate was the major net product of lactate conversion by human fecal microflora. Other SCFAs produced were: propionate, acetate and valerate. Inter-individual differences between the three microfloras were observed. Similar studies performed using ^{2}H-labelled acetate and ^{13}C-labelled lactate and gas chromatography-mass spectrometry (GC-MS) analysis showed that acetic and lactic acids are important precursors of butyrate production in human fecal samples (Morrison et al., 2006).

The metabolic pathway of lactate and acetate utilization to produce butyrate proposed for *Eubacterium hallii* and *Anaerostipes caccae* is shown on Figure 6 (Duncan, et al., 2004; Munoz-Tamayo, et al., 2011). The butyrate produced (in moles) is approximately equal to the sum of half of the acetate and lactate coming from the medium. Lactate is converted to pyruvate by lactate dehydrogenase. The next steps are analogous to those ones presented on Figure 1. Pyruvate is oxidized to acetyl coenzyme A (acetyl-CoA), which is further routed to acetate and butyrate. Acetate is produced via acetate kinase, the pathway generating energy in the form of ATP. For butyrate formation, two molecules of acetyl-CoA are condensed to one molecule of acetoacetyl-CoA, and subsequently reduced to butyryl-CoA. Butyrate can be synthesized from two metabolic pathways: phosphotransbutyrylase and butyrate kinase as shown on Figure 1, and butyryl CoA:acetate CoA transferase as shown on Figure 6. The latter mechanism seems to be the dominant in the human colonic ecosystem. Butyryl-CoA:acetate CoA-transferase transports the CoA component to exterior of acetate releasing butyrate and acetyl-CoA (Duncan et al., 2004; Munoz-Tamayo, et al., 2011). Hydrogen can be produced by both PFOR and NFOR complexes and hydrogenases, as described in section 2. The reaction catalyzed by NFOR is assumed to be the main route for H_2 production by intestinal microflora (Bourriaud et al., 2005). Similar pathway is proposed for clostridial species (eg. *C. acetobutylicum*; Diez-Gonzales et al., 1995) and other hydrogen and butyrate producing bacteria (eg. *B. methylotrophicum*; Shen et al., 1996), as mentioned in section 3.4. Conversion of lactate and acetate to butyrate and hydrogen is an energetically favorable process (Duncan et al., 2004; Jo et al., 2008).

The known lactate-utilizing butyrate-producing bacteria belong to the *Firmicutes* phylum, which includes the following species: *Megasphaera elsdenii, Anaerostipes caccae, Anaerostipes coli, E. hallii* and species distantly related to *Clostridium indolis*. *A. coli* is a dominant member of the human colonic microbiota recognized for its importance in butyrate production. *M. elsdenii* is one of the main butyrate producers from lactate in ruminants as well as monogastric animals, such as pigs or rodents. *A. caccae, A. coli, E. hallii* and species distantly related to *Clostridium indolis* belong to the clostridial cluster XIVa (*Lachnospiraceae*), known butyrate-producing bacteria of gastrointestinal tracts in mammals. However, only a few butyrate-producing species within the clostridial cluster XIVa are capable of converting lactate to butyrate (Duncan, 2004; Hashizume et al., 2003; Munoz-Tamayo, et al., 2011).

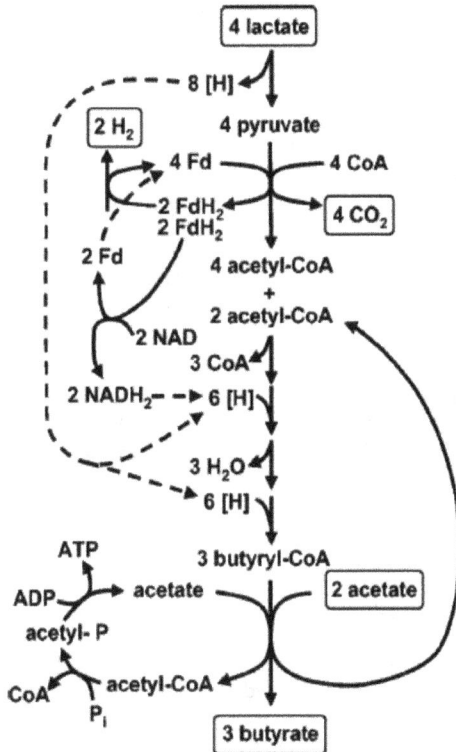

Figure 6. Scheme for butyrate production from lactate in *E. hallii* and *A. caccae*, adapted from Duncan et al., 2004.

The issue of stereospecificity of lactate utilization was addressed in the study of Duncan et al. (2004). Three *E. hallii*-related strains (SL6/1/1, SM6/1 and L2-7) and two *A. caccae* strains (L1-92 and P2) were able to use both D and L isomers of lactate during incubation on DL-lactate-containing medium. Interestingly, the addition of glucose to the medium almost completely inhibited lactate utilization by the tested strains. Additional studies showed that *E. hallii* L2-7, when grown with DL-lactate, used all of the supplied lactic acid together with some acetate, producing more than 20 mM of butyrate. Less butyrate, but a noteworthy amount of formate, was produced during growth on glucose or on glucose plus lactate. Interestingly, the highest level of hydrogen production was observed when strains were grown on lactate and the lowest for growth on glucose plus lactate. However, the *Clostridium indolis*-related strain SS2/1 was able to use D-lactate, but not L-lactate, during growth on DL-lactate containing media, which suggests that it lacks both an L-lactate dehydrogenase, capable of producing pyruvate from L-lactate, and a racemase, capable of converting L-lactate into D-lactate. According to Bourriaud and colleagues (Bourriaud et al., 2005), both lactate enantiomers are equally utilized by human intestinal microflora, treated as a whole consortium, not as pure strains.

Conversion of lactate to butyrate is one of the important factors for maintaining homeostasis in gastrointestinal tracts. Accumulation of lactate leads to different intestinal disorders (Hashizume et al., 2003). A number of studies have been performed to confirm the symbiotic interaction between lactic acid bacteria and butyric acid bacteria, mainly the *Clostridiales* representatives isolated from animal gastrointestinal tracts. Co-culture experiments that simulated the relations occurring *in vivo* were carried out. Symbiotic interactions were described to rely on the phenomenon of cross-feeding of lactate and involve conversion of lactate to butyrate by butyrate-producing bacteria stimulated by LAB.

It is noteworthy that results from studies of the gastrointestinal microflora indicate that acidity seems to be a key regulatory factor in lactate metabolism. The pH values may influence both bacterial growth and development of specific groups of bacteria as well as fermentation processes affecting the relative proportions of SCFAs (Belenguer et al., 2006; Meimandipour, et al., 2009; Belenguer et al., 2011). These observations are in agreement with our position concerning the potentially important role of pH in hydrogen-producing consortia discussed in section 3.4.

We postulate that the phenomenon analogous to cross-feeding observed in the gastrointestinal tract might take place in hydrogen-producing bioreactors. Although LAB may seem to be undesirable in such processes as they use some of H_2 to produce lactate, their stimulatory effects on hydrogen producers seem to exceed the potentially unbeneficial features. In many studies, it has been explicitly proven that the presence of LAB positively affects the production of butyrate. Most probably, hydrogen producers, mainly species belonging to the *Clostridiales* order, are capable of utilizing lactate as the main precursor of butyrate formation. Further investigations are required.

5. Conclusions

Lactic acid bacteria are detected in almost all biohydrogen-producing microbial communities of dark fermentation. Many studies indicate that LAB inhibit hydrogen production due to substrate competition and replacement of hydrogen fermentation by lactic acid and ethanol fermentations, and/or excretion of bacteriocines. On the other hand, some positive interactions between LAB and clostridial species have also been noted. They include hydrogen production from lactate by many clostridial species and symbiotic interactions, called lactate cross-feeding, occurring between LAB and clostridia.

These phenomena rely on the conversion of lactate and acetate to butyrate and hydrogen. Symbiotic interactions between LAB and butyrate-producing bacteria involving clostridia have been described in the gastrointestinal tract. We postulate that similar relations exist in biohydrogen-producing bioreactors. According to our hypothesis, pH may be a critical factor affecting bacterial growth, development of specific groups building hydrogen-producing microbial communities and fermentation processes. Acidity changes in bioreactors might be either the reason or the results of disturbances in the balance between microorganisms constituting hydrogen-producing microbial communities in bioreactors. Still, there are no data on symbiotic interactions between LAB and enterobacteria in hydrogen-producing microbial consortia. All these issues require further investigations.

Author details

Anna Sikora and Urszula Zielenkiewicz
Institute of Biochemistry and Biophysics, Polish Academy of Sciences, Poland

Mieczysław Błaszczyk and Marcin Jurkowski
Faculty of Agriculture and Biology, Warsaw University of Life Sciences, Poland

Acknowledgement

We would like to thank Dominika Brodowska for performance of FISH analyses and Dr. Agnieszka Szczepankowska for editorial assistance.

6. References

Abbas, K.A. (2010). The synergistic effects of probiotic microorganisms on the microbial production of butyrate in vitro, *McNair Scholars Research Journal* 2(1): 103-113.

Angenent, L.T., Karim, K., Al-Dahhan, M.H., Wrenn, B.A. & Domiguez-Espinoza, R. (2004). Production of bioenergy and biohemicals from industrial and agricultural wastewater. *TRENDS Biotechnol.* 22 (9): 477-485.

Antoni, D., Zverlov, V.V. & Schwarz, W.H. (2007). Biofuels from microbes. *Appl. Microbiol. Biotechnol.* 77(1): 23–35.

Baghchehsaraee, B., Nakla, G., Karamanev, D., Margaritis, A. & Reid, G. (2008). The effect of heat pretreatment temperature on fermentative hydrogen production using mixed cultures. *Int. J. Hydrogen Energy* 33(15): 4064-4073.

Baghchehsaraee, B., Nakhla, G., Karamanev, D. & Margaritis, A. (2009). Effect of extrinsic lactic acid on fermentative hydrogen production. *Int. J. Hydrogen Energy* 34 (6): 2573–2579.

Belenguer, A., Duncan, S.H., Calder, A.G., Holtrop, G., Louis, P., Lobley, G.E. & Flint H.J. (2006). Two Routes of Metabolic Cross-Feeding between *Bifidobacterium adolescentis* and Butyrate-Producing Anaerobes from the Human Gut. *Appl. Environ. Microbiol* 72(5): 3593–3599.

Belenguer, A., Holtrop, G., Duncan, S.H., Anderson, S.H., Calder, G.A, Harry, J., Flint H.J. & Lobley G.E. (2011). Rates of productionand utilization of lactate by microbial communities fromthe human colon. *FEMS Microbiol Ecol* 77(1): 107–119.

Bourriaud, C., Robins R.J., Martin, L., Kozlowski, F., Tenailleau, E., Cherbut, C. & Michel, C. (2005). Lactate is mainly fermented to butyrate by human intestinal microfloras but inter-individual variation is evident. *J. Appl. Microbiol.* 99(1): 201–212.

Campos, J.L., Arrojo, B., Franco, A., Belmonte, M., Mosquera-Corral, A., Roca, E. & Méndez, R. (2009). Effect of shear stress on wastewater treatment systems performance, *in* F.E. Dumont & J.A. Sacco (eds.), *Biochemical Engineering*, Nova Science Publishers, Inc., New York, pp. 227-244.

Chojnacka, A., Błaszczyk, M.K., Szczęsny, P., Latoszek, K., Sumińska, M., Tomczyk, K., Zielenkiewicz, U. & Sikora, A. (2011). Comparative analysis of hydrogen-producing bacterial biofilms and granular sludge formed in continuous cultures of fermentative bacteria. *Bioresour. Technol.* 102 (21): 10057-10064.

Das, D. & Veziroglu, T.N. (2008). Advances in biological hydrogen production processes. *Int. J. Hydrogen Energy* 33(21): 6046-6057.

de Nadra, M. (2007). Nitrogen metabolism in lactic acid bacteria from fruits: a review, *in* A. Méndez-Vilas (ed.), *Communicating Current Research and Educational Topics and Trends in Applied Microbiology* vol. 1, Microbiology Book Series, Communicating Current Microbiology with an Educational Vocation, pp. 500-510.

de Vos, P., Garrity, G., Jones, D., Krieg, N.R., Ludwig, W., Rainey, F.A., Schleifer, K.H. & Whitman, W.B. (2009). Bergey's Manual of Systematic Bacteriology, second edition, vol. 3 (The *Firmicutes*), Springer, Dordrecht, Heidelberg, London, New York.

Diez-Gonzalez, F., Russell, J.B. & Hunter, J.B. (1995) The role of an NAD-independent lactate dehydrogenase and acetate in the utilizationof lactate by *Clostridium acetobutylicum* P262. *Arch. Microbiol.* 164(1): 36–42.

Duncan, S.H., Petra Louis, P. & Flint, H.J. (2004). Lactate-utilizing bacteria, isolated from human feces, that produce butyrate as a major fermentation product. *Appl. Environ. Microbiol.* 70(10): 5185-5190.

Fang, H.H.P., Liu, H. & Zhang, T. (2002). Characterization of hydrogen-producing granular sludge. *Biotechnol. Bioeng.* 78 (1): 45-52.

Girbal, L., Croux, C., Vasconcelos, I. & Soucaille, P. (1995). Regulation of metabolic shifts in *Clostridium acetobutylicum* ATCC 824. *FEMS Microbial. Rev.* 17 (3): 287-297.

Hallenbeck, P.C. (2005). Fundamentals of the fermentative production of hydrogen. *Water Sci. Technol.* 52(1-2): 21-29.

Hallenbeck, P.C. & Ghosh, D. (2009). Advances in fermentative biohydrogen production: the way forward? *Trends Biotechnol.* 27 (5): 287–297.

Hallenbeck, P.C. (2011). Microbial paths to renewable hydrogen production. *Biofuels* 2(3): 285–302.

Hashizume, K., Tsukahara, T., Yamada, K., Koyama, H. & Ushida K. (2003) *Megasphaera elsdenii* JCM1772 normalizes hyperlactate production in the large intestine of fructooligosaccharide-fed rats by stimulating butyrate production. *J. Nutr.* 133 (10): 3187-3190.

Hung, C.-H., Lee, K.-S., Cheng, L.-H., Huang, Y.-H., Lin, P.-J. & Chang, J.-S. (2007). Quantitative analysis of a high-rate hydrogen-producing microbial community in anaerobic agitated granular sludge bed bioreactors using glucose as substrate. *Appl. Microbiol. Biotechnol.* 75(3): 693–701.

Ishibashi, N., Yaeshima, T. & Hayasawa, H. (1997). Bifidobacteria: their significance in human intestinal health. *Malaysian J. Nutrition* 3 (2): 149-159.

Ishikawa, M., Nakajima, K., Itamiya, Y., Furukawa, S., Yamamoto, Y. & Yamasato, K. (2005). *Halolactibacillus halophilus* gen. nov., sp. nov. and *Halolactibacillus miurensis* sp. nov., halophilic and alkaliphilic marine lactic acid bacteria constituting a phylogenetic lineage in *Bacillus* rRNA group 1. *Intern. J. System. Evolution. Microbiol.* 55 (Pt6): 2427–2439.

Jo, J.H., Jeon, C.O., Lee, D.S., Park & J.M. (2007). Process stability and microbial community structure in anaerobic hydrogen-producing microflora from food waste containing kimchi. *J. Biotechnol.* 131 (3): 300-308.

Jo, J.H., Lee, D.S., Park, D. & Park, J.M. (2008). Biological hydrogen production by immobilized cells of *Clostridium tyrobutyricum* JM1 isolated from a food waste treatment process. *Bioresour. Technol.* 99(14): 6666-6672.

Juang, C.-P., Whang, L.-M. & Cheng, H.-H. (2011). Evaluation of bioenergy recovery processes treating organic residues from ethanol fermentation process. *Bioresour. Technol.* 102 (9): 5394-5399.

Kalia, V.C. & Purohit, H.J. (2008) Microbial diversity and genomics in aid of bioenergy. *J. Ind. Microbiol. Biotechnol.* 35(5): 403-419.

Kim, T.-H., Lee, Y., Chang, K.-H. & Hwang, S.-J. (2012). Effects of initial lactic acid concentration, HRTs, and OLRs on bio-hydrogen production from lactate-type fermentation. *Bioresour. Technol.* 103 (1):, 136-141.

Kim, S.-H., Han, S.-K. & Shin, H.-S. (2006) Effect of substrate concentration on hydrogen production and 16S rDNA-based analysis of the microbial community in a continuous fermenter . *Process Bioch.* 41 (1): 199–207.

Korhonen, J. (2010). Antibiotic resistance of lactic acid bacteria. Dissertations in Forestry and Natural Sciences, Publications of the University of Eastern Finland, Kuopio.

Kraemer, J.T. & Bagley, D.M. (2007) Improving the yield from fermentative hydrogen production. *Biotechnol. Lett.* 29 (5): 685-695.

Lee, H.-S., Vermaas, W.F.J. & Rittmann, B.E. (2010). Biological hydrogen production: prospects and challenges. *Trends Biotechnol.* 28 (5): 262–271.

Lee, D.-J., Show, K.-Y. & Su, A. (2011). Dark fermentation on biohydrogen production: Pure cultures. *Bioresour. Technol.* 102(18): 8393-8402.

Li, C., Zhang, T. & Fang, H.H. (2006). Fermentative hydrogen production in packed-bed and packing-free upflow reactors. *Water Sci. Technol.* 54 (9): 95-103.

Li, C. & Fang, H.H.P. (2007). Fermentative hydrogen production from wastewater and solid wastes by mixed cultures. *Crit. Rev. Env. Sci. Tech.* 37 (1): 1–39.

Lo, Y.-C., Chen, W.-M., Hung, C.-H., Chen, S.-D. & Chang, J.-S. (2008). Dark H_2 fermentation from sucrose and xylose using H_2-producing indigenous bacteria: Feasibility and kinetic studies. *Water Res.* 42(1-2): 827-842.

Logan, B.E.: Extracting hydrogen and electricity from renewable resources. (2004). *Environ Sci Technol* 38(9): 160A-167A.

Maeda, T., Sanchez-Torres, V. & Wood, T.K. (2007). Enhanced hydrogen production from glucose by metabolically engineered *Escherichia coli*. *Appl. Microbiol. Biotechnol.* 77(4): 879-90.

Makarova, K., Slesarev, A., Wolf, Y., Sorokin, A., Mirkin, B., Koonin, E., Pavlov, A., Pavlova, N., Karamychev, V., Polouchine, N., Shakhova, V., Grigoriev, I., Lou, Y., Rohksa,r Y., Lucas, S., Huang, K., Goodstein, D. & Hawkins, T. (2006). Comparative genomics of the lactic acid bacteria. *PNAS* 103 (42): 15611–15616.

Matsumoto, M. & Nishimura, Y. (2007). Hydrogen production by fermentation using acetic acid and lactic acid. *J. Biosci. Bioeng.* 103(3): 236-241.

Meimandipour, A., Shuhaimi, M., Hair-Bejo, M., Azhar, K., Kabeir, B.M., Rasti, B. & Yazid, A.M. (2009). *In vitro* fermentation of broiler cecal content: the role of lactobacilli and pH value on the composition of microbiota and end products fermentation. *Lett. Appl. Microbiol.* 49(4): 415-420.

Morrison, D.J., Mackay, W.G., Edwards, C.A., Preston T., Dodson B. & Weaver L.T. (2006). Butyrate production from oligofructose fermentation by the human faecal flora: what is the contribution of extracellular acetate and lactate? *British J. Nutr.* 96 (3): 570–577

Munoz-Tamayo, R., Laroche, B., Walter, E., Dore, J., Duncan, S.H., Flint, H.J. & Leclerc, M. (2011). Kinetic modelling of lactate utilization and butyrate production by key human colonic bacterial species. *FEMS Microbiol. Ecol.* 76 (3): 615–624.

Nakashimada, Y., Rachman, M.A., Kakizono, T. & Nishio N. (2002). Hydrogen production of *Enterobacter aerogenes* altered by extracellular and intracellular redox state. *Int. J. Hydrogen Energy* 27(11-12): 1399-1405.

Nath, K. & Das, D. (2003). Hydrogen from biomass. *Current Science* 85 (3): 265-271.

Nath, K. & Das, D. (2004). Improvement of fermentative hydrogen production: various approaches. *Appl. Microbiol. Biotechnol.* 65(5): 520-529.

Noike, T., Takabatake, H., Mizuno, O. & Ohba, M. (2002). Inhibition of hydrogen fermentation of organic wastes by lactic acid bacteria. *Int. J. Hydrogen Energy* 27(11-12): 1367-1371.

Papoutsakis, E.T. (1984). Equation and calculations for fermentations of butyric acid bacteria. *Biotechnol. Bioeng.* 26(2): 174-187.

Piela P. & Zelenay P. (2004). Researchers redefine the DMFC roadmap. *The Fuel Cell Review.* 1(2): 17-23.

Qin, J., Li, R., Raes, J., Arumugam, M., Burgdorf, K.S., Manichanh, C., Nielsen, T., Pons, N., Levenez, F., Yamada, T., Mende, D.R., Li, J., Xu, J., Li, S., Li, D., Cao, J., Wang, B., Liang, H., Zheng, H., Xie, Y., Tap, J., Lepage, P., Bertalan, M., Batto, J.M., Hansen, T., Le Paslier, D., Linneberg, A., Nielsen, H.B., Pelletier, E., Renault, P., Sicheritz-Ponten, T., Turner, K., Zhu, H., Yu, C., Li, S., Jian, M., Zhou, Y., Li, Y., Zhang, X., Li, S., Qin, N., Yang, H., Wang, J., Brunak, S., Doré, J., Guarner, F., Kristiansen, K., Pedersen, O., Parkhill, J., Weissenbach, J.; MetaHIT Consortium, Bork, P., Ehrlich, S.D. & Wang, J. (2010). A human gut microbial gene catalogue established by metagenomic sequencing. *Nature* 464(4): 59-67

Ren, N., Xing, D., Rittmann, B.E., Zhao, L., Xie, T. & Zhao, X. (2007). Microbial community structure of ethanol type fermentation in bio-hydrogen production. *Environ. Microbiol.* 9 (5): 1112–1125.

Saint-Amans, S., Girbal, L., Andrade, J., Ahrens, K. & Soucaille, P. (2001). Regulation of carbon and electron flow in *Clostridium butyricum* VPI 3266 grown on glucose-glycerol mixtures. *J. Bacteriol.* 183(5): 1748-1754.

Salminen, S., Nurmi, J. & Gueimonde, M. (2005). The genomics of probiotic intestinal microorganisms. *Genome Biology* 6 (7): 225.

Sawers, R.G. (2005). Formate and its role in hydrogen production in *Escherichia coli. Biochem. Soc. Trans.* 33(Pt1): 42-46.

Shen, G.-J., Annous, B.A., Lovitt, R.W., Jain, M.K. & Zeikus, J.G. (1996). Biochemical route and control of butyrate synthesis in *Butyribacterium methylotrophicum*. Appl. Microbiol. Biotechnol. 45(3): 355-362.

Sreela-or, C., Imai, T., Plangklang, P. & Reungsang, A. (2011). Optimization of key factors affecting hydrogen production from food waste by anaerobic mixed cultures. *J. Hydrogen Energy*.36(21): 14120–14133).

Wu, S.-Y., Hung, C.-H., Lin, C.-N., Chen, H.-W., Lee, A.-S. & Chang, J.-S. (2006). Fermentative hydrogen production and bacterial community structure in high-rate anaerobic bioreactors containing silicone-immobilized and self-flocculated sludge. *Biotechnol. Bioeng.* 93(5): 934–946.

Yang, P., Zhang, R., McGarvey, J.A. & Benemann, J.R. (2007). Biohydrogen production from cheese processing wastewater by anaerobic fermentation using mixed microbial communities. *In. J. Hydrogen Energy* 32(18): 4761-4771.

Zhang, Z.G., Ye, Z.Q., Yu, L. & Shi, P. (2011). Phylogenomic reconstruction of lactic acid bacteria: an update. *BMC Evolutin. Biol.* 11(1): 1–12.

Chapter 14

Lactic Acid Bacteria and Mitigation of GHG Emission from Ruminant Livestock

Junichi Takahashi

Additional information is available at the end of the chapter

1. Introduction

The gases which bring greenhouse effect are water vapor and trace gases in atmosphere, carbon dioxide (CO_2), methane (CH_4), nitrous oxide (N_2), sulfur hexafluoride (SF_6), hydrofluorocarbons (HFCs), and perfluorocarbons (PFCs). Global warming due to increases in the atmospheric concentration of greenhouse gases (GHG) is an important issue. The worldwide trends of carbon dioxide have shown an increase in the greenhouse effect on global warming (Houghton, 1994). However, CH_4 is an important greenhouse gas second only to CO_2 in its contribution to global warming due to its high absorption ability of infrared in the radiation from sun (IPCC, 1994). The world population of ruminants is important source of methane, contributing approximately 15-18% of the total atmospheric CH_4 flux. The control of CH_4 emission is a logical option since atmospheric CH_4 concentration is increasing at a faster rate than carbon dioxide (Moss, 1993). CH_4 emitted from ruminants is mainly generated in the rumen by hydrogenotrophic methanogens that utilize hydrogen to reduce carbon dioxide, and is a significant electron sink in the rumen ecosystem (Klieve and Hegarty, 1999), although acetotrophic methanogens may play a limited role for rumen methanogenesis (McAllister, 1996). Methane contains 892.6 kJ combustible energy per molecule at 25ºC and 1013hPa, while not contributing to the total supply of metabolic energy to ruminants (Takahashi *et al.*, 1997). As reported by Leng (1991), methane production from ruminants in the developing countries may be high since the diets are often deficient in critical nutrients for efficient microbial growth in the rumen. So far, a number of inhibitors of methanogenesis have been developed to improve feed conversion efficiency of ruminant feeds claimed to be effective in suppressing methanogens or overall bacterial activities (Chalupa, 1984). Attempts to reduce methanogenesis by the supplementation of chemicals such as ionophores (monensin and lasalocid), have long been made (Chalupa, 1984; Hopgood and Walker, 1967). However, these ionophores may depress fiber digestion and protozoal growths (Chen and Wolin, 1979). In addition, some resistant bacteria will appear in the rumen from the results of long term use of the ionophores. Therefore, development of manipulators to mitigate rumen methanogenesis must pay attention to secure safety for animals, their products and environment as alternatives of ionophores.

Theoretically, methanogenesis can be reduced by either a decrease in the production of H_2, the major substrates for methane formation or an increase in the utilization of H_2 and formate by organisms other than methanogens. However, direct inhibition of H_2-forming reactions may depress fermentation in microorganisms that produce H_2, including main cellulolytic bacteria such as *Ruminococcus albus* and *Ruminococcus flavefaciens* (Belaich *et al.*, 1990; Wolin, 1975). Therefore, a reduction in H_2 production by the enhancement of reactions that accept electrons is desirable (Stewart and Bryant, 1988). In the rumen, metabolic H_2 is produced during the anaerobic fermentation of glucose. This H_2 can be used during the synthesis of volatile fatty acids and microbial organic matter. The excess H_2 from NADH is eliminated primarily by the formation of CH_4 by methanogens, which are microorganisms from the *Archea* group that are normally found in the rumen ecosystem (Baker, 1999). The stoichiometric balance of VFA, CO_2 and CH_4 indicates that acetate and butyrate promote CH_4 production whereas propionate formation conserves H_2, thereby reducing CH_4 production (Wolin, and Miller, 1988). By contrast, reductive methanogenesis might contribute to mitigate methane (Immig *et al.*, 1996). Therefore, a strategy to mitigate ruminal CH_4 emission is to promote alternative metabolic pathway to dispose the reducing power, competing with methanogenesis for H_2 uptake. Oligosaccharides are naturally occurring carbohydrates with a low degree of polymerisation and consequently low molecular weight, being commonly found to perform in the various plant and animal sources. β1-4 Galactooligosaccharides (GOS) are non-digestible carbohydrates, which are resistant to gastrointestinal digestive enzymes, but fermented by specific colonic bacteria. The products of fermentation of GOS in the colon, mainly short chain fatty acids, have a role in the improvement of the colonic environment, energy supply to the colonic epithelium, and calcium and magnesium absorption (Sako, *et al.*, 1999). The indigestibility and stability of GOS to hydrolysis by α-amylase of human saliva, pig pancreas, rat small intestinal contents and human artificial gastric juice has been shown in several *in vitro* experiments (Ohtsuka *et al.*, 1990; Watanuki *et al.*, 1996). This is because GOS have β-configuration, whereas human gastrointestinal digestive enzymes are mostly specific for α-glycosidic bonds. From this point of view, expectedly, GOS will be readily degraded in the rumen as a result of the ruminal enzymes being specific for β-glycosidic bonds. Thus, lactic acid bacteria may consume GOS to promote propionate formation through acrylate pathway, and consequently the competition with methanogens for hydrogen will occur. Thus, the amplifying competition of metabolic H_2 with probiotics may be a key factor in the regulation of rumen methanogenesis. However, direct effects of prebiotics and secondary metabolites such as tannin, saponin and natural resin on methanogens and eubacteria in the rumen remain to be elucidated to secure the safety for animals, their products and environment. The mechanism for accreditation of manipulators must be established to mitigate global CH_4 emission.

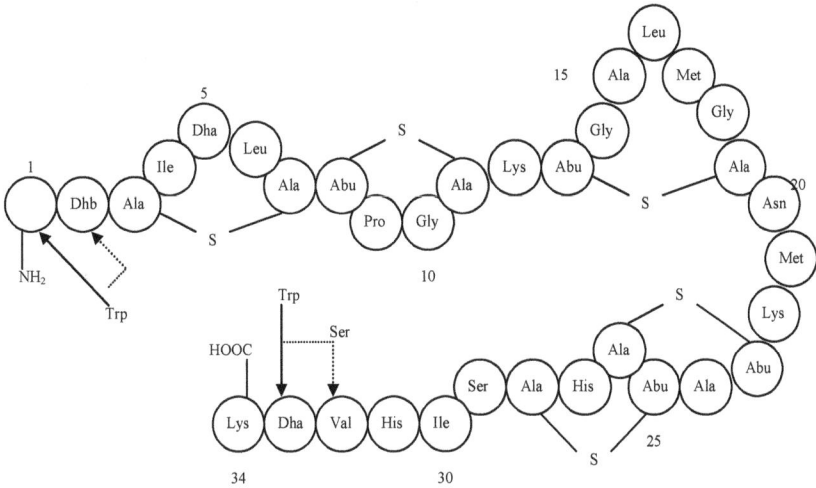

Dha = dehydroalanine, Dhb = dehydrobutyrine, Ala-S-Ala = lanthionine, Abu-S-Ala = β-methyllanthionine. (adapted from Breukink et al., 1998).

Figure 1. Primary structure of nisin.

2. Possible control of indirect action of lactic acid bacteria as probiotics on rumen methanogenesis

Rumen manipulation with ionophores such as monensin has been reported to abate rumen methanogenesis (Mwenya *et al.*, 2005), However, there is an increasing interest in exploiting prebiotics and probiotics as natural feed additives to solve problems in animal nutrition and livestock production as alternatives of the antibiotics due to concerns about incidences of resistant bacteria and environmental pollution by the excreted active-antibacterial substances (Mwenya *et al.*, 2006). Particular interest concerning bacteriocins which produced by lactic acid bacteria has increased recently.

Bacteriocins, antimicrobial proteinaceous polymeric material substances, are ubiquitous in nature being produced by a variety of Gram-negative and Gram-positive bacteria, and typically narrow spectrum antibacterial substances under the control of plasmid. Nisin is produced by *Lactococcus lactis* ssp. *lactis* which is an amphiphilic peptide composed by 34 amino acids with two structural domains that are connected by a flexible hinge (Breukink *et al.*, 1998; Montville and Chen, 1998), and is classified into the group of lantibiotics. Nisin has a mode of action similar to ionophores, which show antimicrobial activity against a broad spectrum of Gram-positive bacteria and is widely used in the food industry as a safe and natural preservative (Delves-Broughton *et al.*, 1996). It is generally recognized as safe (GRAS) and given international acceptance in 1969 by the joint Food and Agriculture Organization/World Health Organization (FAO/WHO) Expert Committee on Food

Additives. Recent works have indicated that *Lactococcus lactis* subsp. *lactis* produce nisin Z, which has been identified from Korean traditional fermented food "Kimchi" besides nisin A (Park, 2003). They have similar antibacterial ability to mitigate methane emission (Mwenya *et al.*, 2004; Santoso *et al.*, 2004; Sar *et al.*, 2006), to inhibit growth both of *Clostridium amoniphilum*, which is obligate amino-acid fermenting bacteria (Callaway *et al.*, 1997) and lactic acid-producing ruminal Staphylococci and Enterococci (Lauková, 1995). *Leuconostoc mesenteroides* ssp. *mesenteroides, Leuconostoc lactis* and *Lactococcus lactis* ssp. *lactis* were isolated from "Laban" which was a traditional fermented milk product in Yemen and determined the mitigating effect on in vitro rumen methane production. These strains isolated from Laban enhanced propionate production and decreased acetate/propionate ratio. In consequence, they reduced methane production remarkably (Gamo *et al.*, 2002). For *Leuconostoc mesenteroides* ssp. *mesenteroides*, in particular, the mitigating effect was amplified with GOS, which was degradable about 80% within 1 hour incubation in the artificial rumen fluid due to the stimulation of reduction reactions consuming metabolic hydrogen. However, direct involvement of bacteriocin or lower molecular substances produced by the strain on rumen methanogenesis remains to be elucidated.

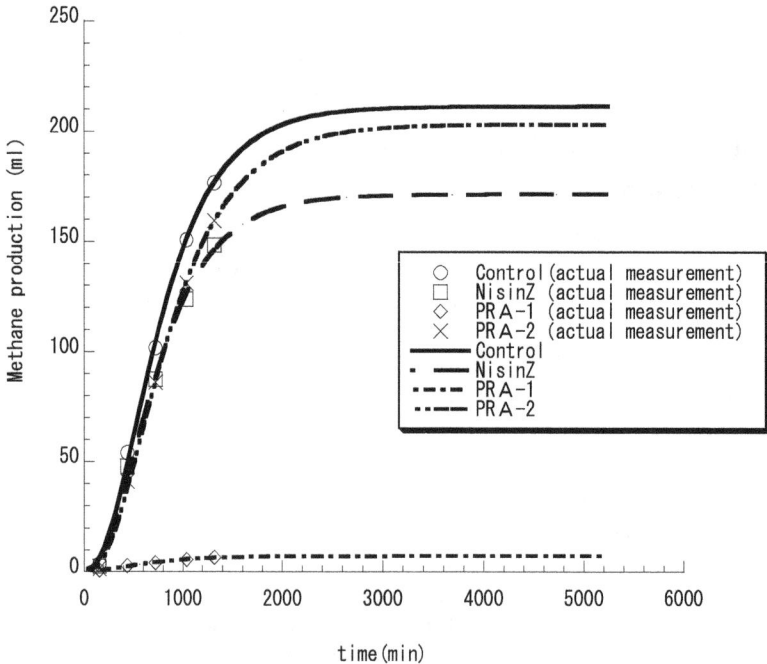

Where, y(ml) =gas produced at time *t* (min), a=first gas production, b=second gas production and c=frctional rate gas production, using Kaleida Graph (Version 3.6, Synergy Software, Reading, PA, USA).

Figure 2. Effect of PRA on the cumulative methane production extrapolated by nonlinear regression analysis; y=a+b $(1-e^{-ct})^3$.

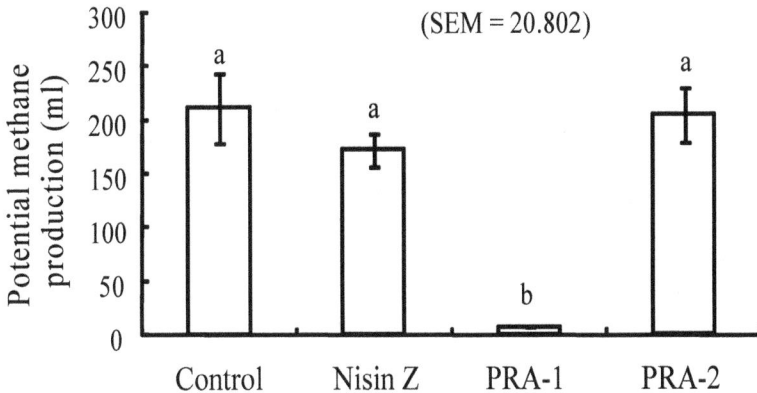

Figure 3. Effect of PRA on potential methane production. Control: *Lactococcus lactis* ATCC19435 (non-antibacterial substances), Nisin-A: *Lactococcus lactis* NCIMB702054, PRA-1: *Lactobacillus plantarum* TUA1490L, and PRA-2: *Leuconostoc citreum* JCM9698. Vertical bars represent standard deviation (n = 4). Means with different letters differ significantly (p<0.01).

3. Abatement of rumen methanogenesis by direct action of lactic acid bacteria as prebiotics producer

For low molecular compounds, small amounts of volatile fatty acids (acetic acid, formic acid), hydrogen peroxide, β-hydroxy-propionaldehyde (reuterin) are produced by   lactic acid bacteria as antibacterial substances in addition to lactic acid. Because lactic acid bacteria themselves don't have a group of catalase, considerable amount of hydrogen peroxide accumulates in the bacterial cells. Many strains of the genus *Lactobacillus* are commonly referred to as having high ability to produce hydrogen peroxide (Jaroni and Brashears, 2000; Aroucheva *et al,* 2001; Gardiner *et al.,* 2002).

Its antimicrobial activity is effective against numerous Gram-positive bacteria. Although it has been reported that nisin suppress rumen methanogenesis, the suppressing efficacy of nisin on rumen methanogenesis may not be sustained, because proteinaceous nisin is degradable in the rumen due to bacterial protease (Sang *et al.,* 2002). Several strains of lactic acid bacteria produce different types of protease resistant antimicrobial substance (PRA). In our research, the strain of lactic acid bacteria that produce PRA were screened on MRS agar plates containing Umamizyme G (protease mixture from *Aspergillus oryzae*, amino Enzyme Inc, Nagoya, Japan) as follows: candidates were inoculated onto MRS agar with or without 1,000 IU ml^{-1} of Umamizyme G and incubated for 24 h at 30 °C. the plates were then overlaid with Bacto Lactobacilli agar AOAC (Becton, Dickinson and Company, NJ. USA) containing an indicator strain, *Lactobacilli sakei* JCM1157T. The agar overlays were incubated for 24 h at 30°C and examined for zones of clearing. Protease degradable anti-microbial substances were decomposed by Umamizyme G, thus a clear zone did not form on the plate with Umamizyme G. Two strains of lactic acid bacteria, *Lactobacillus plantarum* TUA1490L and

Leuconostoc citreum JCM9698 that produced almost the same size of clear zone on a Umamizyme G containing plate as that on a plate without Umamizyme G, were selected as PRA producers. *Lactobacillus plantarum* TUA1490L and *Leuconostoc citeum* JCM9698 were selected as PRA-1 and PRA-2 producers. GYEKP medium to prepare inoculants for PRA-1, PRA-2, nisin Z and control were used for the culture of lactic acid bacteria. Each strain of lactic acid bacteria was inoculated into a shaking flask containing GYEKP, and was cultivated for 20 h at 30°C using SILIKOSEN (Shin-Etsu polymer, Tokyo), which was culture plug for aeration cultivation after confirmation of the stationary phase. The cells were removed by centrifugation at 8,000 × g at 4°C and filtration with 0.45 μm membrane filter. The supernatants were used as PRA inoculants in the in vitro gaseous quantification trials. Methane mitigating effects of PRA-1 from *Lactobacillus plantarum* TUA1490L and PRA-2 from *Leuconostoc citeum* JCM9698 isolated from foods were determined in comparison with *Lactococcus lactis* ATCC19435 which did not produce any antibacterial substances as a negative control and *Lactococcus lactis* NCIMB702954 which produced nisin-Z as a positive control using in vitro continuous incubation system equipped with automated infra-red quantification apparatus (Takahashi *et al.*, 2005). Fig.2 shows effects of PRA-1 and PRA-2 produced by *Lactobacillus plantarum* and *Leuconostoc citreum* on cumulative methanogenesis extrapolated by nonlinear regression analyses. PRA-1 remarkably decreased cumulative methane production. For PRA-2, there were no effects on CH_4 and CO_2 production and fermentation characteristics in mixed rumen cultures. Fig. 3 shows the effect of PRA on potential methane production which estimated from non-linier regression analysis of cumulated methane production. It has been suggested that PRA-1 significantly decreases potential methane production by rumen methanogens (Asa *et al.*, 2010). The PRA maintained their antimicrobial effects after incubation with proteases, while nisin lost its activity. Therefore, the PRA was hypothesized to be a more sustained agent than nisin for the mitigation of rumen methane emission. Fig. 4 shows DGGE band patterns of archaea and eubacteria. All fluorescence brightness of methanogens bands of PRA-1 were remarkably light in color compared with control. Band No. 1 to No.3 in archaea might be *Methanobrevibacter sp.* which is a Gram positive or parasitic methanogens sticking on protozoan surface (Fig.5). PRA-1 increased the fluorescence brightness of the band of the Gram positive bacteria and declined the fluorescence brightness of the band of the Gram negative bacteria. For Gram positive bacteria, *Streptococcus sp., Clostridium* sp., *Butyrivibrio* sp. and *Clostridium aminophilum* were increased, whereas *Prevotella* sp., *Prevotella ruminicola, Pseudobutyrivibrio* sp, *Prevotella* sp, *Succinivibrio dextrinosolvens and Schwartzia succinivorans* in Gram negative bacteria were decreased by adding PRA-1.

Natural antimicrobial substances can be used alone or in combination with other novel preservation technologies to facilitate the replacement of traditional approaches (Brijesh, 2009). *Lactobacillus plantarum* produces bacteriocin from many foods including meat and meat products (Garriga *et al.*,1993; Enan *et al.*,1996; Aymerich *et al.*, 2000), milk (Rekhif *et al.*, 1995), cheese (Gonzalez *et al.*,1994), fermented cucumber (Daeschel *et al.*,1994), olives (Jimenez-Diaz *et al.*, 1993; Leal *et al.*, 1998), grapefruit juice (Kelly *et al.*,1996), Turkish fermented dairy products (Aslim *et al.*, 2005), and sourdough (Todorov *et al.*, 1999). PRA-1 was the antibacterial substance produced from a strain of *Lactobacillus plantarum* TUA1490L that was isolated from tomato in Japan. However, methane suppressing activity of PRA-1 was not inactivated by treatments Umamizyme G and protease K. Moreover, aeration cultivation is an essential procedure for activation of PRA-1 to abate methanogenesis. Therefore, possible mechanism of PRA-1 produced by *Lactobacillus plantarum* TUA1490L on rumen methane production might be assumed as resulting from the direct involvement of low molecule substance such as hydrogen peroxide due to the requirement of aeration for the preparation.

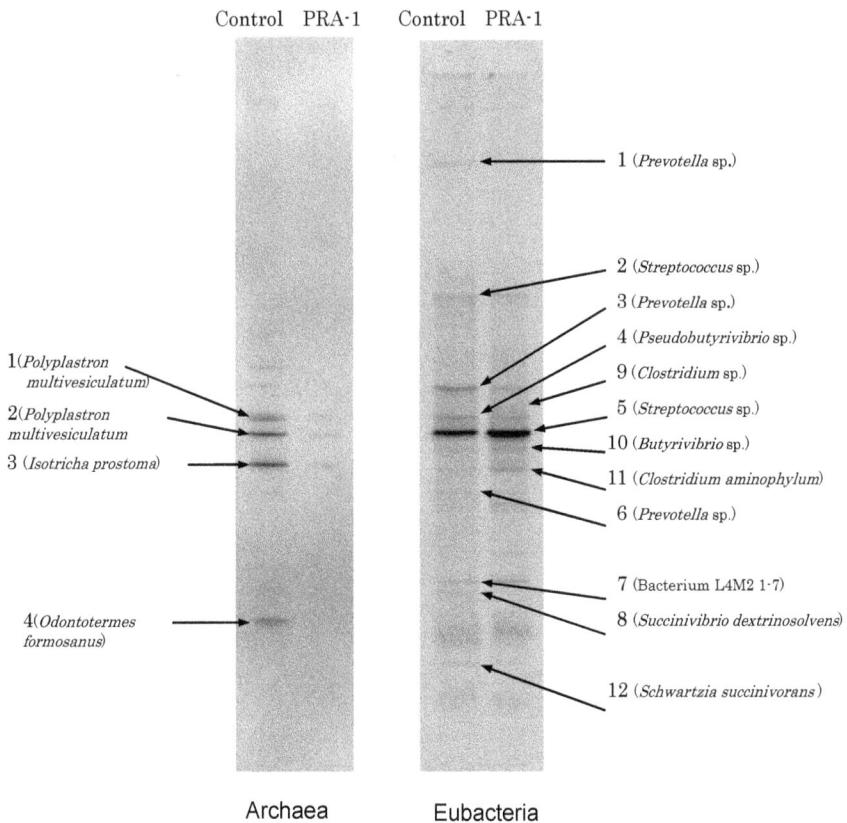

Figure 4. DGGE band patterns

Figure 5. Electric scanning microscopy of symbioses of methanogens on the surface of Ciliate Protozoa.

Author details

Junichi Takahashi

Graduate School of Animal Science, Obihiro University of Agriculture and Veterinary Medicine, Obihiro, Japan

4. References

Asa, R., A. Tanaka, A. Uehara, I. Shinzato, Y. Toride, N. Usui, K. Hirakawa and J. Takahashi., 2010. Effects of protease-resistant antimicrobial substances produced by lactic acid bacteria on rumen methanogenesis. Asian-Aust. J. Anim. Sci., 23:700-707.

Aslim, B., Z. N. Yuksekdag., E. Sarikaya and Y. Beyatli., 2005. Determination of the bacteriocin-like substances produced by some lactic acid bacteria isolated from Turkish dairy products. LWT, 38: 691-694.

Aroucheva, A., D. Gariti and M. Simons., 2001. Defense factors of vaginal Lactobacilli. Am. J. Obstet. Gynecol., 185: 375-379.

Aymerich. M., T., M. Garriga., J. M. Monfort., I. Nes and M. Hugas., 2000. Bacteriocin-producing lactobacilli in Spanish-style fermented sausages: characterization of bacteriocins. Food Microbiol., 17: 33-45.

Baker, S.K., 1999. Rumen methanogens, and inhibition of methanogenesis. Aust. J. Agric. Res., 50: 1293-1298.

Belaich, J.P., M. Bruschi and J. L. Garcia., 1990. Microbiology and Biochemistry of Strict Anaerobes Involved in Interspecies Hydrogen Transfer. Plenum Press, NY. USA.

Breukink, E., C. Van Kraaij, A. Van Dalen, R.A. Demel, R.J. Siezen, B. De Kruijff, and O.P. Kuipers., 1998. The orientation of nisin in membranes. Biochem., 37: 8153–8162.

Brijesh, K. T., P. V. Vasilis, P. O. D. Colm, M. Kasiviswanathan B. Paula and P. J. Cullen., 2009. Applivation of natural antimicrobials for food preservation. J. Agric. Food Chem., 57: 5987-6000.

Callaway, T. R., Alexandra M. S. Carneiro De Melo and J. B. Russell., 1997. The effect of nisin and monensin on ruminal fermentations in vitro. Curr Microbiol., 35:90-96.

Cadieux, P., J. Burton, G. Gardiner, J. Braunstein, A.W. Bruce, G.Y. Kang and G. Reid., 2002. Lactobacilli strains and vaginal ecology. J. Am. Med. Assoc., 287: 1940-1941.

Chalupa, W., 1984. Manipulation of rumen fermentation. In: Recent Advances in Animal Nutrition. W. Haresign and D. Cole, (eds). Butterworths, London, England, pp.143-160.

Chen, M. and M.J. Wolin, M.J., 1979. Effect of monensin and lasalocid-sodium on the growth of methanogenic and rumen saccharolytic bacteria. Appl. Environ. Microbiol., 38: 72-77.

Daeschel, M. A., M. C. Mckenny and L. C. McDonald., 1990. Bacteriocidal activity of Lactobacillus plantarum C11. Food Microbiol. 7: 91-99.

Delves-Broughton, J., P. Blackburn., R. Evans and J. hugenholtz., 1996. Applications of the bacteriocin, nisin. Antonie van Leeuwenhoek. 69: 193-202.

Enan, G., A. A. El-Essawy., M. Uyttendaele and J. Debevere., 1996. Antibacterialactivity of Lactobacillus plantarum UG1 isolated from dry sausage: characterization, production and bactericidal action of plantarcin UG1. Int. J. Food. Microbiol., 30: 189-215.

Gamo, Y., M. Mii, X.G. Zhou, C. Sar, B. Santoso, I. Arai, K. Kimura and J. Takahashi., 2002. Effects of lactic acid bacteria, yeasts and galactooligosaccharides supplementation on in vitro rumen methane production. In: J. Takahashi and B.A. Young, (eds), Greenhouse Gases and Animal Agriculture. ELSEVIER, Amsterdam, Netherland. pp201-204.

Garriga, M., M. Hugas., T. Aymerich and J. M. monfort., 1993. Bacteriocinogenic activity of lactobacilli from fermented sausages. J. Appl. Bacteriol., 75: 142-148.

González, B., P. Arca., B. Mayo and J. E. Suárez., 1994. Detection, purification and partial characterization of plantaricin C, a bacteriocin produced by a Lactobacillus plantarum strain of dairy origin. Appl. Environ. Microbiol., 6: 2158-2163.

Hopgood, M. F. and D. J. Walker., 1967. Succinic acid production by rumen bacteria. II. Radioisotope studies on succinateproduction by *Ruminococcus flavefaciens*. Aust. J. Biol. Sci., 20: 183-192.

Houghton, J., 1994. Global warming. Lion Publishing plc. Oxford, UK, pp: 29-45.

IPCC (Intergovermental Panel on Climate Change), 1994. Houghton, J.H., Meria filho, J. Bruce, L.Hoesung, B.A.Callander, H. Haites, N.Harris and K.Maskell, (eds), Cambridge University Press. New York. pp25-27.

Immig I., D. Demeyer, D. Fiedler, C. Van Nevel and L. Mbanzamihigo., 1996. Attempts to induce reductive acetogenesis into a sheep rumen. Arch. Tierernahr. 49:363-370.

Jaroni, D. and M. M. Brashears., 2000. Production of Hydrogen Peroxide by *Lactobacillus delbrueckii* subsp. *lactis* as Influenced by Media Used for Propagation of Cells . J. Food Sci., 65: 1033-1036.

Jiménez-Díaz R., R. M. Rios-Sánchez, M. Desmazeaud, J. L. Ruiz-Barba and J. C. Piard., 1993. Plantaricin S and T, two new bacteriocins produced by Lactobacillus plantarum LPCO10 isolated from a green olive fermentation. Appl. Environ. Microbiol. 59: 1416-1424.

Kelly, W. J., R. V. Asmundson and C. M. Huang., 1996. Characterization of plantaricin KW30, a bacteriocin produced by Lactobacillus plantarum. J. Appl. Bacteriol. 81: 657-662.

Klieve, A.V. and R. S. Hegarty., 1999. Opportunities of biological control of ruminant methanogenesis. Aust. J. Agric. Res., 50: 1315-19.

Leal, M. V., M. Baras, J. L. Ruiz-Barba, B. Floriano and R. Jimenez-Diaz., 1998. Bacteriocin production and competitiveness of Lactobacillus plantarum LPCO10 in olive juice broth, a culture medium obtained from olives. Int. J. Food. Microbiol. 43: 129-134.

Leng, R.A., 1991. Improving ruminant production and reducing methane emissions from ruminants by strategic supplementation. EPA/400/1-91/004, US Environmental Protection Agency, Washington, DC, pp6-10.

Lauková, A., 1995. Inhibition of ruminal staphylococci and enterococci by nisin in vitro. Lett. Appl. Microbiol., 20:34-36.

Moss, A.R., 1993. Methane: Global Warming and Production by Animals. Chalcombe, Canterbury, UK, p.105.

McAllister, T. A., E.K. Okine, G.W. Mathison, K.-J. Cheng., 1996. Dietary, environmental and microbiological aspects of methane production in ruminants. Can. J. Anim. Sci. 76: 231-243

Montville, T.J and Y. Chen., 1998. Mechanistic action of pediocin and nisin: recent progress and unresolved questions. Appl. Microbiol. Biotechnol., 50: 511-519.

Mwenya, B., C. Sar, Y. Gamo, T. Kobayashi, R. Morikawa, K. Kimura, H. Mizukoshi, and J. Takahashi., 2004. Effects of *Yucca schidigera* with or without nisin on ruminal fermentation and microbial protein synthesis in sheep fed silage- and hay- based diets. Anim. Sci. J., 75: 525-531.

Mwenya, B. C. Sar, B. Santoso, T. Kobayashi, R. Morikawa, K. Takaura, K. Umetsu, S. Kogawa, K. Kimura, H. Mizukoshi and J. Takahashi., 2005. Comparing the effects of β1-4 galacto-oligosaccharides and L-cysteine to monensin on energy and nitrogen utilization in steers fed a very high concentrate diet. Anim. Feed Sci. Technol. 118: 19-30.

Mwenya, B., C. Sar, B. Pen, R. Morikawa, K. Takaura, S. Kogawa, K. Kimura, K. Umetsu and J. Takahashi., 2006. Effects of feed additives on ruminal methanogenesis and anaerobic fermentation of manure in cows and steers. In: C. Soliva, J. Takahashi and M. Kreutzer (eds), Greenhouse Gases on Animal Agriculture update. International Congress Series. Elsevier, Amsterdam. 1293: 209-212.

Ohtsuka, K., K.Tsuji, Y. Nakagawa, H. Ueda, O. Ozawa, T. Uchida and T. Ichikawa., 1990. Availability of 4`-galactosyllactose (o-beta-D-galactopyranosyl -(1-4)-o- beta-D-galactospyranosyl-(1-4)-D-glucopyranose) in rat. J. Nutr. Sci. Vitaminol., 36: 265-276.

Park, S.E., K. Itoh, E. Kikuchi, H. Niwa and T. Fujisawa., 2003. Identification and characteristics of nisin Z-producing *Lactococcus lactis* subsp. *lactis* isolated from Kimchi. Curr. Microbiol., 46: 385-358.

Rekhif, N, A. Atrih and G Lefebvre., 1995. Activity of plantaricin SA6, a bacteriocin produced by Lactobacillus plantarum SA6 isolated from fermented sausage. J. Appl. Bacteriol., 78: 349-358.

Sako, T., K. Matsumoto and R. Tanaka., 1999. Recent progress on research and applications of non-digestible galacto-oligosaccharides. Inter. Dairy J., 9: 69-80.

Sang, S.L., H. C. Mantovani, and J.B. Russell., 2002. The binding and degradation of nisin by mixed ruminal bacteria. FEMS Microbiol. Ecol., 42:339-345.

Santoso, B., B. Mwenya, C. Sar, Y. Gamo, T. Kobayashi, R. Morikawa and J. Takahashi., 2004. Effect of Yucca schidigera with or without nisis on ruminal fermentation and microbial protein synthesis in sheep fed silage- and hay-based diets. Anim. Sci. J., 75: 525-531.

Sar, C., B. Mwenya, B. Pen, R. Morikawa, K. Takaura, T. Kobayashi and J. Takahashi., 2006. Effect of nisin on ruminal methane production and nitrate/nitrite reduction *in vitro*. Aust. J. Agric. Res., 56: 803-810.

Stewart, C.S., and M.P. Bryant., 1988. The rumen bacteria. In: The Rumen Microbial Ecosystem. P.N. Hobson, (ed). Elsevier Appl. Sci., New York, NY, pp.21-75.

Takahashi, J., A.S. Chaudhry, R.G. Beneke, Suhubdy and B.A. Young., 1997. Modification of methane emission in sheep by cysteine and a microbial preparation. Sci. Total Environ., 204: 117-123

Takahashi, J., B. Mwenya, B. Santoso, C. Sar., K. Umetsu, T. Kishimoto, K. Nishizaki, K. Kimura and O. Hamamoto., 2005. Mitigation of methane emission and energy recycling in animal agricultural systems. Aisan-Aust. J. Anim. Sci., 18:1199-1208.

Todorov, S., B. Onno., O. Sorokine., J. M. Chobert., I. Ivanova and X. Dousset., 1999. Detection and characterization of a novel antibacterial substance produced by *Lactobacillus plantarum* ST 31 isolated from sourdough. Int. J. Food. Microbiol. 48:167-177.

Watanuki, M., Y. Wada and K. Matsumoto., 1996. Digestibility and physiological heat of combustions of β1-4 and β1-6 galacto-oligosaccharides *in vitro*. Annu. Rep. Yakult Cent. Inst. Microbiol. Res., 16: 1-12.

Wolin, M. J., 1975. Interactions between the bacterial species of the rumen. In: Digestion and Metabolism in the Ruminant. I.M. McDonald and A.C.I. Warner (eds). Univ. New England Publ. Unit, Sydney, Australia, pp.135-148.

Wolin, M.J. and T.L. Miller, 1988. Microbe-microbe interactions. In: The rumen microbial ecosystem. Elsevier Applied Science, London, UK, pp. 343-359

Permissions

The contributors of this book come from diverse backgrounds, making this book a truly international effort. This book will bring forth new frontiers with its revolutionizing research information and detailed analysis of the nascent developments around the world.

We would like to thank J. Marcelino Kongo, for lending his expertise to make the book truly unique. He has played a crucial role in the development of this book. Without his invaluable contribution this book wouldn't have been possible. He has made vital efforts to compile up to date information on the varied aspects of this subject to make this book a valuable addition to the collection of many professionals and students.

This book was conceptualized with the vision of imparting up-to-date information and advanced data in this field. To ensure the same, a matchless editorial board was set up. Every individual on the board went through rigorous rounds of assessment to prove their worth. After which they invested a large part of their time researching and compiling the most relevant data for our readers. Conferences and sessions were held from time to time between the editorial board and the contributing authors to present the data in the most comprehensible form. The editorial team has worked tirelessly to provide valuable and valid information to help people across the globe.

Every chapter published in this book has been scrutinized by our experts. Their significance has been extensively debated. The topics covered herein carry significant findings which will fuel the growth of the discipline. They may even be implemented as practical applications or may be referred to as a beginning point for another development. Chapters in this book were first published by InTech; hereby published with permission under the Creative Commons Attribution License or equivalent.

The editorial board has been involved in producing this book since its inception. They have spent rigorous hours researching and exploring the diverse topics which have resulted in the successful publishing of this book. They have passed on their knowledge of decades through this book. To expedite this challenging task, the publisher supported the team at every step. A small team of assistant editors was also appointed to further simplify the editing procedure and attain best results for the readers.

Our editorial team has been hand-picked from every corner of the world. Their multi-ethnicity adds dynamic inputs to the discussions which result in innovative outcomes. These outcomes are then further discussed with the researchers and contributors who give their valuable feedback and opinion regarding the same. The feedback is then collaborated with the researches and they are edited in a comprehensive manner to aid the understanding of the subject.

Apart from the editorial board, the designing team has also invested a significant amount of their time in understanding the subject and creating the most relevant covers. They scrutinized every image to scout for the most suitable representation of the subject and create an appropriate cover for the book.

The publishing team has been involved in this book since its early stages. They were actively engaged in every process, be it collecting the data, connecting with the contributors or procuring relevant information. The team has been an ardent support to the editorial, designing and production team. Their endless efforts to recruit the best for this project, has resulted in the accomplishment of this book. They are a veteran in the field of academics and their pool of knowledge is as vast as their experience in printing. Their expertise and guidance has proved useful at every step. Their uncompromising quality standards have made this book an exceptional effort. Their encouragement from time to time has been an inspiration for everyone.

The publisher and the editorial board hope that this book will prove to be a valuable piece of knowledge for researchers, students, practitioners and scholars across the globe.

List of Contributors

Yang Cao
College of Animal Science & Veterinary Medicine, Heilongjiang Bayi Agricultural University, Daqing, China

Yimin Cai
Japan International Research Center for Agricultural Sciences, Tsukuba, Japan

Toshiyoshi Takahashi
Faculty of Agriculture, Yamagata University, Tsuruoka, Japan

Edson Mauro Santos and Carlos Henrique Oliveira Macedo
Department of Animal Science, Federal University of Paraiba, Areia, PB, Brazil

Thiago Carvalho da Silva
Department of Animal Science, Federal University of Viçosa, Viçosa, MG, Brazil

Fleming Sena Campos
Department of Animal Science, Federal University of Bahia, Salvador, BA, Brazil

Mahdi Ghanbari and Mansoureh Jami
University of Zabol, Faculty of Natural Resources, Department of Fishery; Zabol, Iran
BOKU -University of Natural Resources and Life Sciences, Department of Food Sciences and Technology, Institute of Food Sciences; Vienna, Austria

Masoud Rezaei
Tarbiat Modares University, Faculty of Marine Science, Department of Seafood Science and Technology, Noor, Mazandaran, Iran

Shirin Tarahomjoo
Department of Biotechnology, Razi Vaccine and Serum Research Institute, Karaj, Iran
Food Industries and Biotechnology Research Center, Amirkabir University of Technology, Tehran, Iran

Tamara Aleksandrzak-Piekarczyk
Institute of Biochemistry and Biochemistry, Polish Academy od Sciences, Warsaw, Poland

Tsuda Harutoshi
National Institute of Health and Nutrition, Tokyo, Japan

Charina Gracia B. Banaay
Institute of Biological Sciences, College of Arts and Sciences, University of the Philippines Los Baños, Laguna, Philippines

Marilen P. Balolong
Department of Biology, University of the Philippines Manila, Padre Faura Street, Ermita, Manila, Philippines
Institute of Molecular Biology and Biotechnology, National Institutes of Health, University of the Philippines Manila, Pedro Gil Street, Ermita, Manila, Philippines

Francisco B. Elegado
National Institute of Molecular Biology and Biotechnology, University of the Philippines Los Baños, Laguna, Philippines

Sanna Taskila and Heikki Ojamo
University of Oulu, Faculty of Technology, Department of Process and Environmental Engineering, Bioprocess Engineering Laboratory, Oulu, Finland

Bertrand Tatsinkou Fossi
Department of Microbiology and Parasitology, University of Buea, Buea, Cameroon

Frédéric Tavea
Department of Biochemistry, University of Douala, Douala, Cameroon

Panagiota Florou-Paneri and Efterpi Christaki
Laboratory of Nutrition, Faculty of Veterinary Medicine, Aristotle University of Thessaloniki, Thessaloniki, Greece

Eleftherios Bonos
Animal Production, Faculty of Technology of Agronomics, Technological Educational Institute of Western Macedonia, Florina, Greece

Diana I. Serrazanetti, Davide Gottardi, Chiara Montanari and Andrea Gianotti
Department of Food Science, Alma Mater Studiorum, University of Bologna, Bologna, Italy
Inter-Departmental Center of Industrial Agri-Food Research (CIRI Agroalimentare), Cesena, Italy

Anna Sikora and Urszula Zielenkiewicz
Institute of Biochemistry and Biophysics, Polish Academy of Sciences, Poland

Mieczysław Błaszczyk and Marcin Jurkowski
Faculty of Agriculture and Biology, Warsaw University of Life Sciences, Poland

Junichi Takahashi
Graduate School of Animal Science, Obihiro University of Agriculture and Veterinary Medicine, Obihiro, Japan